D1126167

Design and Analysis of Large Lithium-Ion Battery Systems

Recent Artech House Titles in Power Engineering

Dr. Jianhui Wang, Series Editor

The Advanced Smart Grid: Edge Power Driving Sustainability, Andres Carvallo and John Cooper

Battery Management Systems for Large Lithium Ion Battery Packs, Davide Andrea

Battery Power Management for Portable Devices, Yevgen Barsukov and Jinrong Qian

Design and Analysis of Large Lithium-Ion Battery Systems, Shriram Santhanagopalan, Kandler Smith, Jeremy Neubauer, Gi-Heon Kim, Matthew Keyser, and Ahmad Pesaran

Designing Control Loops for Linear and Switching Power Supplies: A Tutorial Guide, Christophe Basso

Electric Systems Operations: Evolving to the Modern Grid, Mani Vadari

Energy Harvesting for Autonomous Systems, Stephen Beeby and Neil White

GIS for Enhanced Electric Utility Performance, Bill Meehan

Introduction to Power Electronics, Paul H. Chappell

Power Line Communications in Practice, Xavier Carcelle

Power System State Estimation, Mukhtar Ahmad

A Systems Approach to Lithium-Ion Battery Management, Phil Weicker

Signal Processing for RF Circuit Impairment Mitigation in Wireless Communications, Huang, Zhu, Leung

Synergies for Sustainable Energy, Elvin Yüzügüllü

Design and Analysis of Large Lithium-Ion Battery Systems

Shriram Santhanagopalan
Kandler Smith
Jeremy Neubauer
Gi-Heon Kim
Matthew Keyser
Ahmad Pesaran

ARTECH HOUSE
BOSTON | LONDON
artechhouse.com

Library of Congress Cataloging-in-Publication Data
A catalog record for this book is available from the U.S. Library of Congress.

British Library Cataloguing in Publication Data
A catalog record for this book is available from the British Library.

ISBN-13: 978-1-60807-713-7

Cover design by John Gomes

10 9 8 7 6 5 4 3 2 1

Contents

Preface ix

CHAPTER 1

Types of Batteries 1

1.1 Lead Acid Batteries 3
1.2 Nickel-Based Batteries 4
1.3 Sodium Beta Batteries 6
1.3.1 Sodium Sulfur Batteries 6
1.3.2 Metal Chloride Batteries 7
1.3.3 Challenges and Future Work 7
1.4 Flow Batteries 8
1.4.1 Redox Flow Batteries 8
1.4.2 Hybrid-Flow Batteries 9
1.4.3 Challenges and Future Work 9
1.5 Li-Ion Batteries 10
1.5.1 Lithium-Ion Cathodes 10
1.5.2 Lithium-Ion Anodes 12
1.5.3 Li-Ion Electrolytes 13
1.5.4 Li-Ion Challenges and Future Work 13
1.6 Lithium-Sulfur Batteries 15
1.6.1 Lithium-Sulfur Cathodes 15
1.6.2 Lithium-Sulfur Anode 16
1.6.3 Challenges and Future Work 16
1.7 Metal-Air Batteries 17
1.7.1 Zinc-Air Batteries 17
1.7.2 Lithium-Air Batteries 18
1.7.3 Challenges and Future Work 19
1.8 Emerging Chemistries 19
1.8.1 Sodium-Ion Batteries 19
1.8.2 Liquid Metal 20

CHAPTER 2

Electrical Performance 21

2.1 Thermodynamics Inside a Battery 21
2.2 Assembling a Li-Ion Cell 24

2.3 Voltage Dynamics during Charge/Discharge 28
2.4 Circuit Diagram for a Cell 29
2.5 Electrochemical Models for Cell Design 31
2.5.1 Charge Transport within the Electrode by Electrons 33
2.5.2 Charge Transport in the Electrolyte by Ions 34
2.5.3 Charge Transfer between the Electrodes and the Electrolyte 36
2.5.4 Distribution of Ions 37
2.6 Electrical Characterization of Li-Ion Batteries 39
2.6.1 Capacity Measurement 40
2.6.2 Power Measurement 40
2.6.3 Component Characterization 41
References 46

CHAPTER 3
Thermal Behavior 47
3.1 Heat Generation in a Battery 47
3.1.1 Heat Generation from Joule Heating 47
3.1.2 Heat Generation from Electrode Reactions 49
3.1.3 Entropic Heat Generation 49
3.2 Experimental Measurement of Thermal Parameters 51
3.2.1 Isothermal Battery Calorimeters 51
3.2.2 Basic IBC Operation 51
3.2.3 Typical Applications for an IBC 54
3.3 Differential Scanning Calorimeters 58
3.3.1 Differential Scanning Calorimeters and Batteries 60
3.4 Infrared Imaging 62
3.4.1 Origin of Thermal Energy 62
3.4.2 Calibration and Error 65
3.4.3 Imaging Battery Systems 65
3.5 Desired Attributes of a Thermal Management System 67
3.5.1 Designing a Battery Thermal Management System 68
3.5.2 Optimization 75
3.6 Conclusions 79
References 79

CHAPTER 4
Battery Life 81
4.1 Overview 81
4.1.1 Physics 81
4.1.2 Calendar Life Versus Cycle Life 82
4.1.3 Regions of Performance Fade 83
4.1.4 End of Life 86
4.1.5 Extending Cell Life Prediction to Pack Level 87
4.1.6 Fade Mechanisms in Electrochemical Cells 88
4.1.7 Common Degradation Mechanisms in Li-Ion Cells 89

4.2 Modeling 99
4.2.1 Physics-Based 99
4.2.2 Semiempirical Models 105
4.3 Testing 108
4.3.1 Screening/Benchmarking Tests 110
4.3.2 Design of Experiments 110
4.3.3 RPTs 111
4.3.4 Other Diagnostic Tests 112
References 115

CHAPTER 5
Battery Safety 117
5.1 Safety Concerns in Li-Ion Batteries 117
5.1.1 Electrical Failure 118
5.1.2 Thermal Failure 118
5.1.3 Electrochemical Failure 119
5.1.4 Mechanical Failure 120
5.1.5 Chemical Failure 120
5.2 Modeling Insights on Li-Ion Battery Safety 121
5.2.1 Challenges with Localized Failure 121
5.2.2 Effectiveness of Protective Devices in Multicell Packs 121
5.2.3 Mechanical Considerations 122
5.2.4 Pressure Buildup 124
5.2.5 Designing Protective Circuitry to Combat Short Circuit 126
5.3 Evaluating Battery Safety 128
5.3.1 Measurement of Reaction Heats: Accelerating Rate Calorimeters 128
5.3.2 Thermomechanical Characterization of Passive Components 131
5.3.3 Cell-Level Testing 133
References 137

CHAPTER 6
Applications 139
6.1 Battery Requirements 139
6.1.1 Electrical Requirements 139
6.1.2 Thermal Requirements 141
6.1.3 Mechanical Requirements 141
6.1.4 Safety/Abuse Requirements 142
6.2 Automotive Applications 142
6.2.1 Drive Cycles 142
6.2.2 SLI 143
6.2.3 Start-Stop (Micro) Hybrids 143
6.2.4 Power Assist Hybrids 144
6.2.5 Plug-In Hybrids 145
6.2.6 BEVs 149
6.3 Grid Applications 151
6.3.1 Demand Charge Management and Uninterruptable Power Sources 154

6.3.2 Area Regulation and Transportable Asset Upgrade Deferral 157
6.3.3 Community Energy Storage 161
6.3.4 Other Grid-Connected Applications 162
References 162

CHAPTER 7

System Design 165

7.1 Electrical Design 166
7.1.1 Power/Energy Ratio 166
7.1.2 Series/Parallel Topology 167
7.1.3 Balance of Plant 169
7.2 Thermal Design 171
7.3 Mechanical Design 173
7.4 Electronics and Controls 174
7.4.1 Roles of Battery Management 174
7.4.2 BMS Hardware 174
7.4.3 Cell Balancing 176
7.4.4 State Estimation Algorithms 177
7.4.5 Battery Reference Model 179
7.4.6 State Estimator 180
7.4.7 Current/Power Limits Calculation 182
7.5 Design Process 183
7.6 Design Standards 184
7.7 Case Study 1: Automotive Battery Design 185
7.7.1 Life Predictive Model 186
7.7.2 Fitting Life Parameters to Cell Aging Data 188
7.7.3 Prediction of Battery Temperature in Vehicle 190
7.7.4 Control Trade-Offs Versus Lifetime 194
7.8 Case Study 2: Behind-the-Meter Peak-Shaving of a Large Utility Customer 196
7.8.1 End User Needs and Constraints 197
7.8.2 End User Load Profile and Rate Structure 197
7.8.3 Baseline 207
7.8.4 Increased Cooling 208
7.8.5 Reduced Target *SOC* 208
7.8.6 Decreased Maximum *SOC* 211
7.9 System Specification 212
References 213

CHAPTER 8

Conclusion 217

About the Authors 219

Index 221

Preface

Battery development is hampered by the lack of cross-disciplinary communication between electrochemists, material scientists, and the mechanical and electrical engineers responsible for scaling up basic electrochemical cells to large systems. Until recently, lithium-ion batteries have been used in small-size applications, either as individual cells in cell phones and laptops or as small modules in power tools and other consumer electronics applications. Over the last decade, the battery market has significantly expanded to include applications that demand thousands of times the energy content typical of the traditional small-size batteries: the electrical grid and automobiles are but two applications that will drive the proliferation of electrical energy storage into the future.

This book is intended to serve as an introductory text that provides a solid understanding of the multiple facets of battery engineering. For systems-oriented engineers, this text demystifies electrochemistry; for the electrochemist, this text introduces topics that must be addressed for scale-up, including developing model-based design and control platforms; for the analyst, this text provides a jump-start on the basics of lithium-ion batteries and the challenges in deployment.

Chapters cover an introduction to the different types of batteries and where lithium-ion batteries fit in; fundamentals of electrochemistry, including sections on what criteria one must use and associated limitations in selecting the materials to build a lithium ion battery; design of thermal management systems for lithium ion batteries, from heat generation in a single-cell to design of cooling channels in a multicell module; a detailed account of the state of the art in prognosis methods for battery life; safety challenges faced when scaling up well-established chemistries to larger formats; and an overview of the wide range of applications with specific engineering examples on the technoeconomic evaluation of large-format lithium-ion batteries used in automotive applications and system design for grid-storage applications.

The book targets readers from diverse backgrounds who would like to get on a fast track to understanding battery engineering within a few weeks and go beyond treating the battery as a black box in their line of work. Each chapter takes a hands-on approach based on the years of experience the authors have accumulated in this discipline, and provides relevant context, requisite theoretical background, experimental tools, and real-world examples. New methods are introduced for battery lifetime prediction that may be practically implemented in design optimization, warranty assessment, systems control, and other analyses. The chapter on battery safety puts specific emphasis on large-format designs. The chapters on thermal management and applications provide detailed examples so readers can familiarize themselves with the design calculations. Detailed case studies on systems-level analyses of vehicle batteries and grid storage have been included to enable readers to perform a technoeconomic analysis for a variety of real-world applications. We hope that readers will enjoy the broad introduction to the topic, the nuances in testing, and the practical case studies as much as we did when we originally worked on each of these topics.

For our related work at the National Renewable Energy Laboratory (NREL), we would like to acknowledge support from the Vehicle Technologies Office at the Department of Energy's Office of Energy Efficiency and Renewable Energy (EERE). Special thanks are due to our Program Managers David Howell, Tien Duong, and Brian Cunningham. Constant support from NREL Management (especially Barb Goodman) and incessant encouragement from our ex-Lab Program Manager (Terry Penney) were key enablers in pursuing such a diverse range of projects over the last three decades. Members of the USABC Technical Advisory Committee and the numerous energy storage developers who worked with us providing samples, sharing their technical insights and the needs from the industry deserve our appreciation. We are deeply indebted to our colleagues at NREL (current and previous): Marissa Rusnek, Jeff Gonder, Tony Markel, Aaron Brooker, Mike Simpson, Rob Farrington, Mark Mihalic, John Ireland, Dirk Long, Jon Cosgrove, Myungsoo Jun, Aron Saxon, Ying Shi, Eric Wood, Chuanbo Yang—the list is long and our gratitude endless. Constant support from the team at Artech helped us stay on a tight leash, for the most part. Support from our family, teachers, mentors, and friends throughout the making of this book has been a true gift—we cannot thank you enough.

Last, but not least, we would like to acknowledge the unrelenting support from our families: Matt dedicates this book to his loving wife, Liz, whose support and guidance always brings out the best in him. Gi-Heon acknowledges support from Sivan, Soyan, and Soyi. Ahmad dedicates this book to his wife, Nahid. Jeremy dedicates this book to his wonderful wife, Anne, whose continued support and endurance of poor battery humor has made this work possible. Kandler dedicates the book to his wife, Jessica, and his parents and children for their loving support and equanimity for a battery engineer. Shriram would like to thank Priya for helping him stay charged throughout this effort, and his parents for their limitless encouragement. Without support from all of you, this work would not have been possible!

CHAPTER 1

Types of Batteries

Batteries are devices that store energy chemically and deliver (and accept) energy electrically. There are two principal classifications of batteries: (1) primary batteries, which can only deliver the stored energy once prior to disposal, and (2) secondary batteries, which can be cycled (discharged and recharged) many times before the end of a device's useful life. This book only discusses the latter.

In a secondary battery, the conversion of chemical energy to electrical energy and vice versa occurs via accepting electrons (reduction) or donating them (oxidation). These reactions are generally referred to as redox reactions and happen both at the negative and positive electrodes. The terms anode is used in the battery community to refer to the negative electrode; and cathode to refer to the positive electrode, and the voltage of the cell is defined as the difference between those of the cathode and anode.[1] Note that these electrodes serve as both the site of energy conversion and of energy storage. An electrolyte is included to support the transfer of ions between electrodes while preventing the flow of electrons between electrodes, which are instead routed external to the cell via current collectors, enabling them to perform work. Accordingly, the electrodes must be both good ion and electron conductors. Often a porous, electronically insulating separator is placed between the electrodes to maintain their physical separation as well. Figure 1.1 illustrates this typical construction along with the basic operational mechanisms of a battery. In many sections of this book lithium-ion (Li-ion) batteries are highlighted as the example for secondary batteries, although there are many different types of batteries, incorporating different cell components (such as electrodes, charge carriers, and electrolytes), and construction that function on the same operating basis.
These principles are similar to that of fuel cells, which rely on redox reactions. Fuel cells, however, are most typically distinguished by the storage of energy in an external fuel and oxidant. These reactants are continually pumped past the electrodes, which are relegated solely to the task of energy conversion and charge transfer.

Electrochemical capacitors (often referred to as super- or ultracapacitors) are also similar to batteries. Electrochemical capacitors are classified into two types based on their charge storage mechanism: electric double-layer capacitors use carbon electrodes to store electrostatic charge, whereas pseudocapacitors employ a material such as a metal oxide that undergoes Faradaic redox reactions to enable

1. This terminology is opposite to that used widely in the classical literature on electrochemical synthesis or electroplating, for example. The reader should exercise caution in interpreting these terms in light of this confusion.

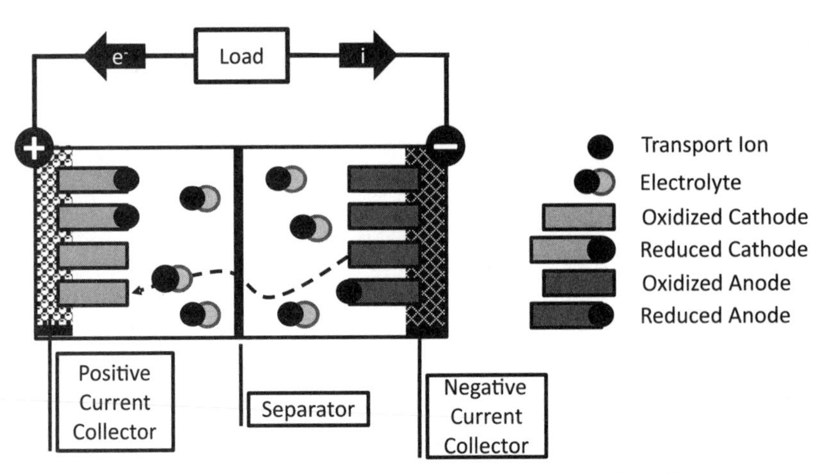

Figure 1.1 Generalized redox battery operation during discharge. When the charge-carrying ion is positively charged, it (or the electrolyte) flows from anode to cathode (right to left) across the separator.

the charge storage. The latter typically has significantly higher energy storage capabilities than the former, although both are much lower in energy and higher in power capability than batteries.

Generally speaking, batteries are best suited to applications that require the supply of rated power for 10 minutes to 10 hours. For shorter durations, battery performance becomes limited by the ability to deliver energy at high rate, and other solutions such as capacitors or flywheels become superior. For longer durations, often associated with very large amounts of stored energy (>1 MWh), the cost and size of the necessary battery installation becomes an issue. However, optimizing a battery for any given application requires looking at many variables, and battery performance can vary dramatically across different chemistries, so these are only rough guidelines.

Batteries find homes in many different applications. Lead acid batteries are ubiquitous in vehicular starting, lighting, and ignition (SLI) roles, as well as commercial and industrial uninterruptible power supply (UPS) applications. As recently as 2010, nickel metal hydride batteries dominated the hybrid electrical vehicle market, providing short-duration engine-off driving and the ability to recapture energy during braking in the Toyota Prius, Ford Escape Hybrid, and others. Lithium-ion batteries have become common place in consumer electronics, as they have enabled smaller and more powerful camcorders, laptop computers, cell phones, portable power tools, and more. At present these batteries are a popular topic as they try to find their way into roles powering fully electric vehicles and supporting our electric grid.

Figure 1.2 shows the relative performance in six key areas of four kinds of batteries selected for their technological maturity and current and near-term relevance to major energy storage applications. It should be noted that terms like "metal-hydride" or "lithium-ion" encompass many different chemistries with varying performances and levels of maturity within each category; the data plotted here represents merely an approximate average of capabilities. Similarly, each individual performance category may mask significant variations within each chemistry. For

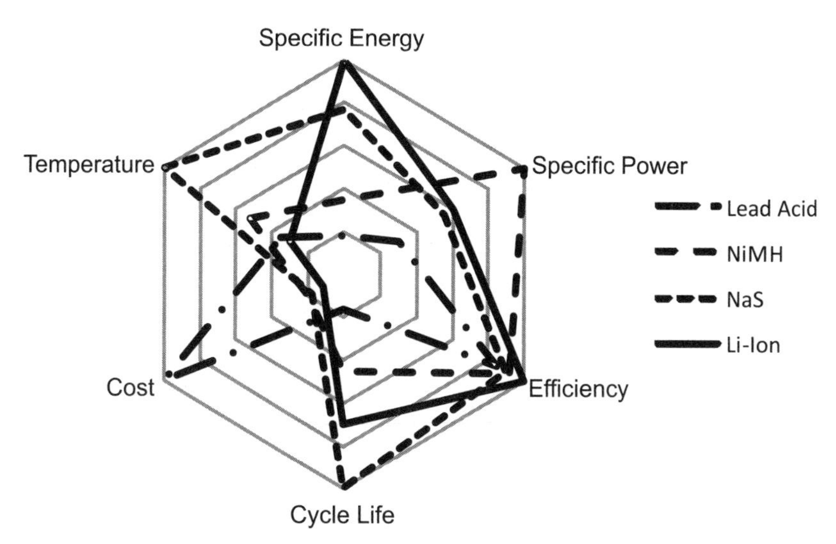

Figure 1.2 Relative performances of select battery types.

example, the simple category "cycle life" does not by itself indicate the sensitivity of cycle life to the depth of discharge, temperature, rate, cycle frequency, and so forth, which for different conditions may alter the relative performance of one chemistry versus another. As such, Figure 1.2 was created in a best effort attempt to capture the general essence of relative battery performance in each of these categories. The plotted values have arbitrary but relative units and are scaled to the best performer in each category.

In the future, not only is it expected that these values will improve (especially so for Li-ion batteries), but other more promising chemistries may also enter the market. For example, batteries comprised of lithium metal anodes and either sulfur (S) or air-based cathodes are being explored for their potential to offer considerably increased energy content per unit weight (specific energy), which is of particular interest to mobile applications. Additionally, flow batteries may provide improved operational flexibility and long cycle life at minimal cost (which is of particular interest to stationary applications). Several battery technologies have been under investigation for decades, and yet still require significant improvements before they become broadly commercially viable.

1.1 Lead Acid Batteries

The lead acid battery consists of a lead dioxide cathode, a lead anode, and an aqueous sulfuric acid electrolyte. During discharge, each electrode is converted to lead sulfate, consuming sulfuric acid from the electrolyte. On charge, the lead sulfate is converted back to sulfuric acid, leaving a layer of lead dioxide on the cathode and metallic lead on the anode.

During the charging process, water in the electrolyte is broken down to hydrogen and oxygen. In a vented design, these gases escape into the atmosphere, requiring the periodic addition of water to the system. In sealed designs, the loss of these gases is prevented and their conversion back to water is possible. This reduces maintenance requirements; however, when the battery is overcharged or charged at too high of a rate, the rate of gas generation can surpass that of recombination, introducing a risk of explosion.

Both valve-regulated gel and absorbed glass mat (AGM) designs thus operate under slight constant pressure, but take different approaches to mitigate the risk of pressure buildup. In the former, silica is added to the electrolyte to cause it to gel, while in the latter the electrolyte is suspended in a fiberglass mat. Both designs reduce or eliminate the risk of electrolyte leakage and offer improved safety by valve regulation of internal pressure, but at a slightly higher cost.

Lead acid is generally the lowest cost secondary battery chemistry on a dollar-per-kilowatt-hour basis, but also exhibits low specific energy and relatively poor cycle life. In practice lead acid batteries only achieve specific energy of ~35 Wh/kg; comparatively, the highest energy Li-ion cells may achieve more than seven times that (~250 Wh/kg), approaching the theoretical maximum value, for the specific energy of the lead acid chemical reaction (252 Wh/kg).

The short cycle life of lead acid batteries, on the order of a few hundred cycles, is primarily attributed to corrosion and active material shedding of the electrode plates. Operation across a wide voltage window, also referred to as high depth of discharge (DOD) operation, intensifies both these issues, in part due to difference in densities of the reactants and products, throughout the cycle, resulting in shedding of the active material from the electrodes. During high rate operations to partially charge or discharge the battery, lead sulfate accumulation on the anode (sulfation) can be the primary cause of degradation. These processes are also sensitive to high temperature, where the rule of thumb is a halving of battery life for every 8°C increase in temperature from the ambient.

The primary disadvantages of lead acid battery-chemistries are their poor energy density and short cycle life, as noted above. Marginal gains to specific energy can be achieved by improving the active material and grid design of the electrodes, but will always be limited by the chemistry's relatively low theoretical boundaries. The replacement of the traditional lead anode with a carbon anode similar to that of an electrochemical capacitor may significantly increase cycle life, but to do so may require that the amount of lead in the cathode be increased several fold, thus raising the cost and mass of the cell. Another approach to improve cycle life is the so-called lead acid flow battery, in which lead is dissolved in an aqueous methane sulfonic acid electrolyte. This system differentiates itself from traditional flow batteries discussed later via the use of just one electrolyte and thus avoids the need for troublesome electrolyte separators. If long deep-discharge cycle life can be proven and costs kept low, this technology may be promising for grid-based bulk electricity storage applications. But given its exceptionally low theoretical specific energy, it is unlikely that lead acid batteries will ever compete with high-energy chemistries such as Li-ion for large mobile applications, which are volume sensitive.

1.2 Nickel-Based Batteries

All nickel-based batteries are related by their use of a common nickel oxyhydroxide cathode and the potassium hydroxide (KOH) electrolyte. The differentiation of nickel chemistries is rooted in the anode. The original anode, cadmium, results in a battery with significantly improved energy density and cycle life over lead acid batteries. Nickel-cadmium batteries have been challenged by the high cost and toxicity of cadmium (nickel-cadmium batteries have largely been banned in the European Union), leading to the rise of nickel metal hydride, and later, Li-ion chemistries.

A nickel-hydrogen cell is essentially a hybrid battery fuel cell, as gaseous, pressurized hydrogen is used as the anode active material. Designed and employed exclusively for aerospace applications, these cells can provide exceptionally long life along with other application-specific benefits, but their extremely high cost essentially prevents their use in most other applications.

Nickel metal hydride (NiMH) systems were originally designed as a method of storing hydrogen within nickel hydrogen cells, although today these find use in hybrid electric vehicles (HEVs) and as a replacement for nickel-cadmium batteries. In this cell, a complex metal alloy is used to store hydrogen on the anode side, consisting of a multitude of alloying agents to tune cell performance. Despite its poor low-temperature performance (below 0°C) and high self-discharge rate (up to 30% per month), this chemistry's advantageous specific energy, cycle life, and high rate capability have driven the trend to replace nickel-cadmium batteries.

The use of metallic zinc as an anode against the nickel oxyhydroxide cathode increases cell voltage, capacity, and improves high rate performance. Furthermore, the relatively high abundance of zinc keeps costs lower than both nickel-cadmium and NiMH batteries. However, the nickel-zinc battery suffers the same drawbacks as other metallic anode systems. In such a system, zinc is plated at the anode during charge. This process can be highly nonuniform, leading to possible formation of metallic dendrites and swelling of the electrodes. Large volume-changes can cause mechanical stress in other cell components, leading to degraded performance. Importantly, dendrites can create internal short circuits or the dendrite can separate from the anode, resulting in the loss of active material and thus irreversible capacity loss.

Iron hydroxide ($Fe(OH)_2$) has also been employed as an anode for nickel-based batteries but is in extremely limited use due to issues with pronounced hydrogen evolution during both charge and discharge.

Progress on nickel-cadmium, nickel-hydrogen, and nickel-iron batteries are severely limited by their toxicity, cost, and obscurity, respectively. NiMH, on the other hand, is commonly employed in HEVs such as the Toyota Prius and Ford Escape. Topics including improving cold temperature performance, reducing self-discharge rates, and extending cycle life have been pursued. Cost is also an issue but is difficult to address through research and development due to the fact that economies of scale in production have already been achieved, and at present 35% of the cost of NiMH batteries is due to the cost of nickel alone.

Nickel-zinc batteries offer many improvements relative to NiMH, but they are currently limited by poor cycle life. Overcoming this obstacle requires mitigation of the zinc dissolution and plating problems. Also the relative abundance of zinc and nickel are similar, so if more zinc is employed for large-scale applications it will likely become more expensive.

1.3 Sodium Beta Batteries

Sodium beta batteries are rechargeable molten salt batteries that use molten sodium for the negative electrode and sodium ions for charge transport. There are two principal types of sodium beta batteries, differentiated by their cathodes: the first employs a liquid sulfur cathode, while the second utilizes a solid metal-chloride cathode. Both include a beta-alumina solid electrolyte (BASE) separating the cathode and anode. This ceramic material offers ionic conductivity similar to that of typical aqueous electrolytes, but only at elevated temperature—thus, sodium beta batteries must typically operate in excess of 300°C.

1.3.1 Sodium Sulfur Batteries

Sodium sulfur batteries operate via the following reactions:

Anode: $Na \leftrightarrow Na^{+} + e^{-}$

Cathode: $xS + 2e^{-} \leftrightarrow Sx^{2-}$

Overall cell: $xS + 2Na^{+} + 2e^{-} \leftrightarrow Na_2Sx$

As noted previously, both the anode and cathode are in a liquid state when the battery is at its operating temperature. The solid BASE electrolyte allows the efficient transfer of sodium ions at high temperature; its impermeability to the liquid electrodes and minimal electrical conductivity supports minimal self-discharge and a near-perfect coulombic efficiency. An illustration of a typical sodium sulfur cell is shown in Figure 1.3.

Sodium sulfur batteries were first developed in the 1960s as rechargeable traction batteries for automotive applications, and they were attractive for their high specific energy exceeding 100 Wh/kg. Although they were never used in automobiles, they are currently employed in limited number for utility load-leveling appli-

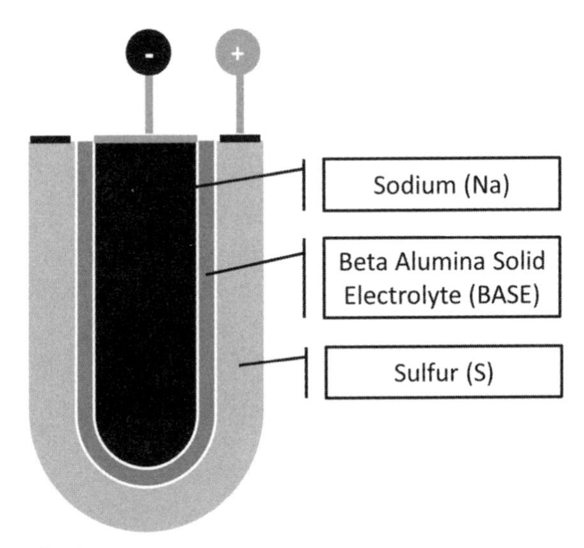

Figure 1.3 Sodium sulfur battery.

cations. The chemistry's long cycle life (up to 5,000 deep discharge cycles) and the applications' relatively low power to energy ratio requirements encourage such use.

Issues with this chemistry generally revolve around the high temperature requirements. Although the ~300°C operating point can be an advantage (given that reasonable changes in ambient temperature are unlikely to impact performance), it has a negative effect on longevity. First, the high temperature exacerbates the corrosive nature of the electrodes. Second, the variances in coefficients of thermal expansion can lead to mechanical stresses that can damage seals and other cell components, including the electrolyte separator, during freeze-thaw cycles. The fragile nature of the electrolyte is also a considerable concern, given that, were a breach to occur and the two liquid cathodes to mix, fire and explosion are possible. Thus, to minimize such risks, freeze-thaw cycles must be avoided, requiring either that the battery be insulated and operated in such a manner as to passively maintain cell temperature above its freezing point or continually be heated from an external source to ensure that this requirement is met.

1.3.2 Metal Chloride Batteries

Metal chloride variants of sodium sulfur batteries most commonly replace the liquid sulfur cathode with a solid nickel chloride cathode (typically called the ZEBRA cell). Also, a secondary sodium chloroaluminate electrolyte is included to provide ionic conductivity between the solid cathode and BASE electrolyte. The reactions that take place in such a cell are provided below:

Anode: $Na \leftrightarrow Na^+ + e^-$

Cathode: $NiCl_2 + 2Na^+ + 2e^- \leftrightarrow Ni + 2NaCl$

Overall cell: $NiCl_2 + 2Na \leftrightarrow 2NaCl + Ni$

Nickel chloride cathodes offer several advantages over sodium beta batteries: they operate at higher voltage, have an increased operational temperature range (due in part to the lower melting point of the secondary electrolyte), the cathode is slightly less corrosive, and cell construction is safer (handling of metallic sodium can be avoided). However, they are likely to offer a slightly reduced energy density.

Such sodium nickel chloride batteries have seen use in several electric vehicles, primarily at the demonstration level. However, their limited power to energy ratios and heating requirements, particularly when the vehicle is parked, are an impediment to broad-scale deployment in the automotive market.

1.3.3 Challenges and Future Work

The high operating temperature of sodium beta batteries is often touted as a benefit, because when ancillary systems are in place to maintain the cells near 300°C, sensitivity to most environmental temperature exposure is nil. However, the principal shortcomings of this technology are also associated with the same requirement of high operating temperatures. This not only presents a problem with respect to safety, but also with efficiency (energy lost to cell heaters), convenience (start-up time), longevity (freeze-thaw cycles can also accelerate degradation), and reliability (the coexistence of highly corrosive sodium metal and high temperatures can lead to electrode containment issues and other failure modes resulting from high resistivity of the corrosion products). Longevity and reliability still have room for marginal improvement via improved cell configurations and designs. There is little ability to improve safety, efficiency, and convenience with modest technological advances—high-temperature operation is largely a matter of fact for these chemistries. New, low-temperature sodium-based chemistries are being researched, however, based on new cathodes and/or sodium ion conductors (e.g., replacing BASE electrolytes with NASICON™).

With respect to the most promising utility-related applications, cost is also a significant issue. Efforts are currently underway to develop stacked planar cell designs that are expected to cut cell costs in half. This departure from the traditional tubular design has the ability to increase specific energy and power (the latter being a limiting factor for the use of these batteries in many applications), improve packing efficiency and modularity, and also presents the opportunity to address long-term corrosion issues. Such efforts will face sealing and material selection challenges.

Two other issues have impeded the use of sodium beta chemistries for mobile applications: low power and fragility. The former issue is largely driven by the cell configuration and low ionic conductivity of the solid electrolyte. The latter is also a property of the brittle ceramic beta alumina electrolyte, compounded by a common mismatch in coefficients of thermal expansion between this component and others within the battery. Thus, use of sodium sulfur cells in an application where exposure to vibration or shock is common (e.g., transportation applications), or under a business strategy where the battery is required to be occasionally relocated (as has been discussed for some utility transmission upgrade deferral applications) is typically impractical.

1.4 Flow Batteries

In a flow battery, energy is stored primarily in active materials dissolved into an electrolyte, which is stored externally and passed through the electrodes when the battery is charged or discharged. The electrodes are separated by an ion exchange membrane that also segregates the cathode and anode side electrolytes (referred to as catholyte and anolyte, respectively). There are two main classifications of flow batteries: redox and hybrid.

Note that per our original definition of batteries and fuel cells, flow batteries may often be classified as a regenerative fuel cell. However, due to the reversibility of the reaction and popular nomenclature, they are discussed here alongside secondary batteries.

1.4.1 Redox Flow Batteries

In a redox flow battery, the active materials are at all times dissolved in the electrolyte. There are several different chemical compositions of redox flow batteries, iron chromium being the first flow battery developed (by the National Aeronautics and Space Administration (NASA)) and vanadium receiving the most attention.

A schematic of the most popularly used vanadium redox battery is shown in Figure 1.4.

The redox configuration offers several benefits to many of the nonflow and hybrid-flow batteries. First, the amount of energy storage available is limited only by the size of the tanks and the amount of electrolyte available. Accordingly, it is also possible to decouple the system's energy and power capabilities, the latter being determined by the number and size of the electrodes. Additional benefits include the absence of a need for cell balancing (allowing a relatively simple construction of higher-voltage batteries) and the ability to mechanically recharge the system by replacing the electrolyte. The disadvantages generally lie with the complexity of the pumping, storage, and control systems, as well as the low specific energy and volumetric energy density (typically less than that of lead acid batteries).

1.4.2 Hybrid-Flow Batteries

In a hybrid flow battery, at least one of the active species is plated onto an electrode; thus, power and energy levels are coupled due to the ability of the electrode(s) to accommodate a finite amount of active material. Several chemistries have been investigated, zinc bromine being the most popular. This configuration benefits from the low cost of the electrolyte and a slightly improved energy density, but issues can arise with dendrite formation during zinc plating.

1.4.3 Challenges and Future Work

The low specific energy of flow battery systems is a major impediment to most markets, restricting this technology primarily to select stationary (mostly, electric utility) applications. Here the low technological maturity level of the product will be a barrier—few flow batteries have been demonstrated to operate at the megawatt scale often desired by utilities, but the system's flexibility and anticipated longevity and low cost are anticipated to be major benefits. To be successful in this market,

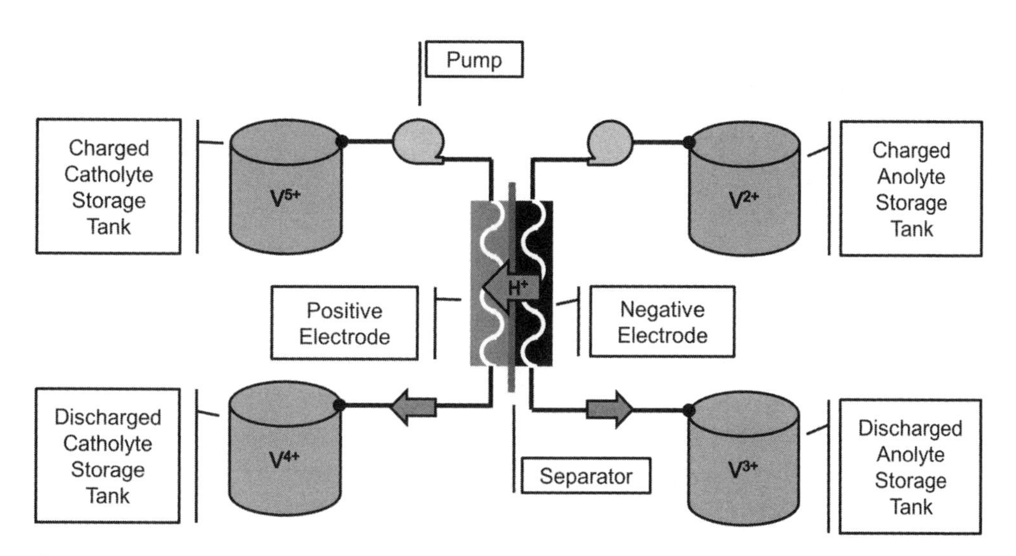

Figure 1.4 Vanadium redox flow battery operating in the discharge mode.

efficiency will likely have to improve beyond present levels (~70%), reliability must increase to attain the acclaimed longevity, and maintenance requirements have to be minimized. Reliability and longevity are complicated by the use of toxic and corrosive electrolytes, which pose significant challenges at the materials level for the hydraulic subsystems and the ion exchange membrane.

There are several pathways to improving flow battery performance and meeting these needs. For one, improved systems design and manufacturing practices can go a long way toward improving reliability and making modest efficiency gains while keeping the costs low. Crucial to this will be advances in the manufacturing of the ion transfer membrane—a major cost contributor and a likely failure point due to frequent leaks. New redox couples that offer higher efficiency, improved specific energy, and/or utilize more cost-effective or less toxic materials are also the subject of recent investigations. These include hydrogen-halogen, hydrogen-bromine, iron-chromium, and others.

1.5 Li-Ion Batteries

Li-ion batteries are currently the batteries of choice for portable consumer electronic devices including cell phones, tablets, laptop computers, digital cameras, power tools, and toys, due primarily to their durability, high specific energy (100 to 200 Wh/kg), and ability to operate at reasonably high power. Recently, Li-ion batteries have begun to enter the automotive market as power packs for hybrid and full electric vehicles. In addition to the high specific energy, the automotive market also benefits from this chemistry's high power, efficiency, and long cycle life capability. As the automotive market drives the expansion of Li-ion production, these batteries may also enter stationary service as well, facilitating the implementation of renewable energy technologies such as solar and wind.

Li-ion batteries operate by shuttling Li^+ ions to the anode host structure on charge, then by transporting the same ions across the porous separator via the electrolyte to the cathode host structure on discharge. The need to maintain charge balance at each electrode in turn drives a current through the external circuit to perform work, as seen in Figure 1.5. Like many other battery chemistries, lithium ion batteries require electrically and ionically conductive anodes and cathodes, and electrically insulative but ionically conductive electrolytes and separators. There are multiple combinations of cathodes, anodes, and electrolytes possible under the category of Li-ion cells, and several of these are aggressively being researched and developed to improve this battery technology.

1.5.1 Lithium-Ion Cathodes

Three classes of cathodes prevail today, including layered transition-metal oxides, spinels, and olivines. Each structure has various advantages and disadvantages. Layered oxides typically offer the highest capacity but can suffer from cost and safety concerns. Spinels can achieve very high power density with fewer cost and safety penalties but can suffer from poor electronic conductivity and structural sta-

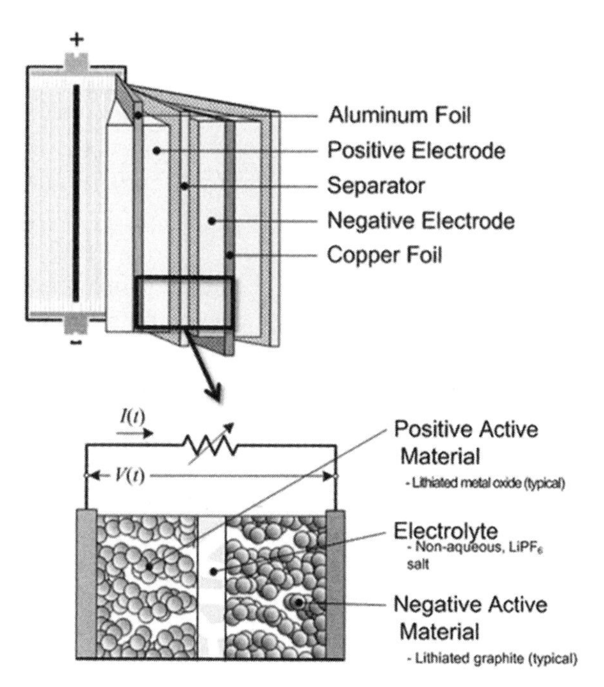

Figure 1.5 A Li-ion battery with a graphite anode and layered lithium metal oxide cathode (e.g., $LiCoO_2$) as the battery is discharged.

bility issues. Olivines, on the other hand, can offer exceptional safety and long life at low cost but sacrifice both capacity and voltage.

Lithium cobalt oxide ($LiCoO_2$) was the first layered oxide cathode material developed and introduced into the commercial market by Sony in the early 1990s and is the most common cathode chemistry in today's Li-ion-powered consumer electronics devices. Originally paired with a hard carbon anode in a small cylindrical consumer cell format, it offered a specific energy of 130 Wh/kg and good cycle life. Since then this cathode material has been paired with a graphite anode, where the specific energy can exceed 200 Wh/kg at the cell level when optimized for energy content. However, the cost and toxicity of cobalt have driven development of other layered oxide cathodes with reduced cobalt content. Most prominently nickel-cobalt-aluminum (NCA) and nickel-manganese-cobalt (NMC) chemistries have since been developed, not only reducing cost but also improving performance based on energy content, power delivery, and longevity.

The largest drawbacks of the layered oxide cathodes are cost and safety. Cost is driven up by cobalt and nickel content. Safety concerns are driven by the evolution of oxygen under abusive conditions (typically high temperature, high voltage, and energy content, where reaction with the electrolyte or dissolution may occur). This process is not only extremely exothermic, but the typical electrolyte is flammable when combined with oxygen, leading to the risk of fire and explosion. Although the newer NCA and NMC cathodes can offer slightly improved safety, they do not sufficiently mitigate the risk.

Spinels (typically lithium manganese oxide) offer further improvements in safety, their structures liberating lower amounts of oxygen and heat when operated under extreme conditions. In addition, the lack of cobalt and nickel, replaced by highly available manganese, significantly reduces cost, while the open structure for Li-ion intercalation allows for higher power and efficiency. The principal challenges are lower specific capacity relative to layered oxides and a tendency for manganese dissolution at high temperature, limiting longevity.

Olivines (typically lithium iron phosphate) offer even further improvements in safety. In some cases the iron phosphate cathode can be even less reactive than a graphite anode. The stability of this chemistry is further demonstrated by its excellent cycle life. Like manganese, iron is cheap and plentiful in comparison to cobalt and nickel. However, the lithium holding capacity (which is measured as coulombs of lithium per gram of the host material, commonly reported as milliampere-hours per gram and referred to as the specific capacity of the host material) is reduced even below that of manganese spinel, and cell voltage is lower as well, resulting in a significant decrease in specific energy relative to the layered oxides. In addition, the rate capability is quite poor without leveraging nanoscale formations in the cathode to maximize the ratio of surface area to volume.

1.5.2 Lithium-Ion Anodes

Development has focused heavily on new cathode technologies over the past several years due to the fact that the common carbon-based anodes (e.g., graphite and hard carbon) are simultaneously less costly, more stable, and of higher specific capacity than their commercially available cathode counterparts. For example, the average capacity of commercial graphite is ~330 mAh/g, whereas the usable capacity of lithium cobalt oxide is only ~140 mAh/g. Thus the cathode is most often the capacity-limiting electrode, marginalizing cell level improvements resulting from increased anode capacity. However, improvements in anode performance still do yield gains at the cell level, and there exists a significant issue with current graphite anode technology—lithium plating.

When the potential of the anode becomes excessively low, most likely due to combinations of low temperature and high charge rate, Li-ions can plate out as metallic lithium on the surface of the anode rather than intercalating within the anode. This leads to dendrite formation, which not only results in irreversible capacity loss (via the loss of active lithium), but also poses a threat of the dendrites creating a short circuit. This is a major safety concern, as heat generated from the internal short circuits may lead to the ignition of the flammable electrolyte and a possible explosion. These issues have long prevented the use of lithium metal as a potential contender as an anode; however, extensive efforts to stablize lithium metal anodes are currently underway.

Lithium titanate spinel structures have been offered as an alternative to graphite, providing high stability by operating at a much higher voltage versus lithium than do carbon anodes, greatly reducing the chance of lithium plating and all but eliminating electrolyte reduction and the formation of the solid electrolyte interface (SEI) layer. Though the use of titanates improves safety, longevity, and efficiency, it also results in a significantly lower cell voltage. With a specific capacity that is about half that of graphite, a lithium titanate cell's specific energy becomes significantly compromised.

1.5.3 Li-Ion Electrolytes

The electrolyte commonly comprises a mixture of organic solvents such as an ethylene carbonate, dimethyl carbonate, and propylene carbonate, which contain a dissolved lithium salt such as lithium hexafluorophosphate ($LiPF_6$). Regardless of the electrode couple, the electrolyte is often exposed to working voltages beyond its stability limits. For carbon anodes operating at a potential close to the plating of lithium, this causes the reduction of the electrolyte at the anode, building a protective coating known as the SEI layer. The SEI layer is typically formed during the first few charge cycles. Though this process leads to some degree of irreversible capacity loss through the consumption of charge-carrying Li-ions as well as negative effects on electrical and ionic transport properties, it is critical to the long-term stability of the cell as the SEI layer prevents further electrolyte reduction. Accordingly, tuning the electrolyte formulation to optimize the SEI layer is an actively pursued means to improve the long-term performance of many electrode couples (notably, however, this does not apply to lithium titanate anodes, which do not form an SEI layer due to their higher voltage versus lithium as mentioned previously).

The stability of the electrolyte is a significant safety issue. When cathodes are operated outside the stable electrolyte voltage window, their reaction with the electrolyte leads to the evolution of oxygen and heat and possibly thermal runaway. This risk is typically exacerbated at higher temperatures. Further, most Li-ion electrolytes are flammable when combined with oxygen and an ignition source. There are several options under development to avert such scenarios, including nonflammable electrolyte additives, inorganic electrolytes, and solid or polymer electrolytes. Although the latter have promise for significant improvements to safety by eliminating concerns of electrolyte breakdown, leaking, and venting, they suffer from significantly low ionic conductivity. For this reason, solid state Li-ion batteries have not penetrated the market in any significant quantity.

The electrolyte also plays an important role in low temperature performance. At low temperatures, typically below 0°C, the ionic transport properties of the electrolyte suffer, which can drastically lower cell performance and efficiency. Electrolyte formulations can be fine-tuned to improve performance at low temperatures, but this may come with penalties at higher temperatures and/or negatively affect long-term degradation and safety.

1.5.4 Li-Ion Challenges and Future Work

Although Li-ion batteries have dominated commercial electronics markets and show promise for vehicular applications, safety remains an extremely important and challenging aspect, especially in the transportation sector. Improvements to cost, specific energy, cold temperature performance, and longevity also require attention, but their importance varies by application. Progress toward each of these goals can be made via development of new cathodes, anodes, and electrolytes, as well as stabilizing additives and coatings. A significant number of research efforts in industry, national laboratories, and academia are presently devoted to these topics.

Safety is the foremost concern of many current and potential Li-ion markets, especially in light of the relatively high number of safety incidents pertaining to laptops, hybrid vehicles, and aircraft that have made the news. Although it is important to keep in mind that the fractions of incidents that have occurred is extremely small in comparison to the quantity of batteries in use in the field, the level of hazard of a single incident (possible fire and explosion)—particularly in the case of large, manned battery installations such as automotive ones—merits this concern. Fortunately, there are multiple avenues from which to address Li-ion battery safety, all offering the capability to make significant improvements, including thermal management and cell management systems at the pack level, cell design and improved manufacturing quality at the cell level, and advanced cathode, anode, and electrolytes at the component level. For example, electrode coatings that can stabilize the electrode-electrolyte interface, benefiting not only safety but also longevity, are currently under investigation. Inorganic electrolyte systems, still in need of further development, hold similar promise.

Cost is often the second largest limitation for Li-ion. In particular for large-format (e.g., automotive) cells, manufacturing at a large scale is often cited as a likely pathway to reduce cost. To this end, $2.4 billion in American Recovery and Reinvestment (ARRA) funds were awarded in late 2009 to create a U.S. manufacturing base capable of supporting the annual production of 500,000 electric vehicles by 2015, targeting a 70% decrease in battery cost. Similar investments have been made by governments across the globe to kick-start large-format Li-ion battery manufacturing. A large portion of such scale-manufacturing-induced cost reductions are attributable to the commoditization of materials, which is reported to make up 60% of current cell costs. The materials alone cost (anode, cathode, and electrolyte) have been shown to make up approximately 10% of pack level costs. Thus, development of anodes and cathodes incorporating lower-cost materials (such as iron rather than cobalt) is another valid route worth pursuing.

Performance, in particular low temperature response, long-term degradation, and specific energy, also show room for improvement, although in many cases it is already superior to competing technologies. It should be noted that long-term degradation and specific energy improvements, achievable via advanced cathodes, anodes, and electrode coatings, also have the potential to reduce the cost of Li-ion on a per-kilowatt-hour basis. Many of these avenues are currently being pursued, including higher voltage and often nanostructured electrodes. However, improvements to specific energy may come at the cost of longevity and/or safety. For example, high-voltage cathodes will certainly improve energy density but necessitate as-yet unidentified electrolytes stable at these high voltages. Alternatively, nanostructured electrodes can simultaneously offer multiple performance improvements, but come with a much higher manufacturing cost.

New material chemistries are also being explored. On the anode side, some new metal oxide anode formulations (such as tin and titanium-based anodes) may offer improved capacity over today's commonplace carbon-based anodes, but generally suffer a significant difference between charge and discharge voltage profiles (resulting in exceedingly poor efficiency). The employment of a silicon-based anode is under extensive study, as it has an extremely high theoretical specific capacity (4200 mAh/g) but is limited by volume expansion on the order of 400% during lithium intercalation. Such extreme volume expansion can lead to particle fracturing and loss of electronic conductivity, leading to high irreversible capacity loss and vastly reduced cycle life. Other possible anode materials including metal sulfides, phosphides, and Li-alloy materials such as tin (Sn) and germanium (Ge) offer similar gains in capacity but also suffer from large volume expansion. Current research efforts are focused on the development of both nanostructured materials to mitigate the effect of volume expansion as well as protective coatings that will minimize the irreversible first-cycle capacity loss as well as provide for better adhesion during volume expansion.

The highest capacity cathode candidates include layered-layered electrodes with Li_2MnO_3 and $LiMO_2$ (M = Ni,Co,Mn) components, layered-spinel electrodes comprised of Li_2MnO_3 and $LiMn_2O_4$, and layered-layered-spinel electrodes comprised of Li2MnO_3, $LiMO_2$, and LiM'^2O^4 (M′ = Ni, Mn) components. These cathode materials deliver a capacity in excess of 200 mAh/g and are currently at the edge of high-capacity cathode technologies. However,because the potential cathode technologies do not come close to the capacity of possible anodes such as silicon (Si), it may be necessary to consider Li-sulfur or Li-air cathodes that are discussed in the next few sections.

1.6 Lithium-Sulfur Batteries

Lithium-sulfur technology currently under development promises extremely high specific energy (theoretical specific energy is ~2500 Wh/kg). Although it is not nearly ready for commercial production, primarily due to cycle life and safety concerns, it has demonstrated its basic performance and energy density potential in niche applications. The basic cell is comprised of a sulfur cathode, generally supported with a porous carbon framework, a liquid electrolyte, and a lithium metal anode. The following reactions occur upon discharge of a Li-S battery as Li+ dissolves in the electrolyte:

$$2Li + S_8 \leftrightarrow Li_2S_8 \text{ (soluble)}$$
$$2Li + Li_2S_8 \leftrightarrow 2Li_2S_4 \text{ (soluble)}$$
$$2Li + Li_2S_4 \leftrightarrow 2Li_2S_2 \text{ (insoluble)}$$
$$2Li + Li_2S_2 \leftrightarrow 2Li_2S \text{ (insoluble)}$$

Long-chain polysulfides are formed in the first step and then dissolve in the electrolyte. Further reduction of the long-chain polysulfide results in reduced soluble polysulfides. Finally, solid Li_2S_2 and Li_2S are formed in the last step. Although Li-S batteries have a high theoretical capacity, the charge reaction is only achieved via electrochemical cleavage and reformation of the sulfur-sulfur bonds in the cathode. Thus the chemistry of Li-S batteries is quite complex.

1.6.1 Lithium-Sulfur Cathodes

Lithium-sulfur cathodes offer the potential for extremely high energy density due in part to the low molecular weight of sulfur. The low cost and high availability of sulfur also aid the cost effectiveness and sustainability of manufacturing these batteries. However, because the electrical conductivity of sulfur is low, it is generally necessary to employ porous carbon supports, the weight of which lowers the theoretical specific energy.

The chemistry also includes a natural sulfur shuttle mechanism that can protect the cell from overcharge. However, tuning the shuttle for optimal performance is complex and it can also promote high self-discharge rates exceeding 10% per month. In addition, the multiple intermediary sulfides formed throughout the charge discharge process make stabilizing the cathode a difficult task, creating longevity concerns. One particular challenge is preventing insoluble products including Li_2S_2 and Li_2S from blocking the porous network of the Li-sulfur cathode.

1.6.2 Lithium-Sulfur Anode

The use of lithium metal as a negative electrode provides extremely high capacity (3860 mAh/g) but comes with many challenges to safety and long-term operation. The anode is utilized by plating lithium on the anode surface during charge and dissolving lithium on discharge as mentioned above. This plating process involves a significant volume change with cycling on the order of 300% (compared to approximately 10% for graphitic anodes), which can induce mechanical stress across all components of the cell and introduce or exacerbate other failure mechanisms. Furthermore, this process can be highly nonuniform, leading to possible dendrite formation that causes the loss of active material and thus irreversible capacity loss under the best of operating conditions. Dendritic growth is a risk as with all metallic anodes, which can lead to an internal short circuit and possible fire (like Li-ion batteries, the electrolyte employed in lithium-sulfur batteries is flammable). However, when combined with the sulfur cathode, the anode surface and newly formed dendrites can become coated by layers of soluble polysulfide chains. Although the increase in inactive lithium and sulfur does lead to irreversible capacity loss and the layering does decrease conductivity, it also reduces the reactivity of the metallic lithium and at least partially suppresses additional dendrite growth. Further, to serve as a reliable protective coating, control over the thickness of the sulfide layers must be achieved. Finally, lithium metal is extremely reactive when exposed to water (H_2O), presenting a safety concern, especially when the cell container is breached.

1.6.3 Challenges and Future Work

The promise of a practical specific energy that is at least twice that of Li-ion batteries is enticing to many applications; however, much work remains to be done to improve lithium-sulfur batteries while addressing capacity fade, self-discharge, and safety. There are many present ongoing efforts to address these matters, such as new cathode structures reliant on different porous carbons and possibly doped or functionalized porous carbons to stabilize the polysulfide products. Surface coatings for increased sulfur utilization, stability, and conductivity, as well as new electrolytes formulated for increased conductivity and shuttle control are also under investigation. Some recent results are encouraging, and future research may find solutions to many of these problems. But it is important to keep in mind that the achievable volumetric energy may eventually be nearly close to that of traditional Li-ion batteries. Thus, where space is at a premium, as is the case with many mobile applications, lithium-sulfur batteries may not offer the level of superior performance over the more mature Li-ion technologies as the differences in specific energy may suggest.

1.7 Metal-Air Batteries

Metal-air batteries have the potential to simultaneously have the highest energy density and the lowest cost of energy storage for many applications. In a metal-air battery, oxygen (ideally supplied freely from the atmosphere) serves as the cathode, combined with a pure metallic anode (see Figure 1.6 for a schematic of a metal-air battery). On discharge, oxygen is combined with the metallic anode to create metal oxides; when charging, these metal oxides are reduced to plate metal back at the anode. Several metals have been considered for such systems, including magnesium, iron, aluminum, zinc, and lithium. Only the latter two have shown any affinity for electrical recharging, and thus the discussion here in will be focused accordingly.

Of note is a variant of metal-air systems; namely, metal-water systems. Here the air cathode is replaced with water. Many of the issues are the same as those of metal-air systems, but the cell voltage is significantly lower and their applications are generally limited; hence, they are omitted from the discussion hereafter.

1.7.1 Zinc-Air Batteries

Zinc-air systems have a theoretical specific energy greater than 3 kWh/kg (when the mass of oxygen is not included) and thus present an opportunity to greatly surpass the energy storage capability of Li-ion batteries. Further, their reliance on a highly abundant and low-cost material—zinc—offers improved sustainability and cost effectiveness.

Historically, zinc-air batteries have been plagued by poor reversibility and long-term performance. This is due to issues concerning zinc plating (volumetric expansion and dendrite formation, common to almost all metal anodes) and the evaporation of the electrolyte (when used in an open system).

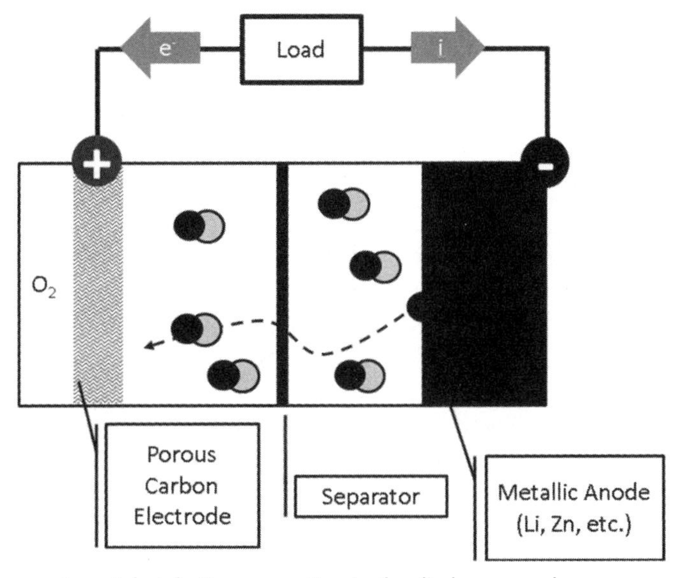

Figure 1.6 A generic metal-air battery operating in the discharge mode.

1.7.2 Lithium-Air Batteries

With a theoretical specific energy exceeding 11 kWh/kg (excluding oxygen), lithium-air is perhaps the most likely candidate battery to come close to matching the specific energy of fossil fuels and thus is of considerable interest to automotive and other mobile applications. However, the air cathode generally requires a porous carbon support. Including the weight of this support structure (~70% porosity), the air cathode, and the electrolyte, the theoretical energy density is then reduced to 2.8 kWh/kg—a far cry from the oft cited 11 kWh/kg, but still far surpassing today's state-of-the-art $LiCoO_2$/graphite-based lithium ion technology (0.25 kWh/kg). Unfortunately lithium-air technology is still in its infancy, and faces several substantial challenges.

For instance, similar to the lithium-sulfur technology, inclusion of a metallic lithium anode comes with considerable volume change during cycling, irregular plating that can lead to dendrite formation, and a high reactivity with H_2O. However, unlike lithium sulfur, there is at present no mechanism to suppress dendrite reactivity, increasing the chance of irreversible capacity loss and short-circuit formation. Furthermore, if an open system is to be used to take advantage of freely available atmospheric oxygen, the lithium anode is at risk of exposure to atmospheric H_2O, especially in environments where humidity is high.

Several other factors are also of concern. At the anode interface, the metal oxides created on discharge can create an insulating layer. Additional oxides can build up with repeated cycling, resulting in considerable capacity loss and reductions in rate capability. The kinetics of both the charge and discharge reactions, of course, impedes the commercial utility of such systems. Not only does this factor limit the achievable specific power, but also results in gradually diminishing efficiencies. Finally, exposure to the atmosphere can cause electrolyte evaporation over time, as well as expose the system to impurities that can react with lithium metal and create effects that remain largely unanalyzed at this stage.

The air cathode also poses very significant challenges. The reactions at the air cathode are shown below:

$$2Li^+ + 2e^- + O_2(g) \leftrightarrow Li_2O_2(s);\ E^o = 3.1V \text{ (equilibrium)}$$
$$4Li^+ + 4e^- + O_2(g) \leftrightarrow 2Li_2O(s);\ E^o = 2.91V \text{ (equilibrium)}$$

Of these, the formation of the peroxide (Li_2O_2) results in a barrier layer formed on the cathode side. This drives the voltage profiles of the Li-air battery, both during discharge and charge, to be far from the equilibrium potential (~3V): implying that we need to supply more energy during the charge process to store energy and are able to retrieve less energy than if the cell were to function at the equilibrium potential. The increase in the voltage gap is a significant barrier to Li-air batteries, as current organic electrolytes are not stable at the voltages encountered during operation of the cell due to reactions occurring at the interface between the electrode and electrolyte.

1.7.3 Challenges and Future Work

The challenges with metal-air chemistries are significant, as discussed above. The core challenges lie with obstruction of the cathode or anode interface with the electrolyte, and those typical of metallic anodes (volume change and dimensional uniformity during recharge). Protection and stabilization of the anode is of particular importance where lithium is employed, as it not only represents a barrier to long-term performance, but also to safety. It is equally important to investigate methods to reduce the large gaps in voltage between the charge and discharge for the reactions at the air cathode. While it may be possible to catalyze this reaction such that the voltage barrier is not as extreme, the difference between the charge and discharge reactions may be due to thermodynamic constraints—and that will hinder any improvements proposed by research focused on improving the rate of the reactions.

Additional challenges include evaporation of the electrolyte and contamination when using air from the environment. The use of ionic and solid electrolytes has the potential to address dryout but typically degrade efficiency and increase the cost. Alternatively, these problems can be approached at the system level by the implementation of closed and/or filtered air systems. Once again, these will bring added cost, complexity, and mass.

Research on metal-air systems is a growing field. At least one company claims to have developed a saleable, rechargeable (up to ~100 cycles) zinc-air system, but much is yet to be learned regarding metal air performance before being successfully mass-marketed for consumer and utility applications. Certainly, the many possible solutions to the problems discussed above are yet to be fully explored. Experimenting with other metals, such as iron, is also in progress and may also prove fruitful. Eventually a viable system may be achieved, but it appears unlikely that metal-air batteries will be widely successful in the near term.

1.8 Emerging Chemistries

Batteries are a lively topic of research at present given the possible transformative effects they could have on the transportation and energy industries; that is, if a battery can meet the necessary performance and cost targets to displace the largely fossil fuel based solutions presently employed. None of the commercialized technologies, and few publicly under development, have the potential to achieve all of these goals. In part, for these reasons, new battery chemistries are quite frequently proposed; and details regarding their technical challenges, present performance, and even basic function are often rigorously protected. In this section, several recently proposed technologies are briefly mentioned. Where possible, their function, performance capabilities, and technical challenges are noted, although in general very little information is currently available.

1.8.1 Sodium-Ion Batteries

Sodium ion batteries function, in principle, like lithium ion batteries do—by shuttling positively charged ions between electrodes. Replacing lithium with sodium could offer several benefits, however. First and foremost is the availability and low cost of sodium as compared to lithium. In addition, some researchers have proposed the use of aqueous electrolytes with lower cell voltages—again reducing costs but also improving safety. The requisite reduction in energy density makes this technology more suited for stationary applications rather than for mobile ones.

Note that sodium batteries classified under this section differ from the aforementioned sodium beta batteries via operation at ambient temperature, since they function via an intercalation mechanism of the sodium ions and do not require the anode to be molten.

1.8.2 Liquid Metal

A novel liquid-metal battery is also currently under development. The system is extremely simple: a liquid antimony anode is poured into a container first and then a liquid sodium-sulfide molten salt electrolyte is added, followed by a liquid magnesium cathode. The material densities are such that they naturally segregate to keep the cathode on top, the anode on bottom, and the electrolyte in-between, greatly simplifying the overall construction and minimizing manufacturing cost. On discharge the antimony and magnesium electrodes dissolve and combine to form magnesium antimonide in the electrolyte; on charge this species separates and the metals return to their respective electrodes.

The previous sections outlined the state-of-the-art batteries and where the lithium-ion technology is placed in terms of its energy and power capabilities. The following chapters elaborate on the underlying principles of designing a battery and commonly encountered issues and factors to be considered when finding an optimal construct suited for a specific application. Most of the examples utilize Li-ion batteries as the chemistry of choice, but the proposed solution invariably is independent of this restriction.

CHAPTER 2

Electrical Performance

The operating principles of a battery are very similar to those of conventional energy sources such as a hydro turbine or compressed air. There are thermodynamic and kinetic limitations not very different from those that govern the combustion of the traditional fuel. There are also a few distinguishing factors that govern the operation of a battery. This chapter discusses the different aspects that determine the suitability of materials used to assemble a battery and the engineering aspects that go into functionalizing these materials. We begin with the description of a few technical terms often used in the community, introduce the mathematical framework used to describe the phenomena that take place within the battery during charge and discharge, discuss the significance of these terms in building a cell, and conclude with a brief outline of the experiments used to make these assessments. The context-sensitive test procedures are deferred until the subsequent chapters that describe the specifics of the test parameters in more detail in relation to the target application. Although the principles described in this chapter are widely applicable for different kinds of batteries in general, the examples provided herein cater to the choice of Li-ion batteries.

2.1 Thermodynamics Inside a Battery

A simple experiment from high school physics is quite effective in understanding how a battery works. Figure 2.1(a) shows a set of glass tubes of different shapes. The first part of the experiment consists of filling each tube with the same amount of water (e.g., 100 ml) and measuring the height of the air column from the top of the tube to the water surface after adding every milliliter. The plots of the air-column height versus volume of water for the different shapes of glass tubes are shown in Figure 2.1(b).

Two observations from this experiment are relevant for a battery engineer:

1. Even though the volume of water inside each container is the same, the height of the water column is different for the different tubes, and as a result, the potential energy at the base of each column is different.
2. The shape of the curves in Figure 2.1(b) is a function of the properties of the container (in this case, the geometry). Thus, the potential energy available at any given instant is a function of not just the volume of water, but the shape of the container as well.

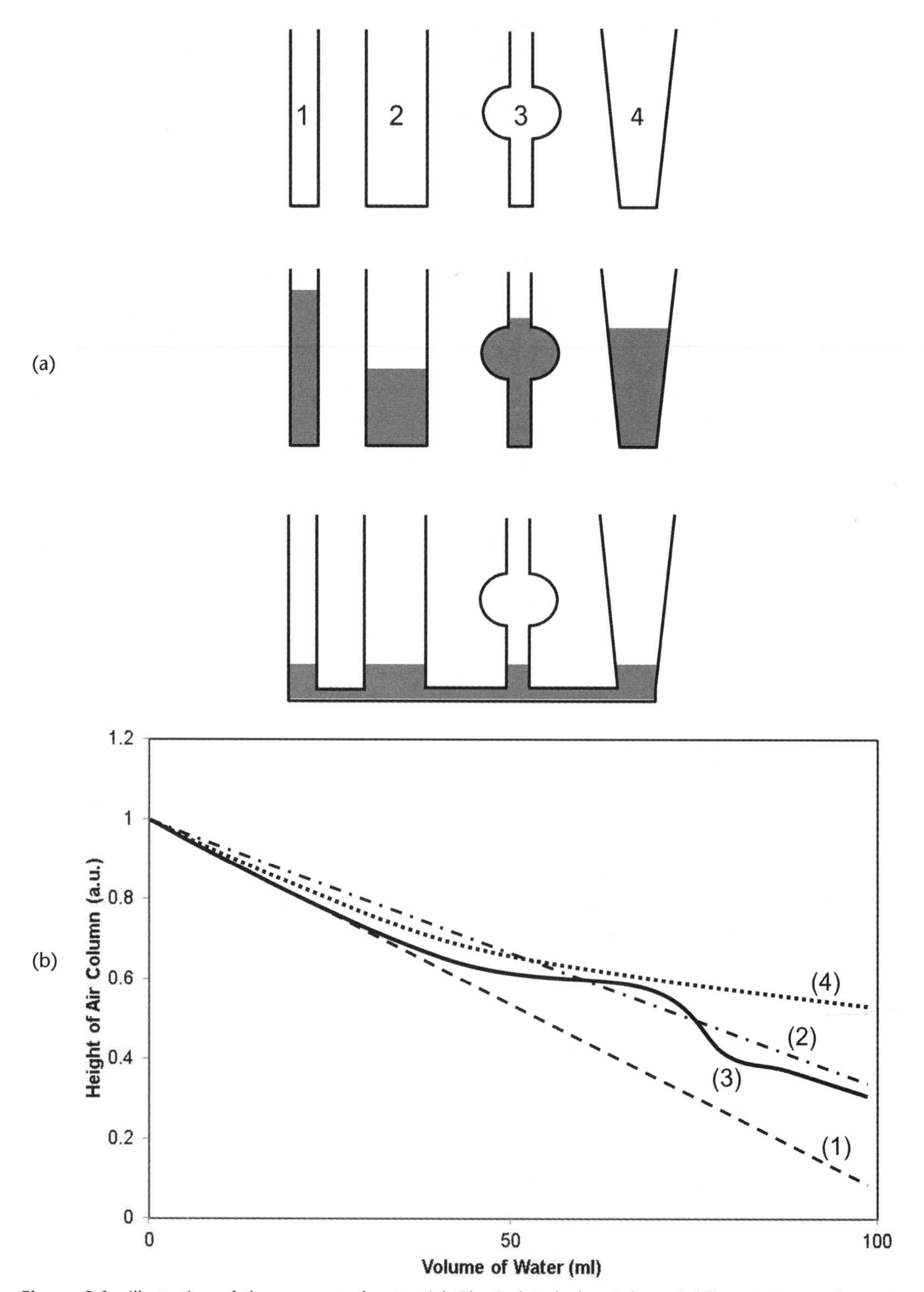

Figure 2.1 Illustration of the concept of potential. The isolated glass tubes of different shapes shown in Figure 2.1(a) all contain the same volume of water despite the difference in the height of the air columns in arbitrary units (shown in Figure 2.1(b)). When the tubes with water at different potential energies are connected, the energy is leveled across the different shapes. Similarly, the positive and negative electrodes of a battery contain the ions at different electrochemical potentials (see Figure 2.2), and by pairing them up suitably, one can make a battery with the desired energy storage and delivery characteristics.

Similarly, battery materials differ in their ability to store and deliver energy: for the same weight of material, the amount of charge you can store within a battery is a strong characteristic of the ability of the material to store charges. This ability is termed the *electrochemical potential*. Like Gibb's free energy, the electrochemical potential is a measure of useful work that can be extracted from a material under ideal conditions.

The electrochemical potential is often measured as the voltage at which one can exchange the working ions from the host material under ideal conditions. Just like we measured the height of the water inside the different columns from the base of the containers, we need a reference value to measure the electrochemical potential of the material. For example, in lithium batteries, this reference is the potential at which Li-ions are converted into metallic lithium. So, at 0V versus Li/Li^+, lithium metal forms Li-ions spontaneously—and no useful work can be extracted from these ions if the material stores them at 0V, since the ions are in equilibrium with lithium metal. This situation is analogous to having 100 ml of water spread infinitesimally thin across the surface of a table—the potential energy of such a film, with respect to the surface of the table, is zero. Thus, a piece of metallic lithium inserted in a container with Li-ions under identical conditions as the battery material under consideration is termed the reference electrode, and all voltage measurements made when this electrode is grounded are listed as being measured with reference to the Li/Li+ electrode. When the voltage measurement is carried out under controlled conditions—constant temperature, pressure, and composition (lithium content) of the material—this voltage is termed the equilibrium voltage. However, it is time-consuming to establish equilibrium conditions. For practical purposes, the equilibrium voltage is approximated with the value measured when the changes to the system are not appreciable: for instance, the temperature of the material is controlled to as much as a tenth of a degree celsius, the pressure is assumed to be constant since the volume changes for these solids and liquids are negligible, and the distribution of lithium in the material is assumed to be uniform if the duration over which the maximum amount of lithium that can be inserted (or extracted) is set to some high value (typically 30 hours or longer). This approximate measure of the voltage as a function of the lithium content at a given temperature and pressure is called the open circuit potential (OCP) of the material.

Just like the height of the water columns, the OCP of the material changes depending on the quantity of Li-ions and the energy levels at which the material stores these ions. The shape of the voltage curve, as a function of the lithium content in the different materials is shown in Figure 2.2. The relationship between the OCP and composition is governed by Gibb's phase rule [1], which states that the number of properties of a material that can change independently (f) is given by:

$$f = c - p + 2 \tag{2.1}$$

where c is the number of chemical species (in our case, this number is 2: one for lithium and one for the material holding it) and p is the number of phases. Temperature and pressure are two of the independent properties; depending on whether the material can hold all the lithium without undergoing a phase transformation, the remaining degrees of freedom are either one or zero. For example, iron phosphate forms two different phases depending on how much lithium it contains. Thus, the

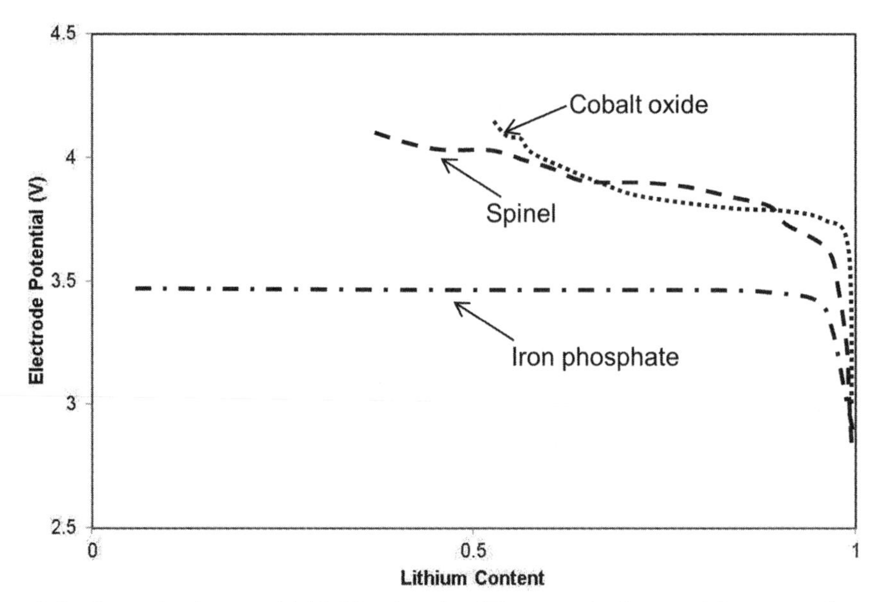

Figure 2.2 Open circuit potential (OCP) values for different cathode materials commonly used in Li-ion batteries. (See the analogy between Figure 2.1(b) showing the potential energy of identical volumes of water inside columns of different shapes and Figure 2.2 showing the electrochemical potential of different electrodes containing identical amounts of Li-ions.)

only independent properties, as read from (2.1) are the temperature and pressure and the chemical potential can be calculated as a function of all the other properties in this system; as seen in Figure 2.2, the OCP of iron phosphate is indeed independent of the composition. On the other hand, cobalt oxide behaves like one homogenous material (i.e., $p = 1$), and thus, the number of degrees of freedom, besides temperature and pressure, for this material is one (i.e., $f = 3$). The chemical potential as represented by the OCP is the additional property that can be varied as a function of lithium content independent of temperature and pressure. This relationship between the OCP and lithium content is one of the fundamental properties governing the choice of materials suitable for use in batteries. Let us explore this idea a bit further, with a specific example of a lithium ion cell.

2.2 Assembling a Li-Ion Cell

A Li-ion cell comprises of a negative electrode (anode) where electrons are accepted from the external electrical circuit when the battery is charged and a positive electrode (cathode) that can supply the positively charged Li+ ions during charge so that the system can remain charge neutral. When the battery is discharged, the Li-ions move back into the cathode and the electrons are released to the external circuit from the anode (resulting in current flow across the load from the positive to the negative electrode).

When assembling a Li-ion cell, the anode and the cathode must be chosen such that each electrode can perform the intended function, as described above, without thermodynamic barriers. This is where relationship between OCP and lithium content comes in handy: for the negative electrode containing a certain amount of

Li-ions to accept more ions, one must set its voltage with respect to Li/Li^+ reference electrode to be lower than its OCP at the same composition. This is accomplished by supplying the battery with electrons from the external circuit, and in turn, shifting the anode from its equilibrium state to a new, energized state. In an attempt to reach equilibrium at the new voltage set point, the anode takes in more Li-ions to bring its composition as close to a value that the electrode would have, if its OCP were equal to the new set point. The same way, when Li-ions must be released from the cathode, one has to raise the voltage to a new set point, at which the electrode would spontaneously release the ions to reach a new equilibrium between its voltage and composition. The reverse process works during discharge of the battery—the voltage values at each electrode shift in the directions opposite to that during charge when the Li-ions are moved back from the anode into the cathode. The continuous change in the voltage of the individual electrodes is illustrated in Figure 2.3(a); the voltage of the cell is calculated as the difference between those of the cathode and the anode when the ions move from the solution into the electrodes or vice versa during a charge or discharge, as is illustrated in Figure 2.3(b).

The movement of the ions is accomplished either by supplying additional incentive to the battery in the form of current or by adjusting the OCP of the individual electrodes as to release or attract excess charges to the electrodes. In either case, the amount of working ions at any given instance in an electrode during a charge or discharge process is related to the amount of charge supplied to the battery (or equivalently, the current flow created due to the shift in the OCP of the individual electrodes) by *Faraday's law*: one Faraday of charge (96,487 C) is required to produce (or consume) one molecular weight of ions (also called one *mole* of ions). In the case of Li-ions, the molecular weight is 6.94 g. Thus, if we charge a Li-ion cell using 1 ampere of current for 1 hour, ideally, we would have extracted 258.9 mg of Li-ions from the cathode. Thus, by varying the amount of current supplied or used from the battery, or by specifying a power demand from the external circuit, the Li-ions can be moved in and out of the electrodes at varying rates, in accordance with Faraday's laws.

In order for the Li-ions to be shuttled efficiently between the two electrodes, the driving force to extract these ions from the anode must be paired up with that used to insert them into the cathode. In other words, the two electrodes must work in sync, with the energetics aligned such that they are additive in favor of transporting lithium in the intended direction and the driving force must be maximized. Correspondingly, the voltage difference between the two electrodes is maximized. On the anode side of the cell, the lowest voltage one can get to without disrupting the flow of ions is 0V versus Li/Li^+, at which point, the energetics favor the plating of lithium rather than movement of Li^+ ions. The first choice for the anode material is thus a sheet of metallic lithium. Practical difficulties, as discussed in Chapter 1, prevent the use of lithium metal as the anode: upon repeated plating and dissolution of lithium, the morphology and mechanical properties of the anode changes drastically resulting in poor cyclability and often safety concerns due to dendrites of lithium that is plated back, inducing a short circuit within the cell. Other materials that can be used as anodes usually host lithium inside a solid matrix. Different types of carbons, tin oxide, titanium dioxide, and silicon are some examples. Once again, it is the voltage at which lithium can be inserted or released, and the

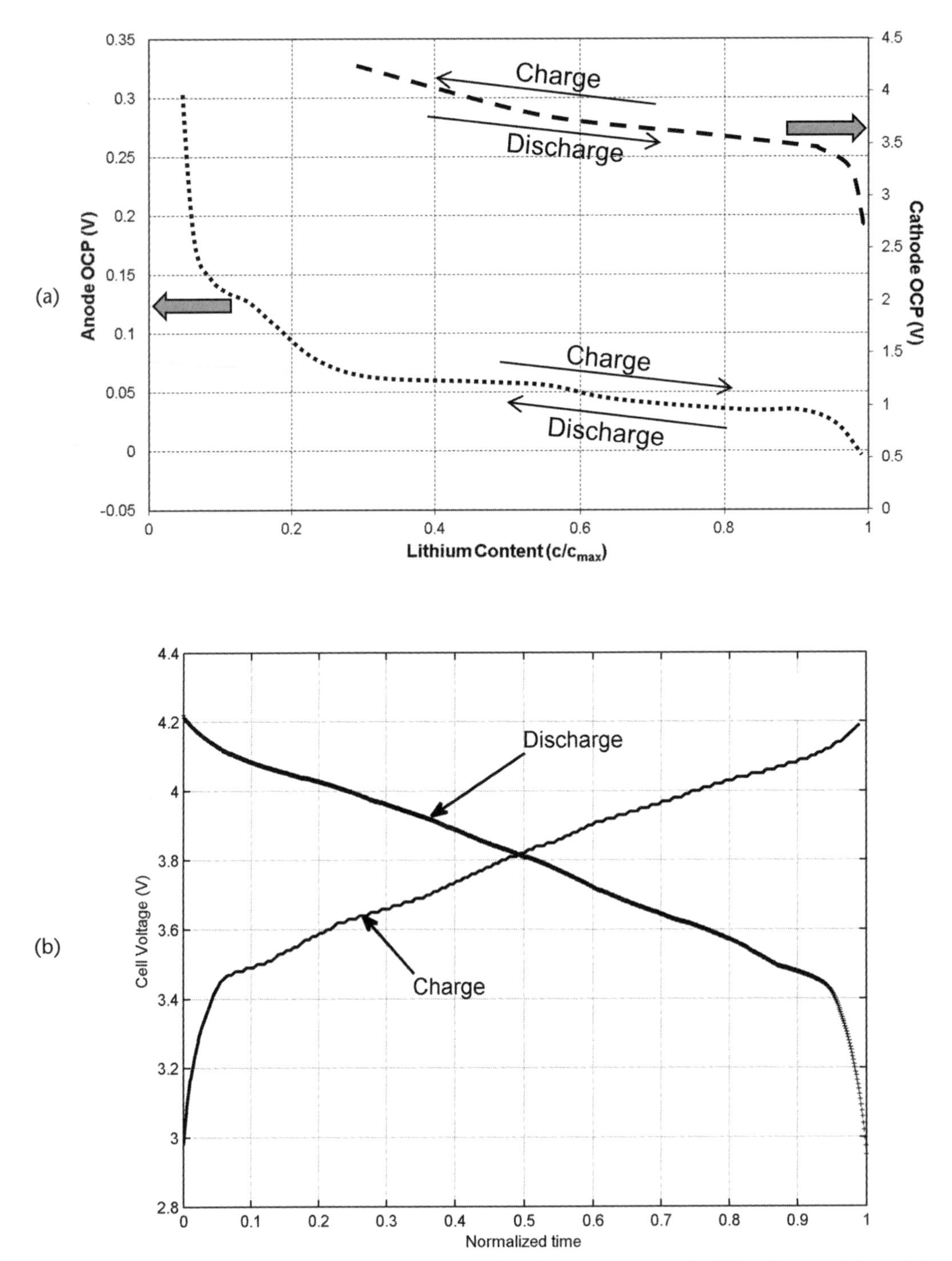

Figure 2.3 Change in OCPs of the individual electrodes during operation of a Li-ion battery (a), and the corresponding cell voltage (b) during charge and discharge.

thermodynamic ability to store electrons from the external circuit that makes these materials good choices as anodes for lithium ion batteries.

On the cathode side, transition metal oxides are the most popular choices for Li-ion batteries, mainly because these materials are durable at very high voltages for prolonged periods of time [2]. Some of these oxides are stable up to 4.6 V versus Li/Li^+. In general, the higher the voltage at which lithium can be inserted or extracted, the greater the driving force for the ions to be shuttled. Also, the objective is to store as much energy by moving as few ions around as possible. One way to accomplish this is by employing materials that store the ions at high voltages: since energy stored equals voltage times the amount of charge, if the voltage values are higher, the amount of charge that has to be stored to accomplish a set target for energy to be stored can be lowered. Thus, in this case, transition metal oxides are good candidates. Another way to accomplish the same goal is to use multivalent ions, which can compensate for the transfer of multiple electrons for every ion that is transported. Magnesium, manganese, lead, and vanadium are some of the ions considered in the past for this purpose; however, it is difficult to shuttle heavier ions and the efficiency drops quickly due to transport limitations. So, we use the Li-ions to transfer the charge from one electrode to the other; but then take advantage of the multiple valence of the transition metals (e.g., Mn^{2+} and Mn^{3+}) inside the host matrix of the cathode material to maximize the amount of energy stored for a given amount of the cathode material. It is worth noting that lithium batteries operate by inserting and extracting ions within electrodes without disrupting the structure of the host lattice, in contrast to alkaline or lead acid batteries, which rely on plating and dissolution of the charge carrying species. The process of insertion of a guest ion into the host matrix without disrupting its structure is called *intercalation.*

There must be an ionic conductor available to transfer the Li-ions from one electrode to the other. This conductive layer must simultaneously be capable of blocking the electrons from being short-circuited within the cell instead of being routed across the external load when called for. This role is played by the *electrolyte*, which is essentially a lithium salt dissolved in a suitable solvent. The choice of electrolytes suitable for use in Li-ion batteries is extremely challenging because the electrolyte must be stable at the extreme voltages (0–6V) it encounters on both the anode as well as at the cathode. The nonavailability of suitable electrolytes has been one challenge that has restricted the usable voltage range of Li-ion batteries to 0–4.5 V. $LiPF_6$ dissolved in organic carbonates is a popular choice of electrolyte. Other solvents like tetrahydrofuran (THF), lactones, and some ethers have been proposed, but they are usually used as additives to address specific problems (e.g., to improve stability of the electrolyte at the anode surface or to improve conductivity of the ions at low temperatures).

Simplistically, a battery looks like two blocks of electrodes with compatible properties as described above, immersed in a bath of electrolyte. However, in a practical system, there are transport limitations in addition to thermodynamic barriers described above. The ions have to be moved from their designated positions within the host matrices to the electrode-electrolyte interface, where they are transferred by means of one or more electrochemical reactions into the electrolyte, and subsequently carried in the electrolyte phase and the reverse process takes place at the other electrode. In order to maximize the efficiency of the cell, the transport of ions must be kept to a minimum. This task is accomplished by placing the electrodes as close to one another as possible without making actual physical contact (which would induce an electronic short circuit). A porous polymer film about 10 to 30 μm thick is used as a spacer between the electrodes. This film is called the *separator* and is usually made of a polymer that is chemically compatible with the electrolyte (e.g., polyethylene or polypropylene membranes are used in the case of Li-ion batteries). In some cell designs, the separator is also designed to conduct ions, and thus, the role of the separator and the electrolyte are integrated. These cells are referred to as polymer cells. In general, the transport of ions within the host matrix is about two to four orders of magnitude slower than moving them in the liquid electrolyte. In order to take advantage of this fact, the electrodes are usually designed as porous films soaked in the electrolyte. They are assembled by coating a thin (~100 μm for Li-ion batteries versus ~100 mils for metal hydride batteries) porous layer of the corresponding active material (carbon, transition metal oxide, etc.) on to a thin metal foil or mesh, which distributes the electrons (or collect the current) across the electrode to ensure a uniform utilization of the active material. The current collectors themselves are chosen by their ability to resist chemical and electrochemical reactions under the environment they operate in. For example, copper is stable in the operating voltage window that the anode of a Li-ion cell is exposed to, and is used as the current collector on the anode, whereas it will oxidize at high voltages on the cathode side to form soluble copper oxide. Hence, an aluminum foil works well at the cathode. A detailed discussion on the choice of materials can be found in any of the standard lithium battery handbooks [1, 2].

The cell components described above can be packaged in different ways: stacking alternating layers of anodes and cathodes with a separator in between is one popular approach, especially with large-format Li-ion cells. Winding long coils of the electrodes on to a central core is another. This assembly of the electrodes and separator is termed a *jellyroll*. The jellyroll is then packaged into a flexible pouch or a metal casing, from which electrical connections to the rest of the circuitry is facilitated using metallic tabs that are attached to the electrodes.

2.3 Voltage Dynamics during Charge/Discharge

Designing a battery for a specific application can considerably benefit from a good understanding of how the different components within the cell work under the stipulated operating conditions and how the electrical response is closely connected to the thermal, mechanical, and chemical behavior of the battery materials. Mathematical models often simplify such analyses and provide a basis for methodic design of the materials as well as engineering aspects. This section provides an overview of the various mathematical frameworks used to describe the electrical response of a Li-ion cell and lays the foundation for the more elaborate discussions of the thermal and chemical aspects that follow in subsequent chapters.

A transition metal oxide (say, $LiCoO_2$) exhibits an OCP of ~ 2.5V versus Li/Li^+ when it contains the maximum amount of lithium it can practically hold. As described earlier, this corresponds to the fully discharged state of the battery. As and when the cell is charged, lithium is gradually removed from the cathode, and the OCP rises toward 4.2 to 4.5V. On the other hand, at the anode (carbon, to be specific), the OCP of the electrode corresponds to ~ 2V versus Li/Li^+ when there is no lithium present. When lithium is inserted the voltage gradually drops until the electrode is fully saturated with lithium, at which point the ions begin to plate out as metallic lithium and the OCP of the anode drops to 0V versus Li/Li^+. The measured voltage of the cell, at any given instant, is the difference between the voltage at the cathode and that at the anode. Thus, at the beginning of the first charge, the cell voltage is around 500 mV. At the end of charge, the cell voltage reaches the 4.2 to 4.5V range (depending on the choice of the cathode material). Similarly, during discharge, the cell voltage starts to drop from this value to around 2V. The anode, after cycling lithium across for the first few times, is seasoned to a blend of lithium in carbon, which has a characteristic voltage versus composition signal that equilibrates to an OCP of 300 mV at full discharge. These break-in cycles that stabilize the interface between the anode and the electrolyte are referred to as *cell formation cycles*. The interfacial film that helps prevent rapid degradation of the electrolyte at such harsh voltage conditions is referred to as the solid electrolyte interface (SEI) layer. Some cyclable lithium is lost in the process of cell formation, but usually the cell manufacturers compensate for this effect by including an excess of lithium when building the cell.

2.4 Circuit Diagram for a Cell

It is common practice for electrical engineers to represent circuit elements such as diodes and operational amplifiers (op amps) as a combination of resistors, inductors, and capacitors. This process helps one to understand performance limitations and to make design calculations. Such a representation of a cell is shown in Figure 2.4. As previously shown in Figure 2.2, each electrode has a characteristic OCP versus composition relationship. The cell voltage at open circuit (V_0) is then calculated as the difference between the OCP of the cathode and that of the anode (U_c and U_a, respectively):

$$V_0 = U_c - U_a \tag{2.2}$$

Thus, the open circuit voltage of the lithium ion cell at any given instant, for example, is a function of the amount of lithium inside each electrode. A resistor Rs is connected in series with the voltage source to represent all contact resistances and ohmic impedance to the flow of charge across the cell. At the interface between the electrode and the electrolyte, the charge is either transferred to the ions, across the

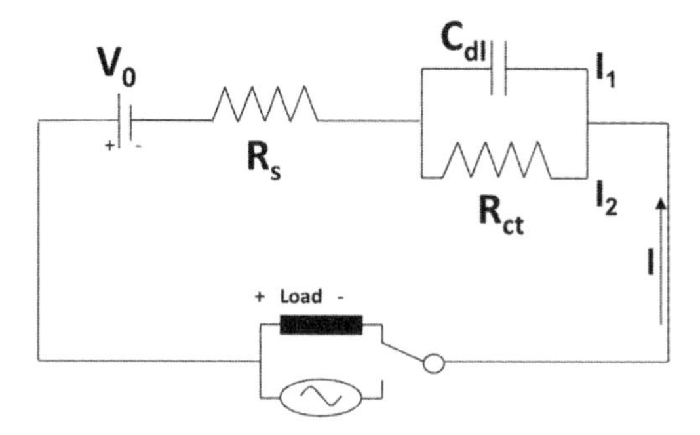

Figure 2.4 Circuit diagram for a cell showing a popularly used equivalent circuit to model electrical performance of the cell for different types of chemistries. The parameters R_s, C_{dl}, and R_{ct} are usually obtained by fitting the circuit equation (see (2.9)) to the current-voltage response measured from the cell. Figure 2.5 shows a comparison between the simulation results using parameters from Table 2.1 and experimental data.

interface into the liquid phase, or tends to accumulate across a double layer, more like a capacitor. To represent the possibility of these two physical phenomena a circuit branch containing a resistor (R_{ct}) and a capacitor (C_{dl}) in a parallel branch is added on to the ohmic part. The resistance R_{ct} represents the barrier at the interface to the charge transfer reaction, and the capacitance C_{dl} corresponds to the buildup of charges at the interfacial double layer (with electrons on the electrode side and the ions in the solution side of the interface to compensate for the charge-balance).

Understanding the dynamics inside the cell is then reduced to solving Kirchhoff's current and voltage equations for the circuit. The voltage drop across R_s follows Ohm's law:

$$V = I\mathrm{R}_s \tag{2.3}$$

where I is the net current flow across the cell, which in turn branches out into I_1 and I_2 across R_{ct} and C_{dl}, respectively. Kirchhoff's node-and-loop rules relate the currents that flow across the different branches of the circuit and the voltage across each branch. Thus we have:

$$I = I_1 + I_2 \tag{2.4}$$

The rate of charge buildup in the capacitor equals the current that flows through the capacitor. This is expressed mathematically as follows:

$$I_2 = \frac{dq}{dt} \tag{2.5}$$

Kirchhoff's loop rule states that the voltage across any branch is the sum of the voltage drop along the branch and that the sum of all currents that enter or exit a node on the circuit is zero. The constraint on the voltage across each branch yields the following equations:

$$V_{Cell} = V_0 + I\mathrm{R}_s + I_1\mathrm{R}_{ct} \tag{2.6}$$

$$V_{Cell} = V_0 + I\mathrm{R}_s + \frac{q}{\mathrm{C}_{dl}} \tag{2.7}$$

Equations (2.4) through (2.7) can be rearranged to obtain the relationship between change in the applied current and the resultant voltage drop:

$$\mathrm{R}_s\frac{dI}{dt} + \frac{1}{\mathrm{C}_{dl}}\left(1 + \frac{\mathrm{R}_s}{\mathrm{R}_{ct}}\right)I = \frac{dV}{dt} + \frac{1}{\mathrm{R}_{ct}\mathrm{C}_{dl}}\left(V_{Cell} - V_0\right) \tag{2.8}$$

The mathematical solution for (2.8) can be found in standard textbooks [3]. For the case of constant current the solution takes the following form:

$$V_{Cell} = V_0 + \frac{Q_0}{C_{dl}} e^{-t/R_{ct}C_{dl}} + IR_s + IR_{ct}\left(1 - e^{-t/R_{ct}C_{dl}}\right) \tag{2.9}$$

This model equation now relates the input current to the change in cell voltage. The parameter Q_0 refers to the total capacity (which, in turn is related to the energy) available in the cell. The change in cell capacity during a charge or discharge is calculated by integrating the current passed as follows:

$$Q = Q_0 - \int_0^t I dt \tag{2.10}$$

Similar results can be obtained for constant power loads by replacing the current (I) in (2.8) with P/V_{Cell} where P is the power demand set by the external load. Voltage versus capacity data for various input signals are used to calibrate the circuit elements (R_s, R_{ct}, etc.) under the operating conditions of interest. A lookup table can then be constructed to characterize the performance of the cell under load. Figure 2.5 shows a comparison of model versus experimental data for the parameter set shown in Table 2.1 for a Li-ion cell.

The circuit-diagram approach to characterize a battery is simplistic, and the upfront investment in terms of time and resources used to obtain a model is minimal. This approach is also extremely effective in system design, wherein engineers of various disciplines treat the battery as a black box with a set of defined electrical characteristics. These models are readily scalable to design of battery packs in which several cells are arranged in series-parallel configurations to meet the load requirements. Once a set of calibrated components is established using the above procedure, commercial software used in circuit programming (e.g., SPICE™) can be readily adapted to include Li-ion batteries as part of the circuit components. The downside alludes to oversimplification of the different physical phenomena taking place inside the cell. For example, when designing large-format Li-ion cells, the use of a single resistance term or capacitance term to characterize the charge-transfer reaction taking place across the entire electrode surface is not sufficiently accurate to capture the localized response: among other limitations, the distribution of the current across the entire surface of the electrode is not uniform, different parts within the cell tend to heat up and dissipate heat at different rates, some parts of the electrode are overutilized while others are underutilized. In such instances, one must modify the circuit diagram to accommodate for such additional complications involving the cell design and the choice of material parameters. Making such ad hoc modifications on a case-by-case basis can be tedious, and the experimental data set used to obtain the relevant calibration curves is usually large.

2.5 Electrochemical Models for Cell Design

A methodic formulation that will help extend the performance metrics observed under one set of operating conditions at one scale or a given design of cells to a different set of situations using known laws of physics has been suggested as the

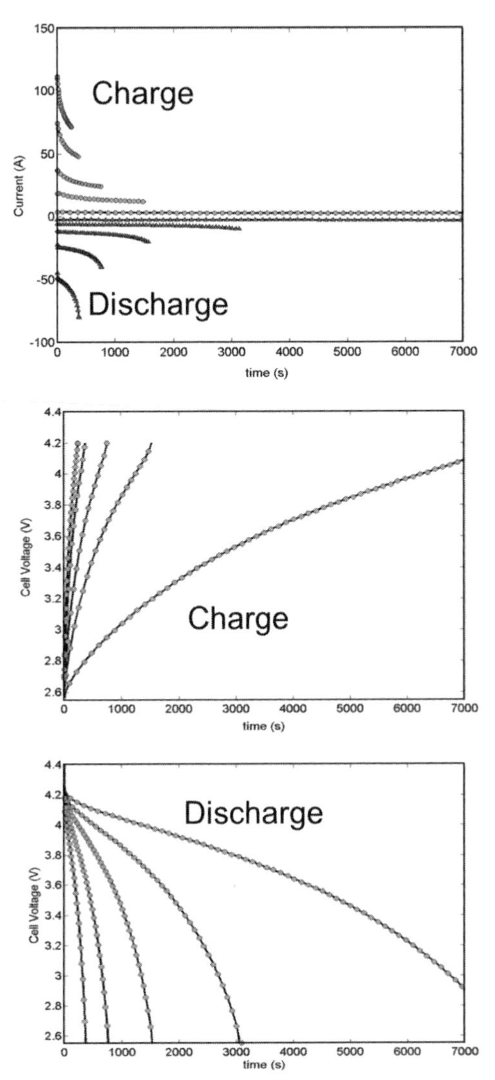

Figure 2.5 Model predictions versus experimental data from the equivalent circuit model under constant power discharge for a Li-ion cell at 0°C. The power was set to 10, 25, 50, 100, and 200W for the curves shown above. The symbols represent the experimental data and the solid lines represent the model predictions. (Figure generated after Verbrugge and Connel [4].)

Table 2.1 Parameters Used in the Equivalent Circuit Model for the Cell Shown in Figure 2.5

Parameter	*Discharge*	*Charge*
τ (s)	5	5
C_{dl} (F)	12500	16667
R_s, mΩ	1.637	1.637
R_{ct}, mΩ	0.4	0.3

From: [4].

alternate choice to battery design. For example, by measuring the porosity of the electrode, its electronic conductivity and the ionic conductivity of the electrolyte as a function of temperature, this process eliminates the need to measure the circuit resistances for electrodes of different porosities at different temperatures if one can formulate a relationship between these independent parameters and the conductivity of the composite electrode. Such phenomenological models are often referred to as physics-based representations of Li-ion cells. For example, a mechanistic model for the Ohm's law equation shown in (2.3) can be built by describing the resistance parameter R_s in terms of physically measurable properties. If we choose R_s to represent the potential drop across the copper bus bar, the properties of interest are the electronic conductivity of the metal (Cu), the cross-sectional area (A_{Cu}), and the length of the bus bar (σ_{Cu}). Each of these properties is characteristic of the bus bar. The resistance $R_{s,Cu}$ is related to these parameters as follows:

$$R_{s,Cu} = \frac{L_{Cu}}{\sigma_{Cu} A_{Cu}} \tag{2.11}$$

and hence (2.3) can be rewritten as:

$$V = I \frac{L_{Cu}}{\sigma_{Cu} A_{Cu}} \tag{2.12}$$

Note that (2.12) can be used for a cable of any given dimension made up of any material whose electronic conductivity is known. However, the use of (2.3) requires that we measure the resistance parameter R_s every time the bus bar is replaced. Also, the use of a mechanistic model enables one to identify better materials (e.g., a bus bar with higher electronic conductivity) suited for an application. We now proceed to develop such mechanistic models to describe the other physical processes commonly encountered within a Li-ion battery, such as the movement of electrons within the electrodes and movement of ions in the electrolyte, chemical, and electrochemical reactions.

2.5.1 Charge Transport within the Electrode by Electrons

The total voltage across a cell (V_{Cell}) can be approximated as the sum of the potential drop across the electrodes that across the electrolytes, the electrolyte, and other losses arising from contact resistances. In the following sections the subscript 1 will be used to denote the properties/variables in the electrodes and 2 to represent the corresponding variables in the electrolyte. We already considered voltage drop due to the flow of electrons across metal cables in (2.12). In order to be able to represent properties such as conductivity as a function of the local environment we use the differential formulation; that is, we use the derivative forms of the equations. Thus, the voltage drop across the electrodes due to the movement of electrons is now measured per unit length ($\nabla \phi_{1,j}$) using an alternate representation of Ohm's law:

$$\nabla \phi_{1,j} = -\frac{i_1}{\hat{\sigma}_j}, \quad j = n \, or \, p \tag{2.13}$$

where i_1 is the current per unit area (called the current density) and $\hat{\sigma}_j$ is the composite electronic conductivity of the electrode material within electrode j ($j = n$ refers to the negative electrode or the anode and $j = p$ refers to the positive electrode or the cathode). Usually a battery electrode is comprised of several components such as solid solutions of different metals or composites of active material that hosts the Li-ions, binders that hold the particles of the active material together, and other components. The composite conductivity corrects for these additional components within the electrode, and is calculated as the sum of the conductivities of the individual components, scaled in proportion to the composition of the electrode:

$$\hat{\sigma}_j = \sum_k x_k \sigma_k \tag{2.14}$$

Here x_k are the mole fractions of the individual components k that constitute the electrode and σ_k refer to the electronic conductivities of the pure components (e.g., σ_{Cu} represents the conductivity of metallic copper). Alternatively, $\hat{\sigma}_j$ can be measured directly after the electrode is assembled, but the utility of the model developed with this parameter will be limited to the specific electrode design used to measure the conductivity value.

2.5.2 Charge Transport in the Electrolyte by Ions

A unique feature of electrochemical devices is the transport of the current by ions. Once the current moves past the electrodes and undergoes the electrochemical reaction, the charge transport from one electrode to another is facilitated by ions. This transport of charge by the movement of ions is more complicated than the current-carrying mechanism by movement of electrons. Usually, there are several species of ions present in the electrolyte. The total current carried by the electrolyte (i_2) across a unit normal area is the sum of the currents carried by each of the ionic species k:

$$i_2 = \sum_k i_k \tag{2.15}$$

The current carried by each species i_k is proportional to how fast the ion can move in the electrolyte. The rate of movement of the ions is termed its flux (N_k) and is related to the current by [5]:

$$i_k = F \sum_k N_k \tag{2.16}$$

The flux of a particular type of ions k may be defined as the product of the number of ions or the concentration of species k and the velocity of each ion:

$$N_k = c_k v_k \tag{2.17}$$

The proportionality factor F in (2.16) is the Faraday's constant (96487C), which is the amount of charge carried by each mole of the ion. The concentration of an electrolyte is a readily measurable quantity (the parameters in the mechanistic equations do not necessarily have to be measured within the battery); the velocity of an ion is proportional to the charge it carries (z_k) and the potential gradient in the electrolyte ($\nabla\phi_2$), which is the electrical driving force for the ion to move:

$$v_k = -u_k F z_k \nabla\phi_2 \tag{2.18}$$

The proportionality constant in (2.18) is called the mobility (u_k) of the ion and is obtained from equivalent conductance measurements. The negative sign indicates that the ions move from a region of higher potential to lower. Equations (2.15) to (2.18) can be rearranged to obtain [5]:

$$i_2 = -\left(F^2 \sum_k c_k u_k z_k\right)\nabla\phi_2 \tag{2.19}$$

Equation (2.19) resembles Ohm's law (see (2.13)) closely, and the electrical conductivity for the electrolyte ($\hat{\kappa}_j$) is now given by:

$$\hat{\kappa}_j = \left(F^2 \sum_k c_k u_k z_k\right) \tag{2.20}$$

As stated in the previous section, (2.20) relates the properties of the component ions (k) to the conductivity of the electrolyte within electrode j. Hence, knowing the composition of the electrolyte, one can model the movement of ions in the electrolyte. Alternatively, one can experimentally measure the electrical conductivity ($\hat{\kappa}_j$) in (2.20); but limitations similar to those following (2.14) apply.

In deriving (2.19) an implicit assumption that the concentration of the electrolyte is uniform throughout the cell precluded the effects of concentration gradients present within the cell. However, this assumption is not strictly valid in large-format Li-ion cells and can be easily relaxed by incorporating flux terms arising from concentration differences using Fick's laws of diffusion. Equation (2.17) then becomes:

$$N_k = c_k v_k - D_k \nabla c_k \tag{2.21}$$

where D_k is the diffusion coefficient of ions k. The equation presented above typically represents the case of dilute electrolytic solutions. In general one can get away with solving for just the concentration of Li-ions (i.e., $k = Li^+$). More sophisticated models that consider the mutual interaction of multiple ions within the electrolyte and the effects of temperature on the conductivity of the electrolyte are available [5].

2.5.3 Charge Transfer between the Electrodes and the Electrolyte

As noted above, charge is carried within the electrode by electrons and in the electrolyte by ions. The transfer of charge from the ions to the electrons and vice versa is similar to any chemical reaction. The rate expression for a chemical reaction states that how fast a chemical species appears or disappears is related to the availability of the individual species participating in the reaction at the electrode surface; the proportionality constant in this relationship is the kinetic rate-constant that determines how fast or slow the reaction takes place:

$$\frac{dc_k}{dt} = kf(c_j) \tag{2.22}$$

where c_k is the species of interest and the function f relates the concentration of all the participating species, c_j. Standard textbooks on reaction engineering [6] discuss the procedure to solve these expressions and how to use these models in reactor design.

In the case of Li-ion batteries, there is an electrochemical reaction involved at the electrode/electrolyte interface: in addition to the availability of the individual chemical species at the electrode surface, the difference in voltage between the electrode surface and the electrolyte at the vicinity of the interface (i.e., $\phi_1 - \phi_2$ at the interface) can be used as another knob to control the rate of the reaction. The rate expression (2.22) should be modified accordingly to reflect this change. The rate of reaction depends exponentially on the available energy (which in this case is the voltage difference scaled to suitable units). The most commonly used expression to capture this dependence is the Butler-Volmer equation:

$$i_j = Fkf(c_j)\left[\exp\left\{\frac{\alpha_{a,j}F}{RT}\left(\phi_{1,j} - \phi_{2,j}\right)\right\} - \exp\left\{-\frac{\alpha_{c,j}F}{RT}\left(\phi_{1,j} - \phi_{2,j}\right)\right\}\right] \tag{2.23}$$

where $\alpha_{a,j}F/RT$ is a scaling factor. For a step-by-step derivation of (2.23), see [7]. It is often convenient to express the concentrations and potentials in the Butler-Volmer expression with respect to reference values; for example, the potentials are referenced to the OCP of the respective electrode and the concentrations to values at which properties such as conductivities and diffusivities are known. Equation (2.23) then becomes:

$$i_j = i_{0,j}\left[\exp\left\{\frac{\alpha_{a,j}F}{RT}\left(\phi_{1,j} - \phi_{2,j} - U_j\right)\right\} - \exp\left\{-\frac{\alpha_{c,j}F}{RT}\left(\phi_{1,j} - \phi_{2,j} - U_j\right)\right\}\right] \tag{2.24}$$

where the term $i_{0,j}$ combines all the concentration dependence:

$$i_{0,j} = Fkf\left({}^{c_j}\!/_{c_{ref}}\right) \tag{2.25}$$

and is termed the exchange current density. The term $(\phi_{1,j} - \phi_{2,j} - U_j)$ represents the departure of the energy barrier across the interface from equilibrium conditions and is popularly referred to as the overpotential (η_j). The higher the value of the overpotential, the greater is the deviation in the voltage response of the cell from ideal behavior. For example, at higher rates of discharge, the voltage drop encountered in a battery is more than that seen at lower discharge currents. This is because as the current increases, the rate of transport of ions does not increase linearly (as one would expect out of ohmic conductors). The concentration (and dilution) of ions across specific portions of the cell gives rise to some overpotential, which in turn induces the additional voltage drop. Similarly in battery chemistries that merely involve plating of the working ions (e.g., lead-acid batteries), buildup of overpotential leads to a shift in the potential at which a reaction happens (in the presence of excess of ions or when ion supply is at a premium, the chemical potential in the vicinity of the electrode/electrolyte interface, and thus the energetics of the charge-transfer reactions, are altered).

2.5.4 Distribution of Ions

As noted previously, for the case of large-format Li-ion cells, nonuniformities in the reaction rates across the different regions of the cells must be accounted for. To accomplish this, we use the local values of the ionic concentrations in (2.17) through (2.25). The local values for the concentrations are computed using Fick's laws of diffusion:

$$\frac{\partial c_k}{\partial t} = -\nabla . \boldsymbol{N}_k + R_k \tag{2.26}$$

The flux used in the above material balance equation is consistent with the previous definition (see (2.21)). The term R_k refers to the change in concentration of Li-ions if they were consumed or generated in a separate chemical reaction taking place within the electrodes or the electrolyte other than the normal functioning of the battery. At the electrode-electrolyte interface, the change in concentration of the ions is because of the electrochemical reaction, and hence the Butler-Volmer equation (2.24) is used.

In the case of battery electrolytes, the interactions between the working ions (e.g., the Li^+ ions in a lithium battery) and that with the other supporting ions (e.g., the PF_6^- anions in the electrolyte) must be considered. These interactions become particularly important to predicting the performance of the cells over a wide range of temperatures and during high power operations. Complexities such as this are usually handled by defining a composite property that considers such interactions. In this case, (2.21) for the flux must be modified accordingly:

$$\hat{N} = c\hat{v} - \hat{D}\nabla c \tag{2.27}$$

In (2.27) the diffusion coefficient is interpreted as an effective property in which the interaction terms has been accounted for. The velocity term $\{\hat{v}\}$ now relates to an effective field within the electrolyte and is usually expressed in terms of the transport number (t_+^0), which is a measure of the fraction of the total current carried by the working ions:

$$c\hat{v} = \left(1 - t_+^0\right)\frac{i_2}{F} \tag{2.28}$$

The diffusivity and conductivity values must be corrected for geometry effects since the cell components are porous blocks of solid immersed into the electrolyte. These corrections introduce effective transport properties that include shape factors such as the porosity (ε) and tortuosity (τ) of the electrodes and the separator:

$$\theta_{\text{eff}} = \varepsilon\theta^{\left(\varepsilon/\tau\right)} \tag{2.29}$$

Similar corrections for temperature dependence are discussed in Chapter 3. The use of effective properties as described in (2.29) and composite properties as shown in (2.14) is what makes the mechanistic models amenable to use in conjunction with easily measureable experimental data. Figure 2.6 shows a comparison between the model predictions and experimental data obtained using a physics-based model over a wide range of operating temperatures. The ability to use one set of parameters under a variety of design conditions is also a direct result of the decoupling of the geometry effects from the composition of the different components.

There are limitations to the mechanistic modeling approach as well: a solution of the set of equations (2.11) to (2.29) is more time-consuming compared to the circuit-diagram approach. The number of parameters to be experimentally measured is large. Most importantly, the use of effective properties and approximations to the kinetic and transport equations limit the utility of these models in designing materials with a known set of properties—for example, in conjunction with quantum mechanical calculations.

The choice of models is often determined by the desired application. Typically, fine-tuning of the parameters such as the conductivity of the electrolyte or the porosity of the electrodes is carried out at the cell-design phase. Hence a mechanistic model is invaluable at the design phase—for example, when determining the chemistry of the cell or to maximize the power available from the cell. Manufacturers of battery packs and systems integrators tend to use circuit-based models for two reasons: (1) the amount of information available in such cases is limited and obtaining parameters for a detailed mechanistic evaluation is both time-consuming and redundant, and (2) once a cell of desired performance has been designed, during the subsequent scale-up of the battery response to a given load or when building packs from the batteries, the cells are treated analogous to other electrical components involved in the application. In any case, the assumptions behind a mathematical model must be carefully explored before employing the conclusions drawn from simulations.

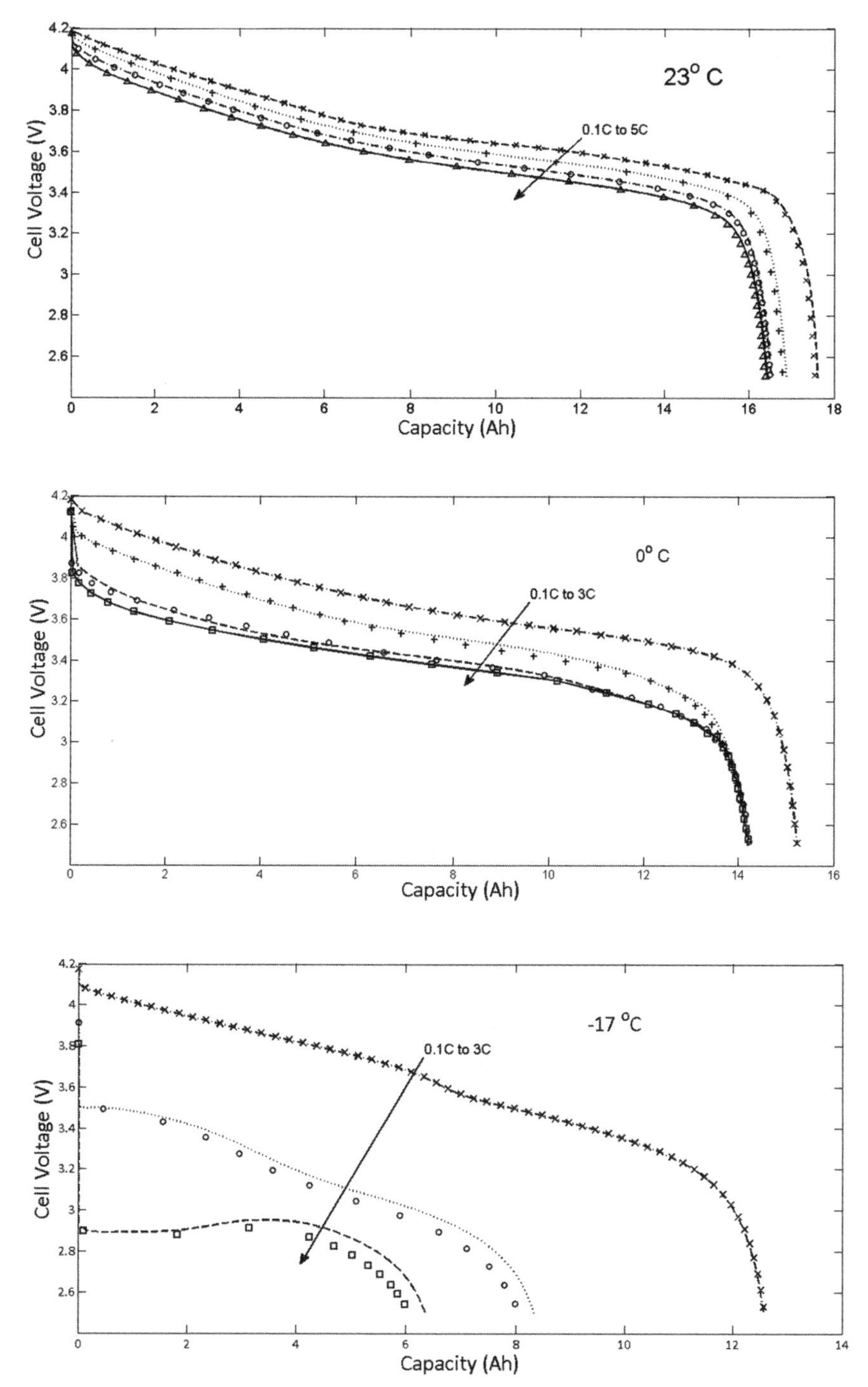

Figure 2.6 Comparison of simulation results from a physics-based model versus experimental data. The symbols represent experimental data collected at three different temperatures from a cell of 16-Ah nameplate capacity.

2.6 Electrical Characterization of Li-Ion Batteries

After cell assembly, the manufacturer subjects the cells to a set of break-in or formation cycles as described earlier to allow for the system to stabilize. Then the cells are charged to about 50% of the nameplate capacity before being shipped to the end user. The following is a basic set of electrical tests carried out on the cells before use in an application. The protocols outlined in this section are usual guidelines practiced in the industry; however, the cutoff values for the voltages, the maximum allowable currents during charge or discharge at different temperatures, as well as the duration for which the cells can be exposed to a given current or voltage are specified by the cell supplier in good detail.

2.6.1 Capacity Measurement

Cells are rated by the capacity specified in ampere hours; with battery packs, the default specification is the energy content in kilowatt-hours together with a listing of the pack voltage. Measurement of the available energy from the cells is carried out by first discharging the cells to the minimum voltage recommended by the manufacturer. Next, the cells are charged to the maximum voltage specified by the manufacturer followed by a 100% discharge. The charge step carried out using a constant value for the current over 2 hours, using the nameplate capacity specified by the manufacturer as the reference point, followed by holding the cell at the maximum charge voltage until the value of the current drops to 10% of the initial charging rate. For instance, if the cell is rated at 40 Ah, the charge is carried out at 20A, followed by a hold at the maximum voltage until the current drops to 2A. The discharge is carried out using the initial charge current as well. In general, the capacity measured from a fresh cell at low discharge rates is higher than the nameplate capacity, and hence the discharge step is likely to continue for a little over 2 hours. The available capacity is then calculated by integrating the current over the duration of the discharge. These steps are repeated until the value of the available capacity for the cells stabilizes, which usually happens over 5 to 10 cycles after manufacture for a well-made Li-ion cell.

The state of charge (SOC) is the percentage of the total capacity measured using the procedure outlined above that is available at a given instant. Some manufacturers and OEMs specify measured capacity across specific voltage windows for use as 100% and 0% SOC. The current at which the cell can be completely discharged from 100% SOC to 0 % SOC over a period of 1 hour is termed the 1C rate. The currents at other *C-rates* are determined as the product between the specified rate and the current at the 1C rate: for example, at 2C, the current is twice that at 1C rate, and ideally, the battery can be fully discharged at half the time. The next series of tests involve measuring the available energy at various C-rates. These measurements give an idea of the power capability of the cells. The maximum charge and discharge rates at a given temperature are specified by the cell manufacturer. A typical set of test data is shown in Figure 2.7. In addition to these constant current measurements, the cells are discharged at a prespecified power, as shown earlier in Figure 2.5.

2.6.2 Power Measurement

The capability of the cells to deliver high power over short periods of time is particularly important for many applications. Traditionally, the power density of the

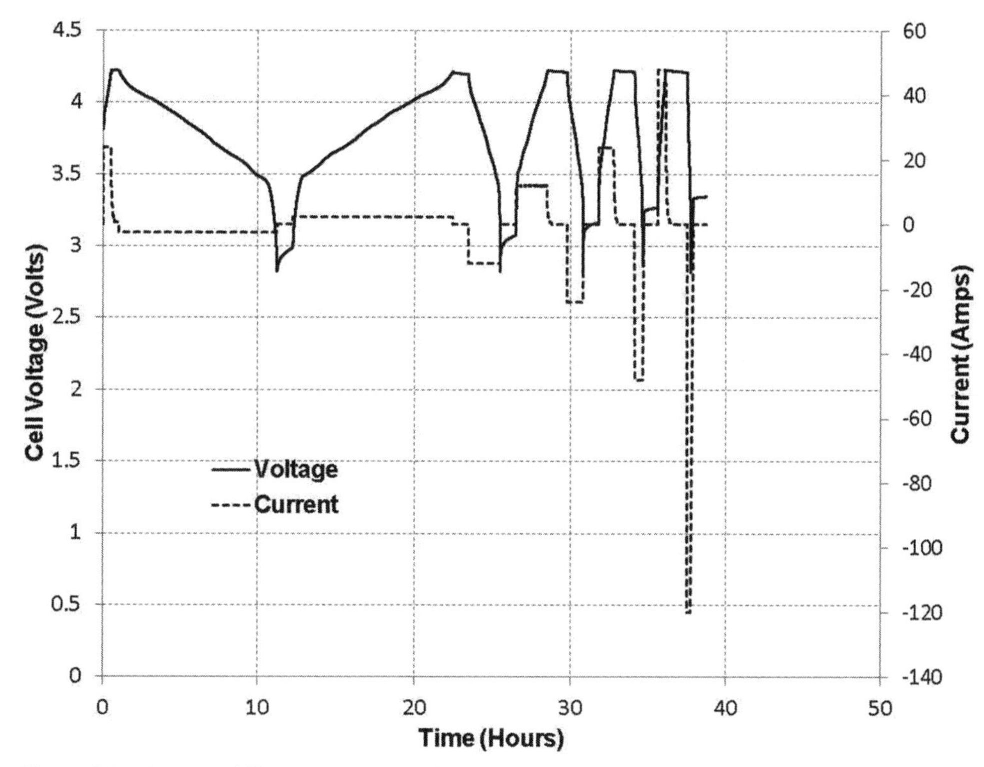

Figure 2.7 Rate capability measurements for a Li-ion; the voltage response to several current inputs to the cell is used to calibrate the circuit-parameters and to understand the energy/power capabilities of the cell.

cells is plotted versus the available energy density in a Ragone plot (see Figure 2.8), which is then used in power-to-energy trade-off decisions. The batteries are subjected to charge/discharge cycles where the constant current discharge is interspersed with short-duration, high-current pulses. These tests are used to determine the ability of the cells to provide the requisite power at a specific state of charge. The hybrid pulse power capability (HPPC) test and the dynamic stress test (DST) are some common examples. A complete set of test procedures for these tests can be found in [8]. Specific tests that combine a string of power demands/input to the cell based on statistical analyses of drive patterns are available. For example, a power cycle that mimics the power requirements from the battery when the vehicle is subjected to a urban dyanmometer driving schedule (UDDS) is available. Such application-specific test schedules are discussed in Chapters 6 and 7.

2.6.3 Component Characterization

In order to utilize the mathematical models discussed earlier in the chapter, one must obtain the relevant parameter set from testing of the individual components. Design parameters like dimensions of the individual layers can be directly measured; whereas porosity and tortuosity have to be estimated from loading (g/cm^2) measurements and the use of standard correlations for porous solids (see for example [9]). In order to measure the properties of materials at the individual electrodes, it is standard practice to assemble a three-electrode cell. This test apparatus (see

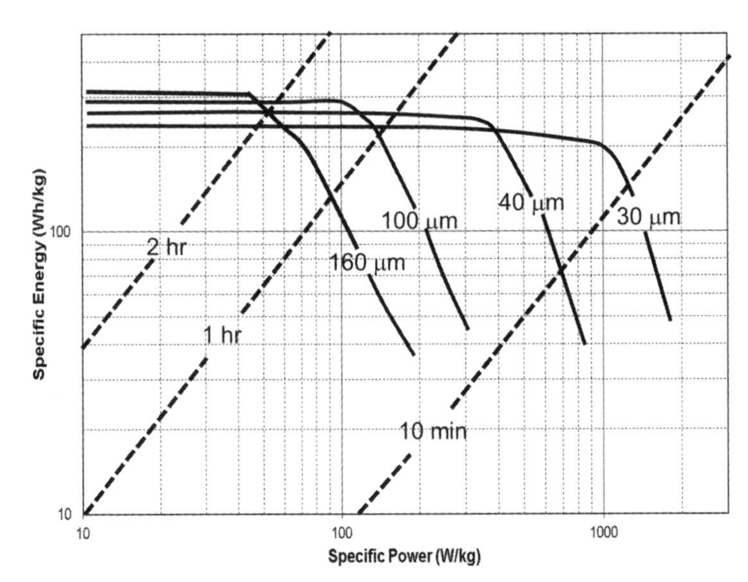

Figure 2.8 A typical Ragone plot used to analyze the energy-versus-power capabilities of Li-ion cells with different design conditions. In this example, the thickness of the cathode is varied as a design parameter while the electrode porosities and the cell capacity are maintained constant. The ability of the cell to deliver prespecified values for power or energy density over a stipulated time duration is measured along the diagonal lines.

Figure 2.9) essentially comprises a small disc of electrode material punched out of the cathode and the anode, which is arranged in a hermetically sealed cell in such a way that each electrode is sufficiently close to a foil of metallic lithium that can serve as the reference electrode, such that the ohmic losses between the reference electrode and the battery electrode of interest are minimal. One popularly used three-electrode cell design using a Swagelok™ fitting is shown in Figure 2.9. The measurements of some of the other electrical parameters discussed in this chapter are briefly described below.

2.6.3.1 Open Circuit Potentials

The OCP of the individual electrodes (see Figure 2.3(a)) are measured by individually monitoring the potential between the anode and the reference electrode as well as that between the cathode and the reference electrode simultaneously. After the first few formation cycles, the cell is subjected to a very slow rate discharge where the current is maintained at C/30 or lower values. Similar measurements are conducted at various temperatures to correct the OCP for changes in the ability of the host material to hold lithium (i.e., the entropic changes) with temperature.

2.6.3.2 Conductivity Measurements

For the electrodes, the electronic conductivity is measured using a standard four-probe method: four equally spaced (~ 1 mm) metal blocks of known dimensions are placed on to the electrode supported by predetermined spring loads to ensure adequate contact. A high-impedance current source is used to supply known values of current across the outer probes and the voltage drop across the inner probes is

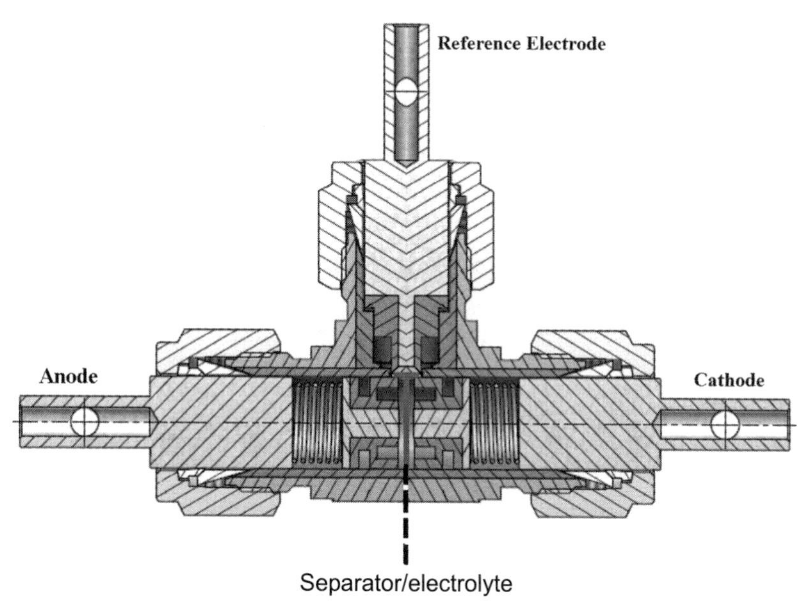

Figure 2.9 Schematic of a three-electrode cell used to measure many of the electrochemical properties described in this chapter. The anode and cathode are discs punched out of the electrodes from the actual Li-ion cell of interest and are positioned as close to each other as possible within a Swagelok-type fitting that is sealed together with a third reference electrode, which is a piece of lithium foil. There is one layer of separator shielding each electrode disc from an electrical short and the space between the electrodes is soaked in the electrolyte. This setup enables the measurement of the individual electrode properties when the cell voltage is set to the desired value between the anode and the cathode.

measured to determine the conductivity. For more details on the instrumentation, consult available standards (e.g., ASTM F390). The electrolyte conductivity is measured using portable electrolytic conductivity meters. Calibration of the test setup, temperature compensation, and a high cell constant (10 or higher) are some of the prerequisites in selecting a suitable conductivity meter. For use with traditional Li-ion battery electrolytes, material compatibility is an additional concern.

2.6.3.3 Diffusivity Measurements

Diffusivity of Li-ions in the electrolyte (D_2) is related to the viscosity of the electrolyte using the Stokes-Einstein relationship:

$$D_2 = \frac{kT}{6\pi r\eta} \tag{2.30}$$

where k is the Boltzmann constant ($1.3806488 \times 10^{-23}$ m^{2}kgs^{-2}K^{-1}), r is the radius of the ion (68 pm), and η is the viscosity at temperature T. Alternatively, solutions with different concentrations (and hence mass densities) can be introduced to create a concentration and the distribution of lithium ions across different distances from the electrode surface can be monitored by using refractive index measurements (see [10, 11]). This data can then be used together with (2.26) and (2.27) to back-calculate the diffusivity within the electrolyte.

The diffusivity of lithium within the electrodes (D_1) can be measured using a current or voltage perturbation technique, wherein the time constant for the relaxation process is a measure of the diffusivity. The following steps summarize the titration technique:

1. The cell with the reference electrode is equilibrated at a predetermined voltage.
2. A current pulse is introduced such that the time-constant R_s^2/D (R_s being the radius of the particles constituting the electrode) is much larger than the duration of the pulse (t^*). Typical amplitude of the pulse is set to be the current at 2C.
3. The voltage response is monitored at a high sampling rate (kHz). A typical response curve is shown in Figure 2.10.
4. The diffusivity is then calculated using the following relationship:

$$D_1 = \frac{4}{\pi t^*}\left(\frac{R_s}{3}\right)^2\left(\frac{\Delta V_a}{\Delta V_b}\right)^2 \tag{2.31}$$

The potential drop values corresponding to ΔV_a and ΔV_b are shown in Figure 2.10. The above process utilizes a fixed current pulse and is referred to as the galvanostatic intermittent titration technique (GITT). A voltage pulse is used in the potentiostatic intermittent titration technique (PITT).

2.6.3.4 Reaction Rate Constants

Frequency domain analysis, which is a popular tool used to analyze electronic circuitry, also finds use in characterizing electrochemical systems. Typically, the circuit representation of a battery (see Figure 2.4) is used in conjunction with a sweep of sinusoidal perturbations around the equilibrium voltage and the corresponding

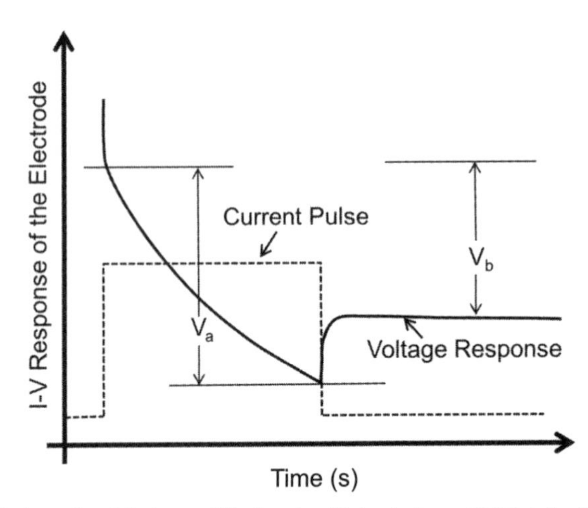

Figure 2.10 Calculation of solid phase diffusion coefficients from GITT data: the cell is subjected to a current pulse of known magnitude and the drop in voltage V_a and V_b are measured. The diffusivity is then calculated using (2.31) shown in the text.

perturbations in the currents are measured using the electrochemical impedance spectroscopy (EIS) technique [12]. A typical response is shown in Figure 2.11. The different processes such as electron and ion transport within the battery have response times that range from microseconds to days. Correspondingly, the frequency of the sinusoidal input is varied from MHz to μHz depending on the process of interest. In general, to measure reaction rate constants, the frequencies of interest lie between 1 MHz and 100 mHz; the Nyquist response corresponds to that of a semicircle for a typical RC circuit, whose intercept on the x-axis is the resistance to charge transfer (R_{ct}). The exchange current density is then calculated using the following expression:

$$i_{0,j} = \frac{RT}{FR_{ct,j}} \tag{2.32}$$

These measurements can be carried out at various OCP values for each electrode and then to deduce the concentration dependence shown in (2.25).

This chapter discussed underlying terminology, some basic principles behind the operation, provided the mathematical framework used in subsequent analyses, and discussed a set of common experimental measurement techniques used to calibrate the electrical performance of a Li-ion cell. More detailed case studies

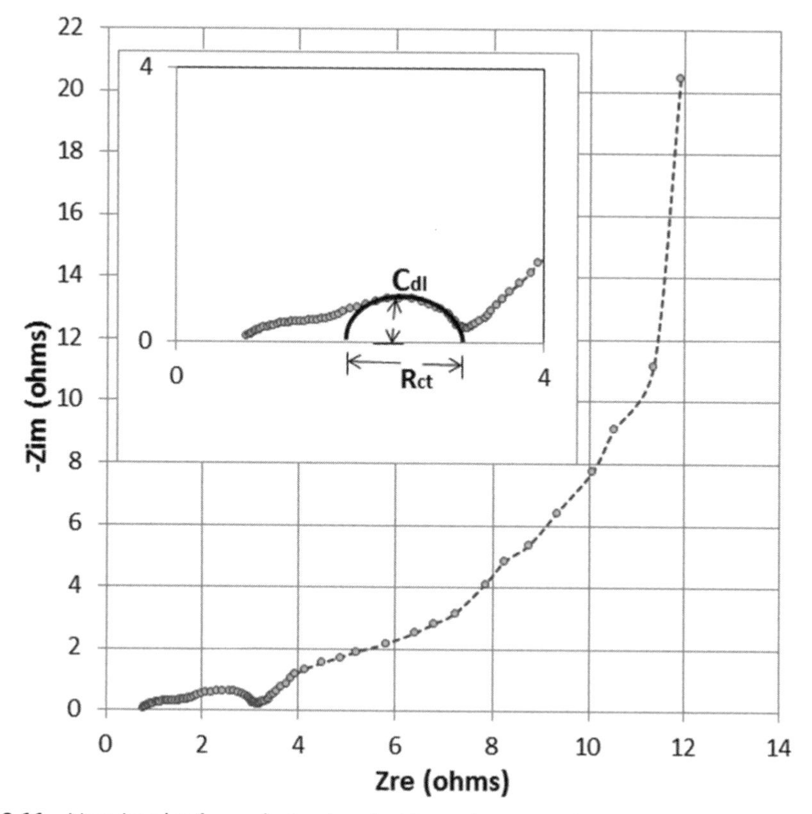

Figure 2.11 Nyquist plot from electrochemical impedance spectroscopy measurements used to calculate the kinetic rate constants for the charge-transfer reactions.

and application related examples are discussed in the next few chapters. Each of these chapters also provides additional discussion on context-specific modeling and measurement techniques.

References

[1] Pistoia, G. (ed.). *Handbook of Lithium-Ion Batteries Applications,* Netherlands: Elsevier, 2013.

[2] Yuan, X., H. Liu, and J. Zhang (eds.). *Lithium Batteries: Advanced Materials and Technologies,* Boca Raton, FL: CRC Press, 2012.

[3] Malvino, A., and D. Bates. *Electronic Principles,* 7th Edition, New York: McGraw-Hill Science and Engineering, 2006.

[4] Verbrugge, M. W., and R. S. Connel. "Electrochemical and Thermal Characterization of Battery Modules Commensurate with Electric Vehicle Integration," *Journal of the Electrochemical Society*, Vol.149, No. 1, 2002, pp. A45–A53.

[5] Newman, J. S., and K. T. Alyea. *Electrochemical Systems,* 3rd Edition, Hoboken, NJ: Wiley Interscience, 2004.

[6] Levenspiel, O. *Chemical Reaction Engineering,* 3rd Edition, Hoboken, NJ: John Wiley and Sons, 1998.

[7] Bockris, J. O'. M., A. K. N. Reddy, and M. E. Gamboa-Aldeco. *Modern Electrochemistry,* Volume 2A: Fundamentals of Electrodics, 2nd Edition, New York: Springer Science, 2001.

[8] USABC Electric Vehicle Battery Test Procedures Manual, http://www.uscar.org/guest/article_view.php?articles_id=74.

[9] Whitaker, S. *The Method of Volume Averaging,* Series: Theory and Applications of Transport in Porous Media, Vol. 13, Dordrect, Netherlands: Kluwer Academic Publications, 1999.

[10] Hafezi, H., and J. Newman. "Verification and Analysis of Transference Number Measurements by the Galvanostatic Polarization Method," *Journal of the Electrochemical Society*, Vol. 147, No. 8, pp. 3036–3042, 2000.

[11] Newman, J., and T. W. Chapman. "Restricted Diffusion in Binary Solutions," *AIChE Journal*, Vol.19, No. 2, 1973, pp. 343–348.

[12] Orazem, M. E., and B. Tribollet (eds.). *Electrochemical Impedance Spectroscopy,* ECS Series of Texts and Monographs Series, Hoboken, NJ: John Wiley & Sons, 2008.

CHAPTER 3

Thermal Behavior

Batteries are comparable to catalytic chemical reactors that are highly sensitive to temperature. The ambient temperature as well as the heat generation rate within the cells has a big impact on the selectivity of the reactions that take place within the battery and the stability of the different species that participate in the functioning of the battery. This is a topic of particular significance, as the size of the individual cells and the battery packs keep increasing to meet demands from high energy (and high power) applications. For example, the variations in temperature across the different cities in the United States have been shown to create a difference of as many as 4 years on the life of batteries used in electric cars. Given the extremes in the operating environment, thermal management of batteries is an area of increasingly active research that has lasting impact on the increasing adoption of batteries for larger-scale applications such as power grids and transportation. In this chapter, we have laid down some of the basic principles of heat generation and heat transfer, applied these in the context of batteries, and illustrated some design examples.

3.1 Heat Generation in a Battery

Even though a battery is seen as a device to convert chemical energy into electrical directly and vice versa, there are a few inevitable energy losses due to mismatch of efficiencies among the different components that constitute the battery. Such energy losses are often manifested in the form of heat. The most common example is the Joule losses—the electrons cannot move across a plastic film as fast as they do across a sheet of copper, and as a result, the kinetic energy is dissipated as heat. Similarly, when the heats of formation of the reactants and the products are different, the difference in the energies is usually given out in the form of heat. Similar concepts apply to the movement of ions as well. As a result, there are several sources of heat within a Li-ion battery. It is important to identify and characterize each source of heat generation because the mitigation strategy for the various sources of heat are different, as described in the remainder of this chapter.

3.1.1 Heat Generation from Joule Heating

The joule heating term, as described earlier, arises due to the inability of the electrons to move as fast as they are required to (due to the presence of the external field or potential gradient). The amount of heat generated under this category is a function of the external field as well as the electronic conductivity of the different components [1]:

$$q = i^2 R = \sigma \left(\nabla \phi_s \right)^2 \tag{3.1}$$

where σ is the electronic conductivity (S/m) and $\nabla \phi_s$ is the voltage gradient (V/m) for a given current density. As observed from the expression above, the higher the gradients in the potential, the greater the rate of heat generation. As a result, the Joule effects are predominant during higher rates of charge/discharge for the battery. When the battery is under storage or if the current drawn out of the cell is meager, other forms of heat generation are dominant. Similarly, if the cell components are good conductors of electrons, this helps in reducing the potential gradient along the different parts of the cell that the electrons traverse. The most commonly used anode material for Li-ion cells is carbon-based, and carbon has an electronic conductivity that is two orders of magnitude higher than that for the ceramic-like cathode active material. As a result, higher-potential gradients tend to develop in the cathode material compared to the anode. In addition to improving the electronic conductivity of the cathode, the carbon-black based additives also serve the purpose of minimizing these gradients.

The movement of ions introduces similar heating effects, only at a larger scale. The ions are bulkier (by about 15,000 times) than that of the electrons, and thus the efficiency of moving the larger-sized charges is much lower. The constitutive property of the electrolyte corresponding to the electronic conductivity (σ) within the electrodes is the ionic conductivity of the electrolyte (κ). One other side effect of the ions being bulkier is that they tend to move at a much slower pace than the electrons. This slow movement often results in the ions being accumulated over selective regions of the cells, resulting in an excess of ion concentrations in some parts and deficiency of the same species in other parts within the electrolyte. These concentration differences introduce additional driving forces for the ions to move from a region of higher occupancy to the deficient zones. There is a corresponding diffusion-related conductivity (κ_D), and the driving forces are the gradients of the ionic concentration. The contributions from these two phenomena are summed up in the following expression for the joule heating that happens within the electrolyte:

$$q = i^2 R = \varepsilon^t \kappa \left(\varepsilon^t \kappa \nabla \phi_e + \kappa_D \nabla \ln c_e \right)^2 \tag{3.2}$$

One difference between (3.1) and (3.2) is the presence of the porosity term in the latter. The implication is that the porosity (ε) of the electrodes and/or the separator can be used as a design knob in order to adjust the amount of heat generated from the ionic contributions. For this reason, batteries designed for high-power applications (e.g., plug-in hybrid electric vehicles (PHEVs)) often use a larger porosity. The other parameter to alter is the electrolyte conductivity—this is accomplished by using the appropriate salt-to-solvent ratio for the electrolyte, or by the use of additives. In either instance, the objective is to make sure the viscosity of the electrolyte is in the appropriate range to minimize excess heat build-up due to viscous dissipation. In optimizing either of these parameters (porosity or the electrolyte conductivity), there is often a trade-off between obtaining the best thermal environment for the cell to operate in versus the other functionalities viz., ensuring that adequate ions are available for transporting the charge in the case of the ionic conductivity and maximizing the available energy within a given volume of the cell in the case of the electrode porosity.

3.1.2 Heat Generation from Electrode Reactions

Current through the cell is carried by electrons within the electrodes and ions within the electrolyte. At each interface between the electrodes and the electrolyte, there is a transfer of charge between the two carriers that takes place during the electrochemical reactions at the interface. Part of the kinetic energy associated with these reactions is lost as heat due to the inefficiency of the charge-transfer process. Every time there is a charge-transfer happening across the electrode-electrolyte interface, there is an associated potential across the interface due to the resulting difference in the free energy in the electrode (and the electrolyte) before and after the charge is transferred. This potential is a measure of the amount of work done in order to carry a unit charge across the interface—it has the unit of volts (which in turn is joules per coulomb). The difference in energies between the two sides of the interface accounts for the inefficiency of the charge-transfer process. Thus the heat generation rate from the reactions is given by:

$$q = \sum_j a_s i_j \eta_j \tag{3.3}$$

where the product of the overpotential (η, joules per coulomb) by the area-scaled reaction rates ($a_s i_j$ amperes per unit volume) is a measure of the total amount of energy released (or consumed) per second per unit volume. As previously discussed in Section 2.5.3, the higher the overpotential, the farther the deviation from equilibrium performance of the battery—and as shown in (3.3), one of the contributors to this deviation is the additional heat generation.

Whereas the kinetic effects depend on the power requirements for the target application akin to the Joule effect, this dependence of the reaction heat changes with the remaining usable energy available in the cell. This allows for the operating window for a given cell to be used as a knob to size the amount of heat generation. Another parameter of interest is the kinetic rate constant; altering the rate constant is comparable to improving the charge-transfer efficiency of the reactions at the interface, thereby minimizing thermal losses. The best way to accomplish such a change is by employing surface modifications to the electrode particles.

3.1.3 Entropic Heat Generation

The insertion and deinsertion of Li-ions in and out of the electrodes results in a change in the arrangement of atoms within the crystal structure, and in turn their mutual interactions. Such changes are reversible from a practical perspective; however, there is some energy loss associated with these phenomena. We call these losses associated with the arrangement of the different species the entropic heat. Changes to the entropic heat are often incorporated into the heat-generation calculations by a correction factor to the overpotential:

$$\eta = \eta_{ref} - \left(\frac{\partial U}{\partial T}\right)_{T_{ref}} \left(T - T_{ref}\right) \tag{3.4}$$

where U is the equilibrium value for the potential at the reference temperature T_{ref} and is calculated from the Gibbs' free energy for a given composition of the material. The entropic term, as represented in (3.4) assumes a linear change with temperature in the equilibrium potential across the interface for each electrode. Given that the entropic contributions are small compared to the kinetic or joule heating under normal operating scenario for the battery, this approximation is appropriate. The entropic heat is a characteristic signature of the material; hence, the knob to control this term lies in the choice of the electrode materials. However, evaluation of this quantity can be used as a screening test during the material selection process.

Figure 3.1 shows a comparison of the heat generated by these different mechanisms within a spinel-based cell simulated for a specific set of design parameters. Joule effects (i.e., ohmic heat generated in the electrodes and from the contact resistances) play a dominant role in the heat generation within the battery typically during high power demands. For the constant-current discharge on a 16 Ah cell shown in the figure, the average joule heating rate is ~20W. This value increases slightly toward the end of discharge (i.e., DOD $\geq$ 80%) since the electrodes become more resistive at those chemical compositions; however, the values stay constant for majority of the discharge duration. As a result, the temperature rise, which

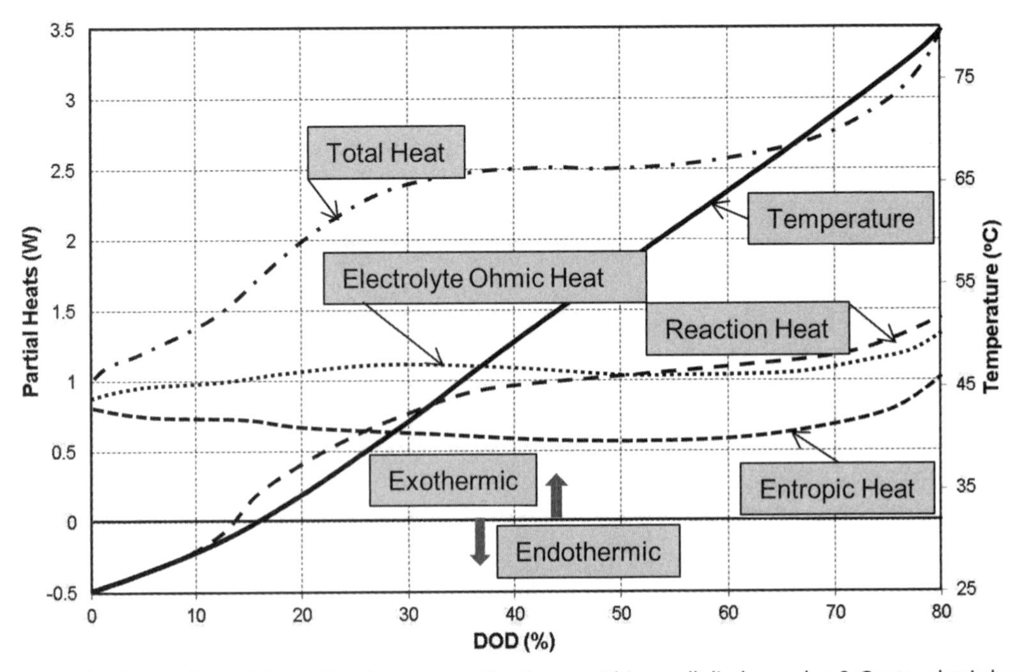

Figure 3.1 Comparison of the various heat generation terms within a cell discharged at 2-C rate; ohmic heat from the electrodes and contacts is not shown on the figure for clarity, since these values are much higher than the other sources of heat for the conditions simulated.

is computed from the integral of the heat generated, has a fairly constant slope. Another noticeable result is that the reaction heats can be negative (endothermic) or positive (exothermic) depending on the standard energy of formation for the reaction products and the reactants. Hence, the net heat generation term for the reaction heats can accordingly be positive or negative.

For efficient cell design, a balance must be struck among the different strategies outlined in this section. A large heat transfer area can help dissipate the ohmic heats, for example. However, in large-format cells the chances for uneven utilization of the active material across the area of the electrodes is quite common; as a result, some parts of the electrode (especially the region in the immediate vicinity of the tabs) end up handling higher currents than the others. At the module or pack level, the heat generation may still not be spatially uniform because of other factors: thermal conductivity of the case, placement of the positive and negative terminals, or size and position of the cell interconnects within the module.

3.2 Experimental Measurement of Thermal Parameters

Thermal parameters associated with battery design are often measured using calorimetric methods. This section outlines the basic operating principles and different types of equipment used commonly in characterization of a Li-ion battery's thermal performance. Aspects related to battery safety are deferred until Chapter 4.

3.2.1 Isothermal Battery Calorimeters

Isothermal battery calorimeters (IBCs) provide critical heat generation and efficiency data for the battery under test. Understanding how much heat is produced by the battery allows manufacturers to design a cooling system that will operate the battery within a range that extends the life and operational safety of the battery. Before IBCs, battery manufacturers could only estimate the round-trip efficiency of a battery through electrical measurements—the battery would be discharged and then charged back to its original state of charge (SOC). This technique is limited by the inability to determine the discharge and charge efficiency independently. By using IBCs to directly measure heat, the efficiency of the battery can be determined independently for both charge and discharge currents rather than a combination of the two.

3.2.2 Basic IBC Operation

IBCs are primarily based on two different designs: temperature control and heat conduction. The temperature control design attempts to regulate the temperature of the battery under test; in essence, it attempts to keep the battery isothermal while the battery is charged and discharged. Figure 3.2 shows a basic schematic of a temperature-controlled IBC. The test chamber in which the battery is located is surrounded by thermal electric devices (TEDs). TEDs are used because they have the ability to cool and heat depending on the direction of the current to the TEDs—batteries will experience endothermic and exothermic conditions during cycling. Essentially, TEDs are used to control the temperature of the test chamber as well as

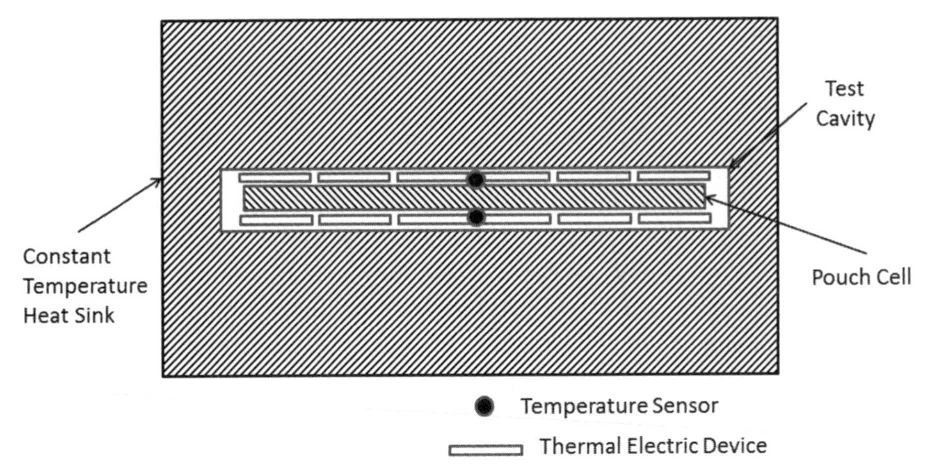

Figure 3.2 Schematic of a temperature-controlled IBC.

the temperature of the battery. As the battery is charged and discharged, the power to the TEDs is modulated to keep the battery at an isothermal temperature. The difference in power required by the TEDs to maintain a temperature under different test conditions is calibrated to a known heat flow into and out of the chamber. The primary limitation of this design is that the battery temperature used for the TED control is limited to a finite number of temperature sensors. Furthermore, only the surface temperature of the battery can be measured so the battery is not truly isothermal; only its surface is considered isothermal. Hence the measurement of reaction heats under normal operating conditions (e.g., within a pack) will need to be carried out after calibrating out the temperature nonuniformity.

The second type of IBC is a heat-conduction IBC. A schematic of a heat-conduction calorimeter is shown in Figure 3.3. The heat conduction calorimeters sense heat flux between the battery and a heat sink. If the battery is hotter or colder than the heat sink, heat flows between the heat sink and the sample. In actual practice, the thermal conductivity of the path between the sample and the heat sink is

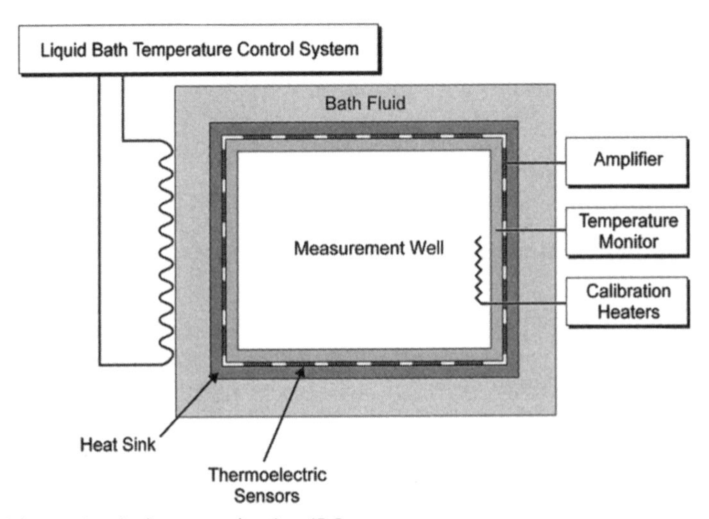

Figure 3.3 Schematic of a heat conduction IBC.

matched to the expected heat flow so that the temperature difference between the battery and the heat sink is minimized. The temperature of the heat sink is kept constant and the entire calorimeter shielded from its surroundings by a constant-temperature bath. The temperature control of the heat sink, together with proper matching of the thermal conductivity of the path between battery and the heat sink, renders a passive isothermal measurement condition. The heat flow between the battery under test and the heat sink is measured with a set of TEDs. A temperature difference across the TEDs generates a voltage proportional to a known heat flow. The limitation of this type of IBC is that the measured heat is dependent on the mass (time constant) of the calorimeter chamber; the larger the chamber, the larger the time difference between when the heat is generated in the battery and when the heat is sensed by the TEDs. The other limitation is that the battery is not isothermal under high charge and discharge currents; coupling the battery to the test chamber walls can mitigate the temperature change but not eliminate it.

Figure 3.4 shows a typical IBC response to a battery being tested over a range of currents; this particular battery is used for PHEV applications. The heat curve generated from the calorimeter is essentially power (heat generated by the battery) versus time; the test should begin and end when an isothermal temperature condition exists for the calorimeter and the battery under test. The thermal power (dQ/dt) generated within the battery upon charge or discharge is integrated to determine how much heat (Q) is produced during the test. Whether the heat comes from the entropic heating or joule heating of a battery, the time integration of the instantaneous thermal power allows for the total heat (Q) to be measured. Equation (3.5) shows how to calculate the heat release from/to the battery under test.

$$Q = \int (P_{thermal})\,dt \tag{3.5}$$

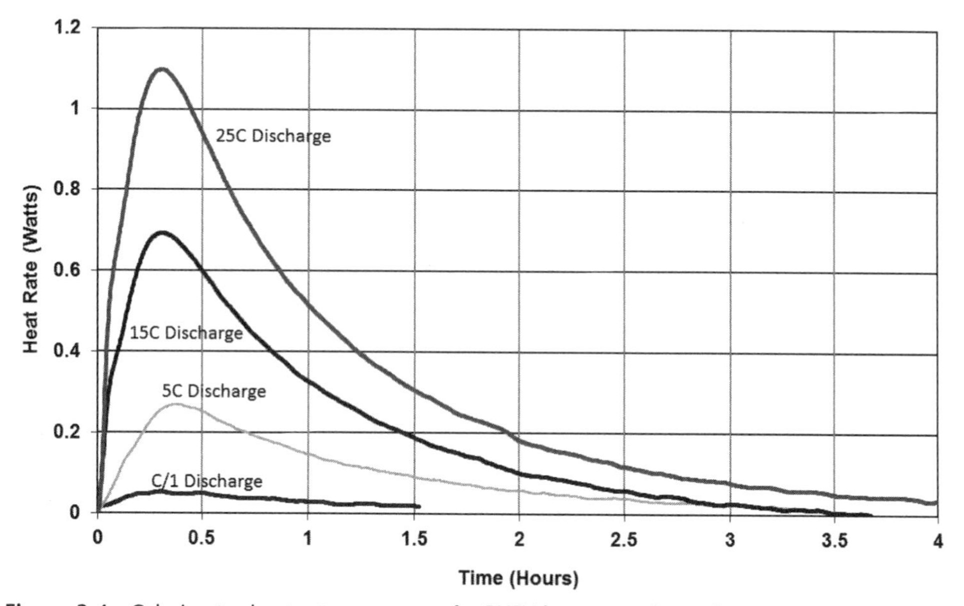

Figure 3.4 Calorimeter heat-rate response of a PHEV battery under various constant current discharges. The discharge is from 100% to 0% SOC for all currents.

The energy efficiency of the battery during constant current charge/discharge cycles can then be determined by the following equation:

$$\eta = 100\left[1 - \frac{Q}{E(Input/Output)}\right] \tag{3.6}$$

The heat generated by the cell is due to the I^2R losses in the cell and the chemical changes within the cell as measured by the calorimeter. The E(input/output) is the electrical energy supplied or taken away from the cell over the testing cycle. Both the heat generated and the electrical energy are measured in joules. The average cell heat rate in watts is determined by the following equation:

$$HeatRate_{Cell} = \frac{Q}{Time_{Cycle}} \tag{3.7}$$

The heat generated is in joules and is divided by the cycle time in seconds. The cycle time is the time over which the charge or discharge of the battery was completed. For instance, a C/1 discharge from 100% to 0% SOC takes approximately 60 minutes (3,600 seconds).

3.2.3 Typical Applications for an IBC

As mentioned earlier, IBCs find use in a variety of applications to characterize the thermal performance of batteries. This section outlines some examples.

3.2.3.1 Efficiency and Heat Generation

A typical heat generation curve produced by the calorimeter for a battery is shown in Figure 3.5(a). The figure shows how the heat generation is dependent on ambient temperature conditions and the magnitude of the current applied to the battery. The associated efficiency losses are summarized in Figure 3.5(b). Battery manufacturers can use such data to design a thermal management system to maintain the battery pack within the desired temperature range, preventing it from becoming too hot. Regulating the operating temperature of a battery pack is essential because it affects performance (power and capacity), charge acceptance (during regenerative braking), life span, safety, and vehicle operating and maintenance expenses.

3.2.3.2 Entropic Heating of Batteries

IBCs are designed to measure the joule heating and entropic heating of a battery under test. As a battery is discharged from 100% to 0% SOC, the battery goes through several crystalline phase transitions as shown in Figure 3.6. The battery in this figure was discharged at a very low current to limit the joule heating of the battery. As shown in the figure, the battery undergoes an endothermic transition at about 3 hours, which is the equivalent of 70% SOC.

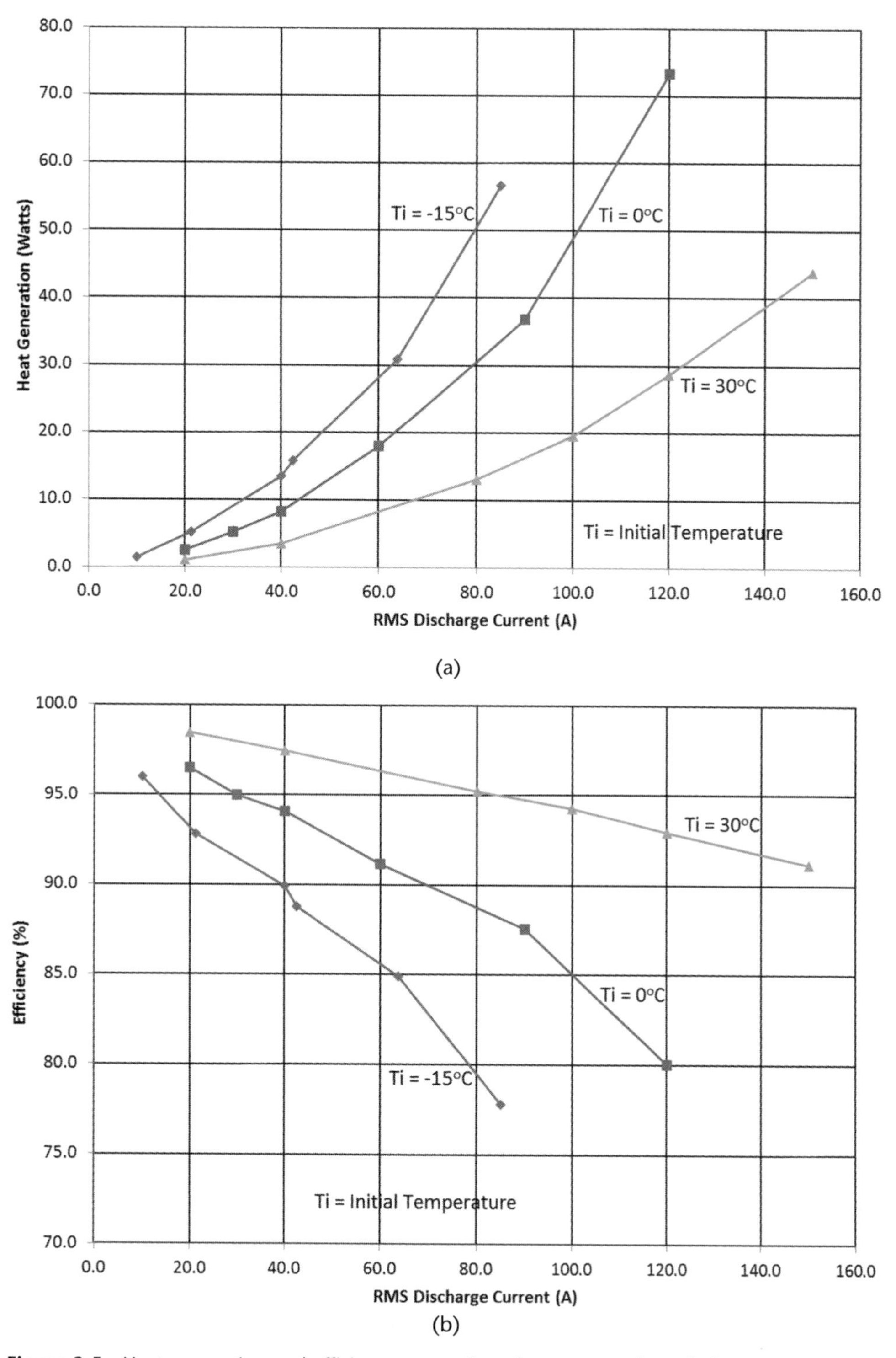

Figure 3.5 Heat generation and efficiency curves for a battery at various discharge currents and temperatures: this critical information is used to regulate the temperature of the battery during use.

With a crystalline or any other phase transition, the material going through the phase transition expands and contracts. For the most part, batteries are engineered

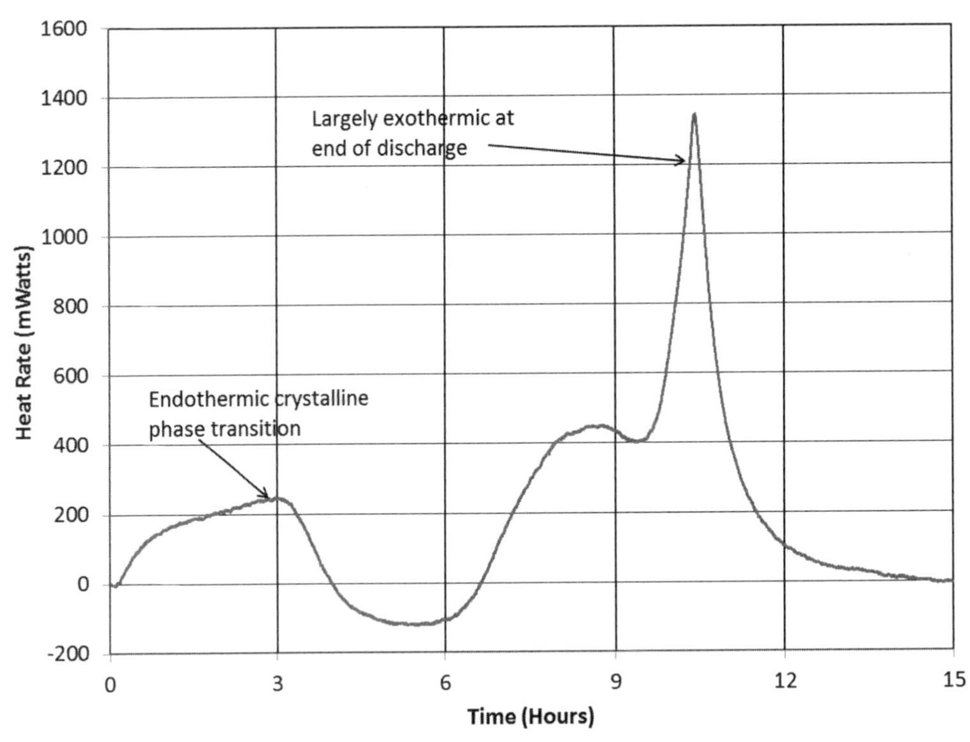

Figure 3.6 Endothermic crystalline phase transition at 70% SOC.

to mitigate damage during these phase transitions for a limited number of cycles. However, batteries typically are not engineered for repeated cycles through the phase transition that may occur in various applications. Knowing where the phase transition occurs within the SOC window allows the battery manufacturer to design a control strategy to prevent repeated cycling at this point. In the end, the control strategy can be adjusted to prevent the formation of cracks in the battery material (cathode and anode), reducing the warranty costs and the life-cycle cost of the battery while increasing the reliability of the energy storage system.

Furthermore, an energy storage device can be an endothermic device. The device requires heat to be pulled from the environment in order to sustain the internal chemical reaction—in essence, it provides mild cooling under certain test conditions. Figure 3.7 shows a Li-ion capacitor (LIC) that is completely endothermic during charge up to currents as high as 70A. Comparing these results to the heat generation rates for Li-ion batteries shown in Figure 3.1 gives us an idea of the energetics associated with the two devices employing the same working ion.

Knowing the entropic nature of the device under test allows for the thermal management system to be designed according to how the device is being cycled or charged. If this device were to be charged by a wall outlet for an electric vehicle application, then a cooling system may not be required—the cooling of the pack would occur when charging the cell, minimizing the efficiency loss. All of the available energy would be used to charge the battery pack, resulting in a quicker charge time for the pack.

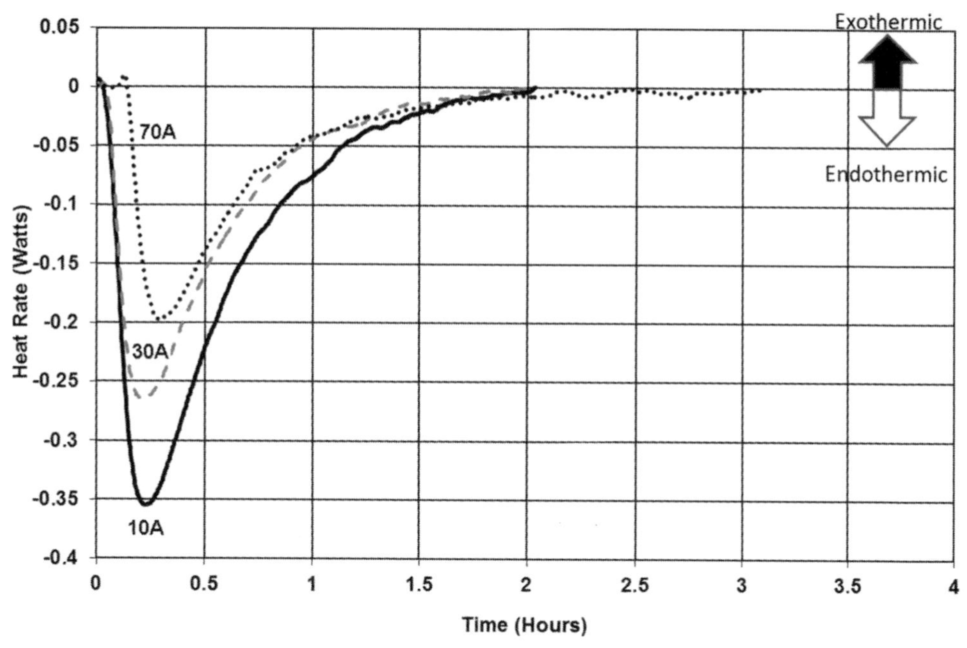

Figure 3.7 Calorimeter response to an LIC: the LIC is completely endothermic under charge currents up to 70A.

3.2.3.3 Efficiency Comparison between Cells

Figure 3.8 compares the efficiency of a second and third generation cell from the same manufacturer. The cells were discharged under a constant current from 100% to 0% SOC. The efficiency of the Gen 3 cell is slightly below the efficiency of its predecessor, the Gen 2 cell, indicating from this snapshot of data that the cell design has not improved from one generation to the next. However, cells are not typically used over their full capacity range due to life-cycle limitations of the cell. In this particular case, the cells will be used in an HEV vehicle application and be cycled from approximately 70% to 30% SOC. Figure 3.9 compares the efficiency of the Gen 2 and Gen 3 cell over this usage range of the cell. As can be seen from the figure, the

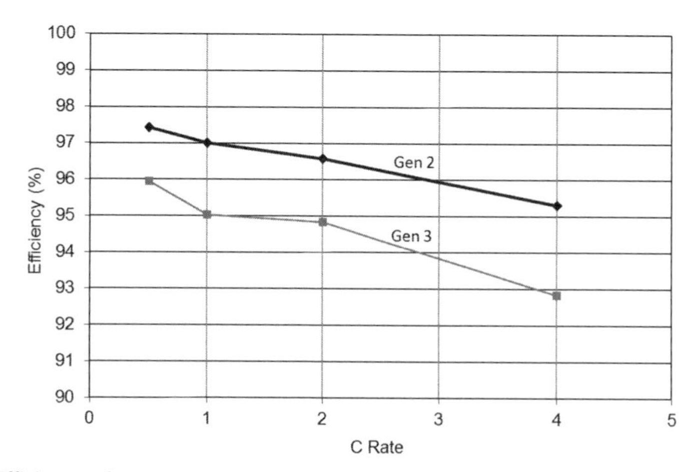

Figure 3.8 Efficiency of two generations of cells tested at 30oC from 100% to 0% SOC.

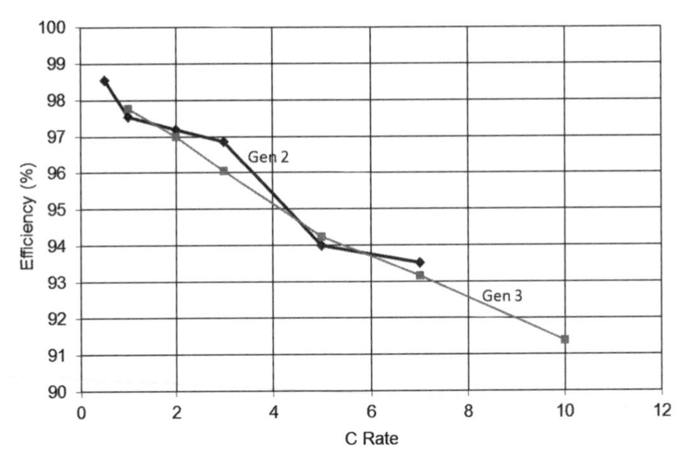

Figure 3.9 Efficiency of two generations of cells tested at 30oC from 70% to 30% SOC.

efficiencies of the two cells are fairly well matched. Battery manufacturers use the data from IBCs to ensure that the cell has the desired efficiency over the usage range while making trade-offs on other aspects of the cell design such as low temperature operation, safety, cost, and ease of manufacturing.

IBCs are crucial tools in assessing the thermal performance of lithium ion cells. IBCs provide key data for the accurate design of a thermal management system, which directly affects the cycle life cost of the battery pack. Faster degradation due to temperature means replacing the expensive batteries sooner, which makes electric drive vehicles (EDVs) less affordable. Early reports on capacity fading in the Nissan Leaf EV batteries and Li-ion battery fires in Boeing Dreamliner airplanes provide dramatic illustrations of how uncontrollable temperature increases can be during system failure. IBCs can help understand how temperature affects a battery's life and how changes made to the battery design affect its thermal signature at different temperatures.

3.3 Differential Scanning Calorimeters

Differential scanning calorimeters (DSC) are used primarily to assess the heat capacity of a material. Heat capacity is a measure of how much energy it takes to increase the temperature of a known mass by 1 Kelvin. The SI units for heat capacity are joules/kg/K whereas the imperial unit for heat capacity is BTU/°F/lbm. There are two types of differential scanning calorimeters: heat flux and heat flow DSCs. The heat flux and heat flow DSCs are shown in Figure 3.10. The heat flux DSC consists of a chamber that contains two sample holders or pans; the sample pan contains the material under study whereas the reference pan is empty. The two pans are identical in their mass and volume and are typically made from high thermal conductivity materials such as aluminum. If a pressure-rated pan is used, then the pan material is made from stainless steel or titanium—a higher tensile strength material but a lower thermal conductivity of these materials also limits the accuracy of the

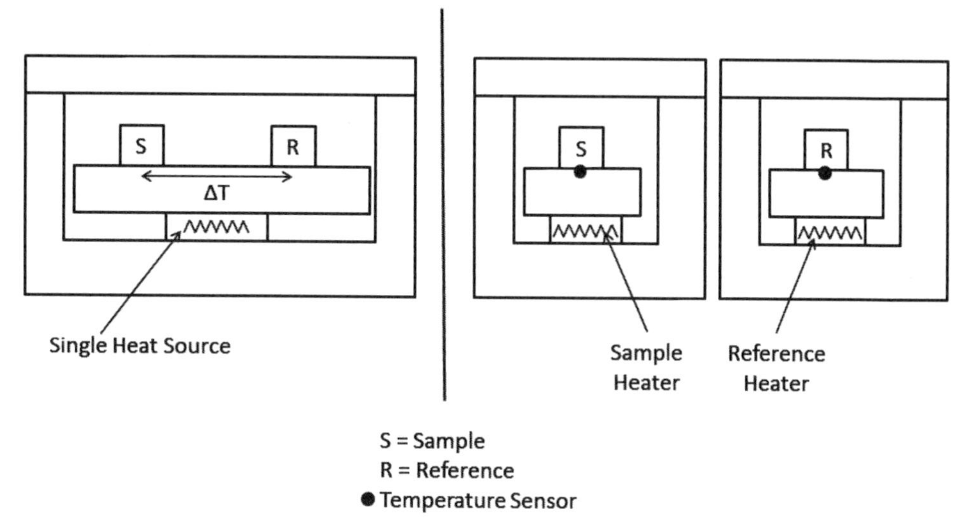

Figure 3.10 Heat flux (left) and heat flow (right) differential scanning schematics.

results. In a heat flux DSC, the temperature difference between the sample pan and the reference pan is measured. Using calibration curves, the temperature difference between the sample and the reference pans allows for the heat flux to be calculated. In general, the basic governing equation for a DSC experiment is shown in (3.8) where the dH/dt is the change in specific energy (J/g) with respect to time (s), C_p is the heat capacity of the material in J/g/°C, and the dT/dt (°C/s) is the programmed heating rate for the DSC:

$$\frac{dH}{dt} = C_p \frac{dT}{dt} \tag{3.8}$$

Different manufacturers have different correction factors associated with (3.8) that relates to the design of their specific unit. Heat flux DSCs are more cost-effective and most of the commercially available calorimeters are based off of this design. The heat flow DSC has two identical chambers that have their own temperature sensors and heaters. The temperature of each chamber is independently controlled and matched to the other. The difference in the heater power between the two chambers directly relates to the heat capacity and/or enthalpy of the sample under test.

The typical heating rate used for DSC experiments is 10°C/min—a matter of tradition only. It should be noted that the sensitivity and productivity are increased by going to higher heat rates but this limits the temperature resolution and accuracy. In essence, if one is primarily interested in the melting temperature of a material, then a lower heating rate would better suit one's needs. As for the benefits of higher heating rates, these suppress kinetically hindered processes such as crystallization; thus, the high heating rates allow for the thermodynamic properties of unstable materials such as electrolytes to be investigated.

3.3.1 Differential Scanning Calorimeters and Batteries

In its intended and basic operation, a differential scanning calorimeter can be used to assess the heat capacity of the different materials associated with a lithium cell—electrolyte, separator, cathode, anode, and electrodes. The heat capacity is very useful and is required for developing accurate electrochemical/thermal models for a lithium cell. Figure 3.11 shows the heat capacity of an electrolyte solution containing different amounts of $LiPF_6$ salt in a 1:1 mixture of EC/EMC. The heat capacity of the electrolyte varies as a function of temperature and composition, indicating that the heat generation from the electrolyte as shown on Figure 3.11, at least in part, originates from poor distribution of the electrolyte (or the ions within the electrolyte) within the large-format cell.

DSCs are being used more and more extensively to assess the safety of the different components within a Li-ion cell. One of the safety features of interest to battery manufacturers is shutdown separators. A shutdown separator consists of multiple polymeric layers and each layer has a different melting temperature. A typical shutdown separator consists of three layers: polypropylene (PP), polyethylene (PE), and another layer of PP. The PE layer has a melting point of around 130°C whereas the PP layers have a melting point of around 155°C. The separator shuts down when the PE melts and fills the pores, preventing Li-ions from transferring between the cathode and anode. DSCs are being used to assess the effectiveness of these shutdown separators by providing critical data such as melting temperatures and the enthalpy of fusion. Figure 3.12 shows a DSC run to assess the melting

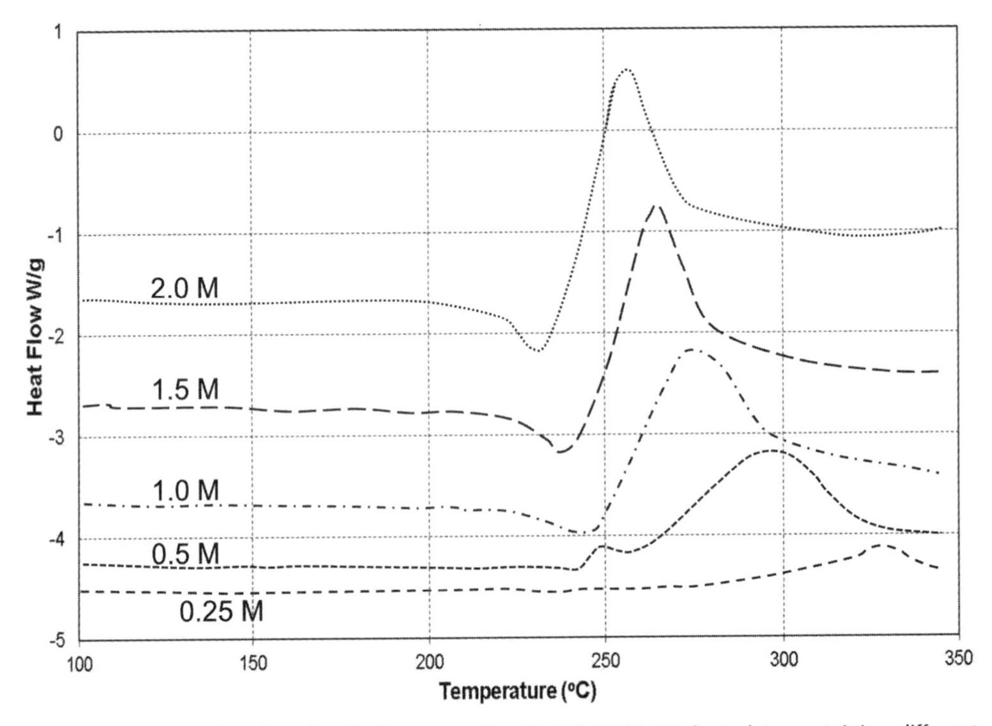

Figure 3.11 DSC data showing the heat capacity of a 1:1 EC/EMC electrolyte containing different amounts of LiPF6. (Based on data from: Botte et al. [2].)

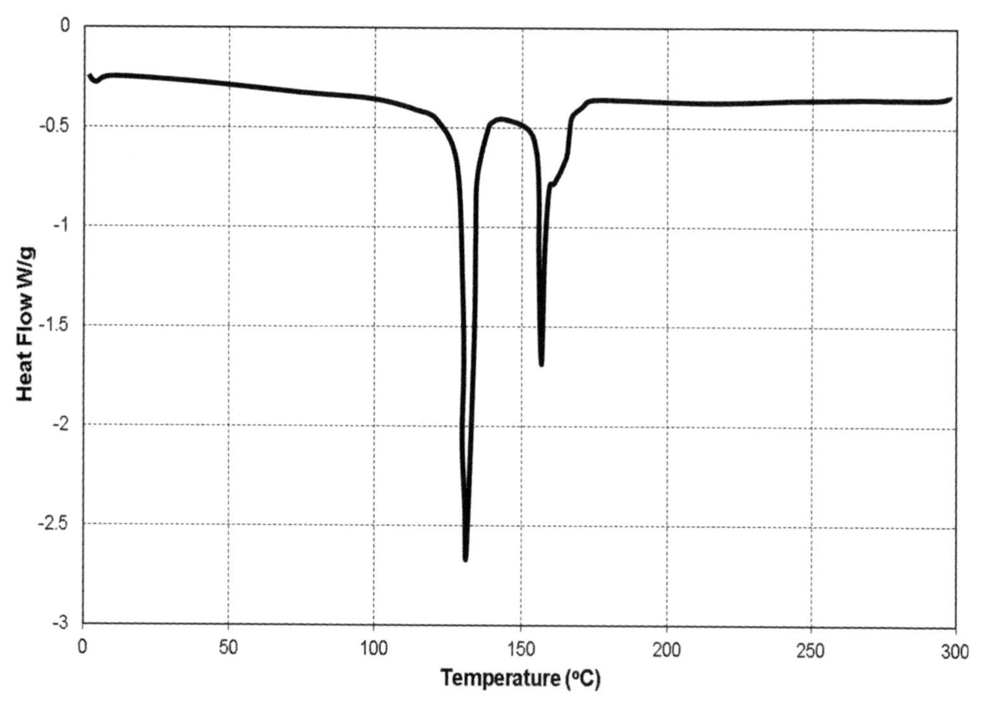

Figure 3.12 DSC run showing the melting point and enthalpy changes in a trilayer separator.

temperature of a typical trilayer (PP/PE/PP) shutdown separator. The gradient of heat generation near the melting point is an indicator of how fast the separator can shut down in case of excessive heat generation within the cell.

Despite the promise of Li-ion batteries, there are problems such as life and safety issues that have been bottlenecks for the introduction of these batteries in mass-produced advanced electrified vehicles. Some of the degradation and capacity loss can be traced to corrosion of the electrodes due to electrode dissolution and reactions that occur between the electrodes and the surrounding electrolyte. To address these problems, an artificial solid electrolyte interface layer can be deposited on the cathode or anode through a process called atomic layer deposition (ALD). ALD can affect the thermal stability of Li-ion electrodes and thermal behaviors of the full cell at high temperatures. Figure 3.13 shows how repeated ALD applications of Al_2O_3 can affect the thermal stability of a NCA cathode. Four different samples were tested with a DSC: (1) bare NCA, 92) NCA with six ALD applications of Al_2O_3, (3) NCA with 10 ALD applications of Al_2O_3, and (4) an aluminum blank with excess electrolyte. The DSC results show that repeated applications of Al_2O_3 reduces the amount of energy released as compared to the bare NCA sample. At the same time, there is not much difference between the cathode surface coated with six ALD layers of Al_2O_3 versus that coated with 10 layers, indicating that six layers of the coating is adequate to mitigate the heat generated due to reaction between NCA and the electrolyte. This is but one example that calorimetry is an essential tool for the development of safe and reliable Li-ion batteries for future vehicle and grid applications.

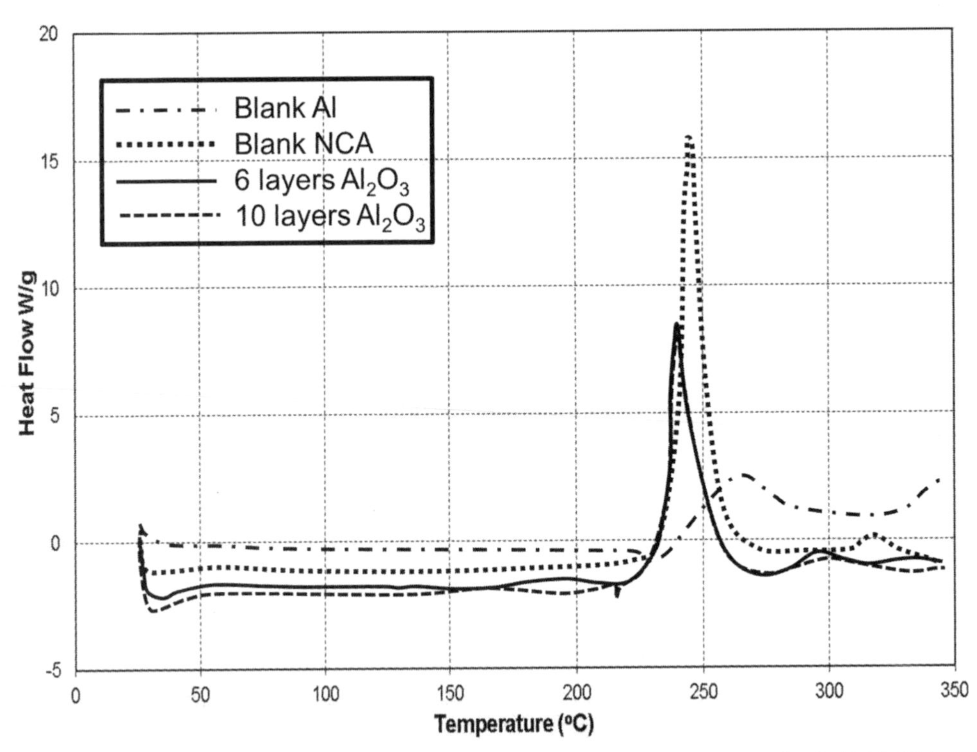

Figure 3.13 DSCs can be used to assess whether safety features such as an artificial Al2O3 SEI layer incorporated into Li-ion cells interfere with the nominal operation of the cell.

3.4 Infrared Imaging

Thermal imaging has been in use for the past 50 years and was originally developed for military applications. Presently, infrared (IR) thermal imaging technology is widely used by industry and is being applied in such fields as health care, firefighting, building maintenance, and construction. Basically, IR thermography or thermal imaging is used to detect radiation in the infrared range of the electromagnetic spectrum—typically, the radiation has a wavelength of 9 to 14 microns. Specialized IR cameras have been developed to detect the radiation and produce thermal images of the object. The thermal images can be used to understand heat flow paths in buildings, the human body, and battery systems.

3.4.1 Origin of Thermal Energy

All objects above a temperature of absolute zero emit infrared or thermal energy. The IR radiation is a measure of the atomic or molecular movement in an object due to its temperature. All thermal energy emitted by an object will be absorbed, reflected, or transmitted. Figure 3.14 shows a schematic of the process.

The absorption energy is a measure of how much incident energy is absorbed by the object. Typically, objects like asphalt, ceramics, and plastics are good at absorbing energy. The reflection energy is a measure of how much energy reflects off of an object. Materials like aluminum, copper, and silver are good reflectors. The transmission energy is a measure of how much energy is transmitted through an

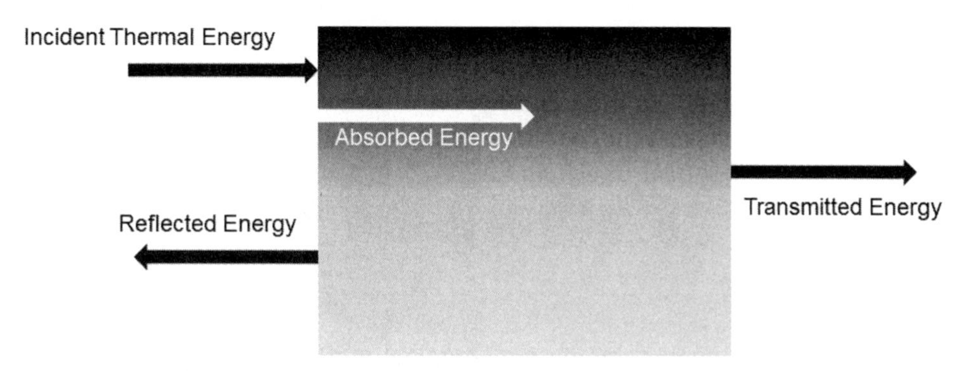

Figure 3.14 Incident thermal energy on an object is either absorbed, reflected, or transmitted.

object—as a general rule, objects, including window glass, are poor transmitters of IR radiation. Barium fluoride, calcium fluoride, and germanium are exceptions to this rule and are therefore used, at times, for the lenses in IR cameras.

If an object is under isothermal steady-state conditions, then 100% of the thermal energy directed at it is reflected, transmitted, or absorbed.

$$\%\text{Absorbed} + \%\text{Transmitted} + \%\text{Reflected} = 100\% \tag{3.9}$$

If the object is assumed to be opaque to IR radiation (and most objects are), then no energy is transmitted through the object and (3.9) becomes:

$$\%\text{Reflected} + \%\text{Absorbed} = 100\% \tag{3.10}$$

Since the object in our example is at an isothermal condition, then the absorbed energy must equal the emitted energy and (3.10) becomes:

$$\%\text{Reflected} + \%\text{Emitted} = 100\% \tag{3.11}$$

Stated in another way, the reflectivity and emissivity for opaque objects add together to be 100%. Thus, we get to the more practical definition of emissivity and reflectivity for opaque objects.

$$\%\text{Reflectivity} + \%\text{Emissivity} = 100\% \tag{3.12}$$

3.4.1.1 How Does the Emissivity and Reflectivity of an Object Affect Its Thermal Image?

The emissivity and reflectivity of materials can have a large impact on an object's thermal signature. To further complicate the matter, the emissivity of a material can vary depending on how the material was fabricated or processed—the emissivity of highly polished copper is approximately 0.02 whereas burnished copper has an emissivity of approximately 0.10. Figure 3.15 shows a thermal image of a battery under isothermal conditions. The thermal image indicates that the battery terminals are hotter than the rest of the battery despite the isothermal conditions.

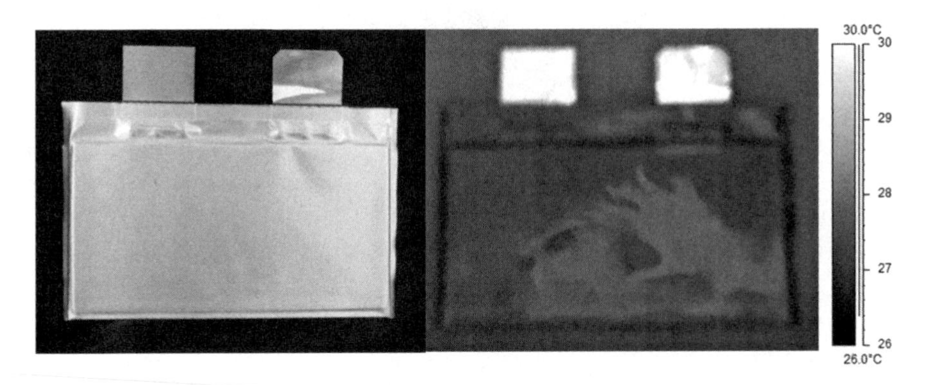

Figure 3.15 Isothermal pouch cell (left) and its corresponding infrared image: the terminals are reflecting the IR radiation in the laboratory and therefore appear hotter than the rest of the cell.

The terminals of the battery are made from aluminum (positive) and nickel-coated copper (negative) materials; these materials are highly reflective and are reflecting the IR radiation within the laboratory instead of emitting a radiative signature that is indicative of its temperature. Upon looking close enough, one can see the IR signature of the camera-shy researcher taking the IR image.

When IR-imaging a battery, it is imperative that the emissivity of the battery be consistent across its entire surface, including the connecting busbars and battery terminals and tabs. The battery can be painted with high-emissivity paint; flat paints typically have a higher emissivity than glossy paints. However, using a paint to adjust the emissivity of a surface requires a fairly thick coating that could artificially affect the heat signature of the object being imaged and paints are not easy to remove. Boron nitride, on the other hand, has a high emissivity and comes in convenient spray can that requires a very thin coating to increase the emissivity of a surface to approximately 0.9. Boron nitride is a ceramic material that is used primarily as a lubricant in vacuum systems due to its low vapor pressure but works very well for adjusting the emissivity of a highly reflective object. Figure 3.16 shows a picture of a battery sprayed with a thin layer of boron nitride. The text "Cell #1" written on the surface of the battery can be read through the sufficiently thin layer of boron nitride. Figure 3.16 also shows a thermal image of the

Figure 3.16 Isothermal pouch cell covered with boron nitride to diminish reflectivity of its surface (left) and its corresponding IR image.

boron nitride coated battery under isothermal conditions. The battery temperature is very uniform, as expected, and the terminals no longer appear to be warmer than the rest of the battery because the reflectivity of these surfaces has been diminished due to the coating of boron nitride.

3.4.2 Calibration and Error

Before imaging a battery system, the IR thermal camera needs to be calibrated for the emissivity of the battery being imaged. The easiest way to perform this calibration is to place several calibrated thermocouples (TCs) on the surface of the battery with a thermally conductive epoxy. The battery and the TCs/epoxy should be coated with boron nitride to have a consistent high emissivity surface before performing the calibration. The temperature of the TCs and the temperature given by the IR camera can be synchronized by adjusting the emissivity setting on the camera. The temperature calibration of the camera should take place over the temperature range that the battery will be experiencing during imaging.

When capturing the IR image of a battery, the largest source of error is due to the variation in emissivity across the battery's surface. The next largest source of error is due to the viewing angle between the camera lens and the battery. A flat and uniform battery, as long as it is limited in size, will have limited temperature error due to viewing angle. However, a curved surface on a cylindrical battery will result in progressively different viewing angles and focal points. The result will be an increasingly blurred surface with less temperature accuracy as the imaged surface approaches a 90° angle to the plane of the camera/lens, as shown in Figure 3.17. If a hot spot or defect is detected on a curved surface, then the camera angle should be adjusted to ensure that the defect is real and not an artifact of the viewing angle.

3.4.3 Imaging Battery Systems

IR imaging is an important tool used to determine areas of thermal concern in battery systems. The thermal design of a cell, module, or pack can affect the cost, cycle life, and safety of a battery system. IR imaging has been used to determine manu-

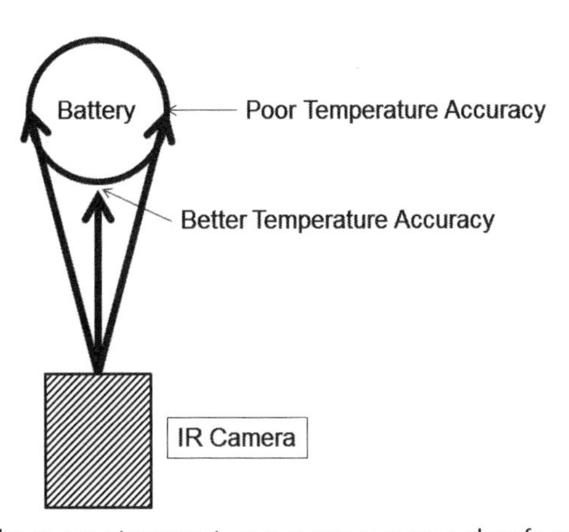

Figure 3.17 IR images have poor temperature accuracy on curved surfaces.

facturing defects, cell thermal uniformity, and how to improve the electro/thermal design of the battery.

Figure 3.18 shows a thermal image of a 12V lead acid battery module with a defect in one of its cells. This particular module made it through formation and quality acceptance but when it was put on test the modules amp-hour capacity decreased quickly and therefore had a short cycle life. To understand the root cause of the short cycle life, the module was thermally imaged. The imaging revealed that the module had apparently been struck with an object, possibly during shipping, causing a soft short in the layers of one of the spiral wound cells. The short could not be detected by visually inspecting the surface of the module and was only detected after thermal imaging the module. Understanding the limitation of the modules external enclosure, the manufacturer of the battery changed the hardness of the plastic so that any force applied to the surface would not be transferred to the cell layers contained within and cause an early failure of the module.

Figures 3.19 and 3.20 show the thermal images of two PHEV cells from different manufacturers at the end of a constant current C/1 discharge; the Ah capacities

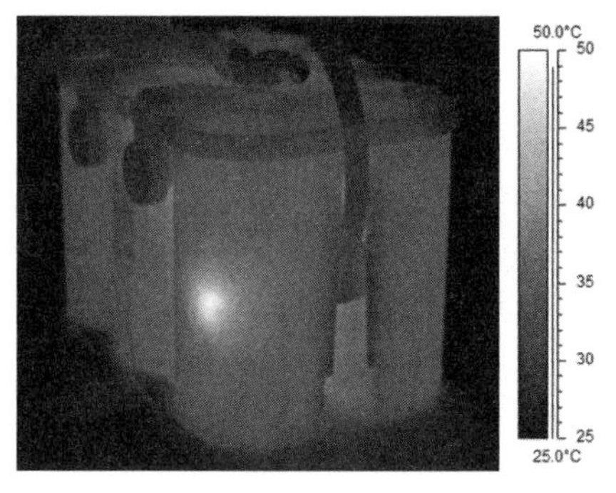

Figure 3.18 IR image of a 12V lead acid battery shows damage that is otherwise undetectable during a visual inspection.

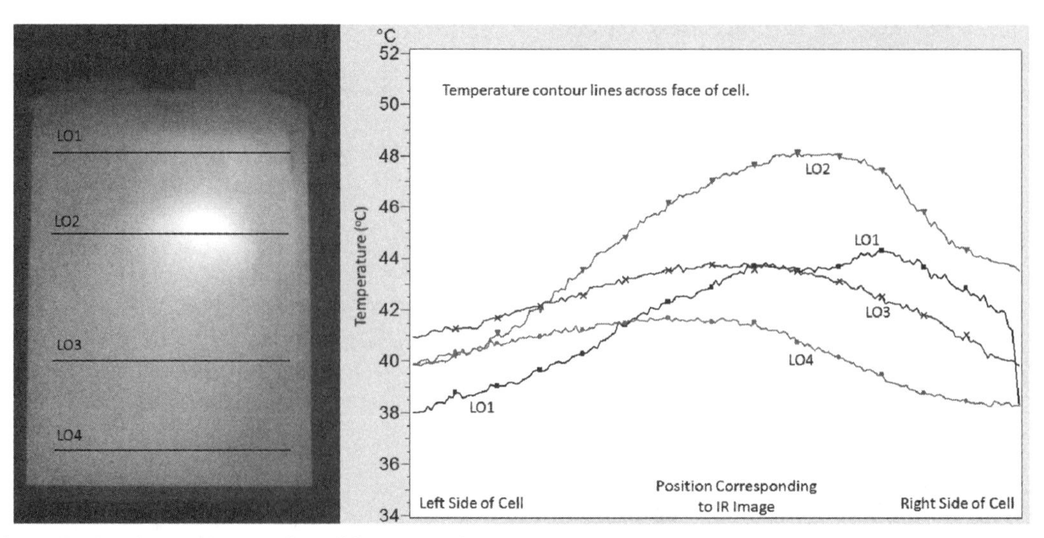

Figure 3.19 Thermal image of a cell from manufacturer A under constant current C/1discharge from 100% to 0% SOC.

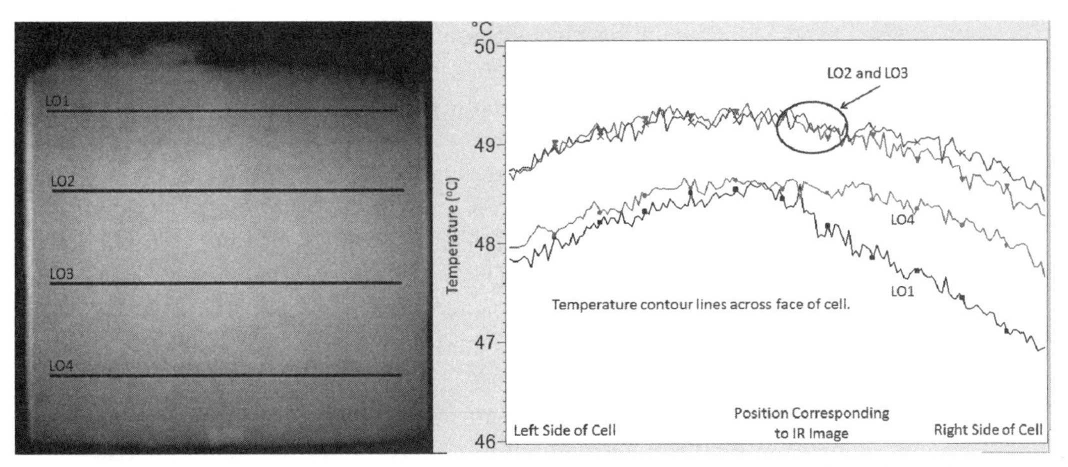

Figure 3.20 Thermal image of a cell from manufacturer B under constant current C/1 discharge from 100% to 0% SOC.

of the cells are within 5% of one another. Each figure contains a thermal image of the cell at the end of the discharge as well as plot indicating horizontal contour lines across the face of the cell: L01, L02, L03, and L04. Figure 3.19 shows a hot spot in the upper right corner of the thermal image of the cell as well as a wide spread in temperature across the face of the cell from top to bottom and left to right. Figure 3.20, on the other hand, shows a very uniform temperature distribution across the face of the cell at the end of discharge. When the cell temperature is uniform and consistent, all areas within the cell age at the same rate, leading to a better cycle life. After discussions with the manufacturer of the less thermally uniform cell, it was determined that the cell was gassing during cycling and that the contact resistance between the different layers in the cell increased, which led to the temperature variations across the surface. The minimal gassing of the cell during cycling was not detectable by visually inspecting the cell and was detected only by thermally imaging the cell.

IR imaging of battery systems is a crucial tool in understanding the thermal inefficiencies associated with a battery design. IR imaging can identify mechanical and thermal defects that are not apparent when visually inspecting the cell. The use of IR imaging continues to expand and will continue to be a useful tool in determining areas of thermal concern with regard to battery cell and system design.

3.5 Desired Attributes of a Thermal Management System

Given the various modes of heat generation within a battery, a thermal management system invariably helps improve battery performance. The goal of a thermal management system is to ensure that the battery pack can deliver the specified load requirements at an optimum average temperature (as dictated by life and performance trade-off) with an even temperature distribution (i.e., only small variations within the cell, between the modules, and within the pack) as identified by the battery manufacturer. However, the pack thermal management system has to meet the requirements of the vehicle as specified by the vehicle manufacturer as well; for example, it must be compact, lightweight, low cost, easily packaged, and compatible with location in the vehicle. In addition, it must be reliable and easily accessible for maintenance. It must also use low parasitic power, allow the pack to operate under a wide range of climate conditions (very cold to very hot), and allow ventilation if the battery generates potentially hazardous gases. There are several approaches to classify thermal management techniques for battery packs: a thermal management system may use air for heat/cooling/ventilation, liquid for cooling/heating, insulation, thermal storage such as phase change materials, or a combination of these methods; the thermal management system may be passive (i.e., only the ambient environment is used) or active (i.e., a built-in source provides heating and/or cooling at cold or hot temperatures). Figure 3.21 shows schematics of an air-cooled and liquid cooled battery system to illustrate the differences in complexity between the different cooling systems. Depending on the type of coolant(s) and the mode of heat transfer, an effective control strategy is designed for thermal management and implemented through the battery pack's electronic control unit.

3.5.1 Designing a Battery Thermal Management System

The design aspects of a battery thermal management system (BTMS) are closely related to the conventional design of heat exchangers. The basic sequence of design involve identification of the design objectives and constraints, measurement of heat generation rate and heat capacity of the modules, perform design calculations based

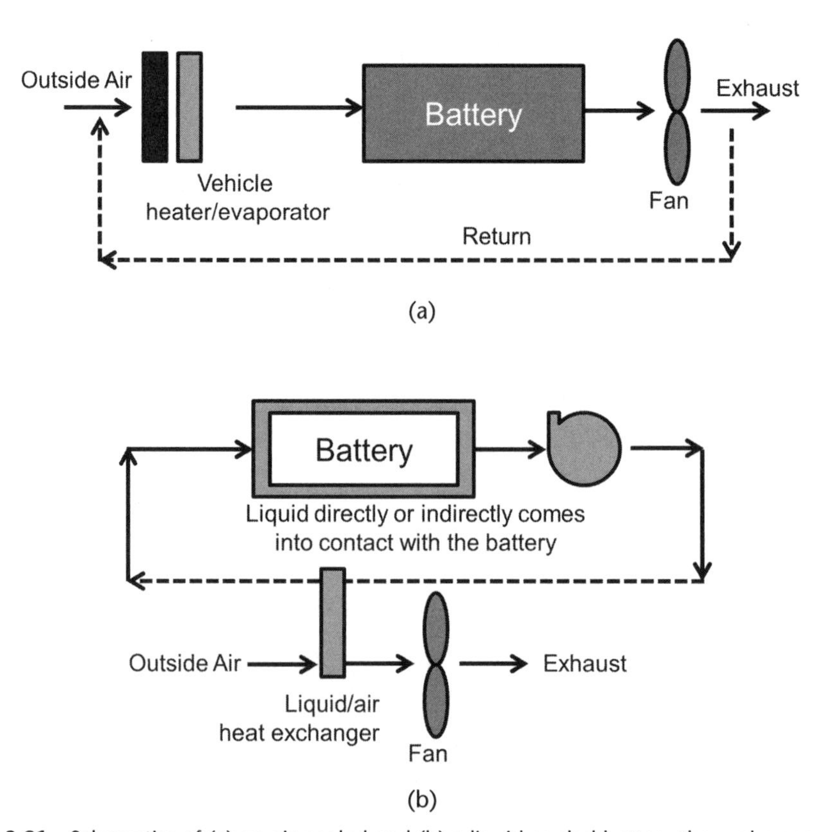

Figure 3.21 Schematics of (a) an air-cooled and (b) a liquid-cooled battery thermal management system. (From: Ahmad A. Pesaran [3].)

on the energy balance equations outlined earlier in the chapter, predict battery performance using the design equations, design, build and test the preliminary battery thermal management system, and finally optimize the unit. Each of these steps is discussed in detail in the following pages.

3.5.1.1 Defining the BTMS Design Objective and Constraints

There are three major constraints that one needs to identify when designing the BTMS: Specification of desired thermal performance for the modules and packs under various operating conditions (i.e., the average temperature, the difference between the maximum and minimum temperatures within a module and between modules). Once the target temperatures are identified, the other key challenge is to determine the potential electrical and mechanical interfaces with the rest of the vehicle:

- Size constraints (geometry, dimensions, number, orientation, and packaging of modules to be managed) as dictated by the vehicle integrator's requirements;
- Safety requirements (e.g., specification of the maximum pressure drop across the cooling channels, need for ventilation);
- Cost requirements.

In practical design scenarios, the three sets of constraints often have contrasting design outcomes that the engineer must grapple with during the optimization phase.

3.5.1.2 Obtaining Module Properties

Once the thermal performance at the individual cell level has been optimized by the cell manufacturer, it is common industrial practice to treat the module as a lumped reactor together with the packaging elements and circuitry in designing the BTMS. Toward this end, the thermal characterization of the module should be carried out. For example, the overall heat generation rate from the module must be measured under the desired charge/discharge rates at various operating temperatures. The magnitude of overall heat generation rate from the modules in a pack affects the size and design of the thermal management system. It depends on the magnitude of module internal resistance and thermodynamic heat of the (electrochemical) reaction. Thus, the heat generation rate depends on the discharge/charge profile and the module's state of charge and temperature. The heat generation could be estimated by measuring the internal resistance and enthalpy of the chemical reaction at the desired conditions. However, direct measurements could lead to more accurate values. One method is to charge/discharge a module with a cycle so the initial and final state of the charge (and temperature) of the battery remains the same. The difference between the electrical energy in and electrical energy out of the module is the heat generated in the module. When using this approach one must ensure that the state of charge of the battery at the end of the test is the same as that from the beginning of the test, but it can be time-consuming to exactly match the SOCs before and after the test. The other difficulty in using this method is that the module temperature changes with heat generation and one should estimate the energy stored in the module due to its thermal mass.

In order to do any reasonable transient or steady-state thermal analysis, the designer must know the heat capacity of the module. Overall or average heat capacity can either be measured or calculated from knowledge of heat capacity of individual components using a mass-weighted average of cell/module components. The effective heat capacity of the module is usually determined at various states of charge and temperature for proper calibration of the heat transfer models. The measurement of heat generation rates and the estimation of the heat capacity are best carried out using calorimeters as described in the previous section.

3.5.1.3 Design Calculations to Perform a First-Order BTMS Evaluation

Using the standard energy balance equation, the transient or steady-state temperature of the module/pack is determined for the anticipated operating conditions as specified in the design requirements. One of the key decisions the designer is faced with at this stage is the choice of the heat transfer medium. The heat transfer medium can be air, liquid, a phase change material, or any combination of these. Heat transfer with air is achieved simply by blowing air across the modules.

However, heat transfer with liquid can be achieved either through discrete tubing around each module, a jacket around the modules, submerging modules in a dielectric fluid for direct contact, or by placing the modules on a liquid heated/cooled plate. If the liquid is not in direct contact with the modules—for example, when a tube or jacket is used—the heat transfer medium can be water-based automotive fluids. If modules are submerged in the heat transfer fluid, the fluid must be dielectric, such as silicon-based oils, to avoid any electrical shorts. Using the air as the heat transfer medium may be the simplest approach, but the heat transfer rates are lower compared to a liquid-cooled system. With liquid systems, an extra heat exchanger for heat rejection/addition is required. In some cases, the side of the modules is wrapped with a phase change material to control the temperature of the module temporarily; however, carrying the heat from the phase change material to the exterior of the pack still relies on another suitable heat-exchange mechanism. The final selection of the heat transfer medium is made upon evaluating other factors such as added volume, mass, complexity, ease of maintenance, and cost. Such evaluations are carried out by repeating the design calculations with the different heating/cooling fluids (air, liquid), different flow paths (direct or indirect, series or parallel), and different flow rates. As part of this step, the designer needs to estimate the thermophysical properties of the fluid for further thermal analysis. The trade-off between the flow rate across the cooling channels to accomplish the requisite heat transfer rate (based on the heat transfer coefficient of the cooling fluid) and the maximum allowable pressure drop across the channels determines the choice of coolants. Figure 3.22 shows the trade-off comparisons for three different coolants, whose properties are listed in Table 1.

The following simple expression was used to relate the pressure drop (ΔP) within a channel to the coolant properties and the flow conditions:[1]

1. Parts of this example were originally published by Kim, G. H., and A. Pesaran, *World Electric Vehicle Association Journal,* Vol. 1, pp. 126–133, 2007.

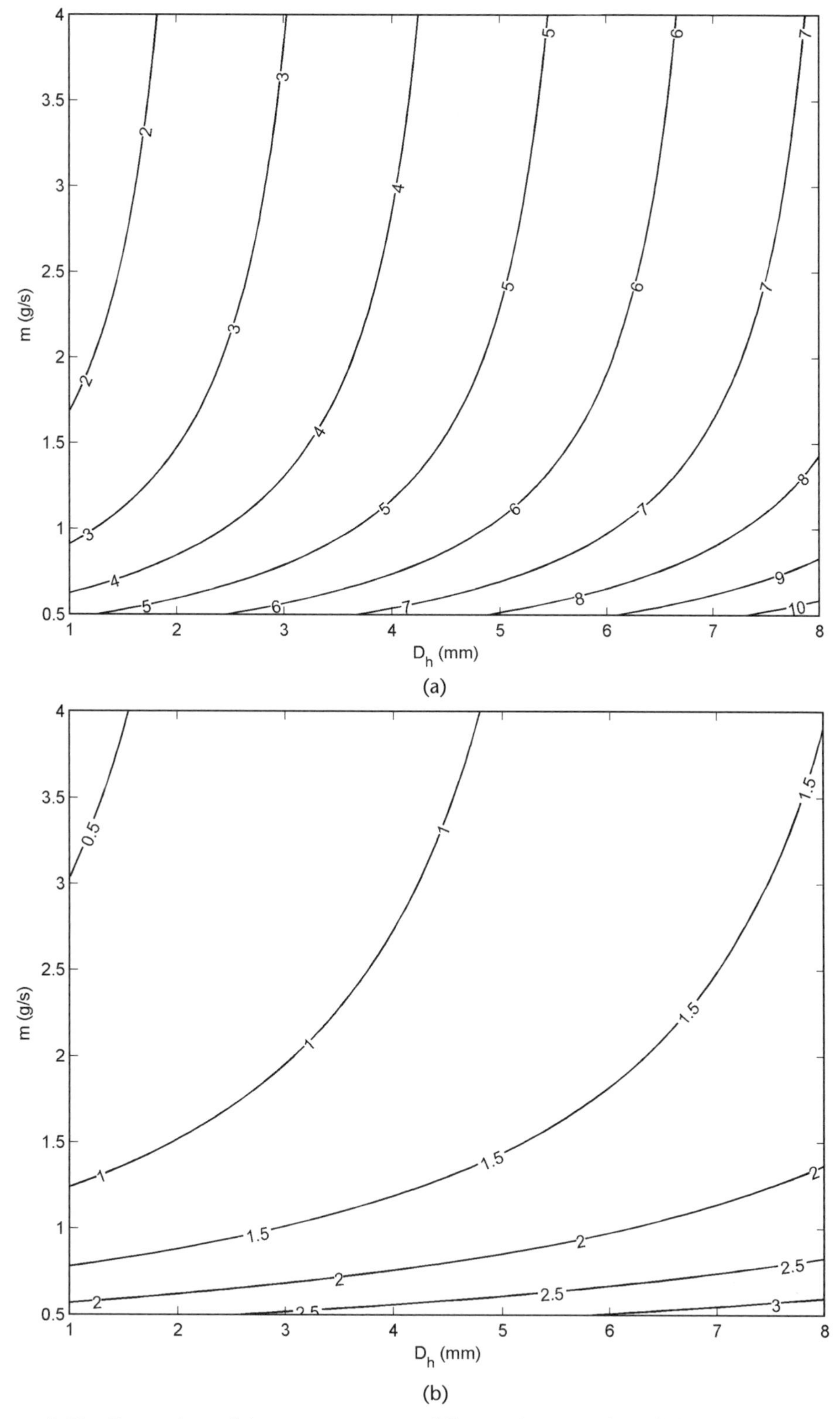

Figure 3.22 Comparison of the net temperature difference between the cell skin temperature and the inlet temperature. The coolants chosen were (a) air, (b) oil, and (c) 1:1 mixture of water/glycerol. The temperature distribution is broader for the air-cooled channel and is more sensitive to the hydraulic diameter compared to the liquid cooled channels.

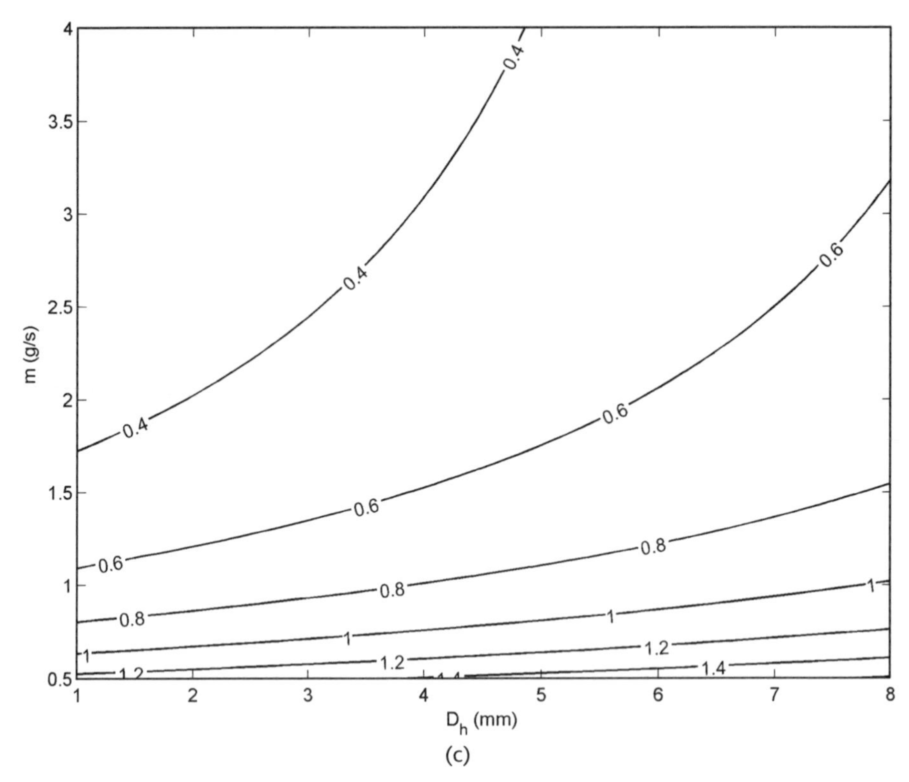

Figure 3.22 (continued)

Table 3.1 Coolant Properties for the Results Shown in Figures 3.22–3.24

Property	*Air*	*Mineral Oil*	*Water/Glycol*
Density, ρ (kg/m^3)	1.225	924.1	1069
Specific heat, C_p (J/kg/K)	1006.43	1900	3323
Thermal conductivity, k (W/m/K)	0.0242	0.13	0.3892
Kinematic viscosity, ν (m^2/s)	1.461e-5	5.6e-5	2.582e-6

$$\Delta P \propto \frac{mv}{D_h^3} \tag{3.13}$$

The temperature difference between the inlet and the exit streams is treated as the sum of two parts, the first component (ΔT_1) being the difference between the inlet stream and the mean temperature of the coolant within the flow channels, and the second part (ΔT_2) the difference between the skin temperature on the cell and the mean temperature of the coolant:

$$\Delta T_1 = \frac{q}{mC_p} \tag{3.14}$$

$$\Delta T_2 = \frac{qD_h}{k} \tag{3.15}$$

Thus, separate constraints may be imposed on the maximum temperature difference across the cell surface and along the flow direction. Simultaneously, a comprehensive bound on the net temperature difference between the cell skin temperature and the inlet temperature for the coolant can be specified as $\Delta T_{net} = \Delta T_1 + \Delta T_2$. Figure 3.22 shows a plot of ΔT_{net} contours as a function of coolant mass flow rate (m) and the hydraulic diameter (D_h) of the coolant channel for three different coolants. The values of ΔT_{net} in the air-cooled channel are much higher compared to the liquid cooled channel due to the small heat capacity and thermal conductivity of air. Another notable difference among the contours for the different coolants is that in the case where air-cooled channels are employed, the flow rate changes significantly even for small changes in the hydraulic diameter, whereas in the case of liquid-cooled channels, the contours on Figure 3.22 (b–c) have lower gradients. This indicates that ΔT_{net} is more sensitive in the case of air-cooling to the flow channel's hydraulic diameter than the mass flow rate for the range of these design variables shown in the figure, whereas, in the same domain, for the water/glycol mix, the hydraulic size of the flow channel is not expected to be the limiting factor in the channel design.

Once a preliminary choice of heating/cooling fluid and the fluid flow rate have been established, based on the flow-rate calculations, a rough estimate of the parasitic power requirements for the fan or pump is obtained. Usually, the choice of the cooling fluid follows a standard subset of industrial coolants that have been approved by regulatory bodies for use in passenger vehicles.

Let us consider another example when heat transfer by radiation is important for designing packaging material for cylindrical cells. For this case, the amount of heat generated is related to the amount of heat transferred from the surface of the cell by Newton's law of cooling:

$$mC_p \frac{dT}{dt} = hA\Delta T_{net} \tag{3.16}$$

where A is the area of the cell surface from which heat is radiated. Substituting the expressions for a cylinder and ignoring the edge effects, the rate of rise of skin temperature (which can be measured using a thermocouple attached to the cell surface) is given by:

$$\frac{dT}{dt} = \frac{h\Delta T_{net}}{\rho C_p (D_h)} \tag{3.17}$$

For the 18,650-size cells, the value of dT/dt is about 0.01 K/s for ΔT_{net} of 10 K and 0.025 for ΔT_{net} of 20K, giving an estimate for the total heat transfer coefficient as ~15 W/m^2–K for a stainless steel cell casing. Now, the total heat transfer coefficient can be interpreted as being composed of two parts: one due to convection and the other due to the radiation (i.e., $h = h_c + h_r$). From standard heat transfer texts, the Stefan-Boltzmann correlation between the skin temperature and the heat transfer coefficient due to radiation is given by:

$$h_r = 4\sigma_T \varepsilon_T T_S^3 \tag{3.18}$$

σ_T is the Boltzmann constant (5.669×10^{-8} W/m^2–K) and ε_T is the emissivity of the surface. For a perfect blackbody, the emissivity is 1, and that for steel is 0.33. These numbers correspond to the heat transfer coefficients (h_r) of 2.33 and 7 when the skin temperature of the cell (T_s) is 40°C, respectively, for steel and a cell with a suitable coating to improve radiation. Thus, there is a potential to transfer as much as 50% of the heat away from the cell by radiation.

3.5.1.4 Predicting Battery Module and Pack Performance

Building a battery pack is a labor-intensive process and it is expensive to subject the battery to several build-and-break cycles. In order to minimize the cost and effort, right before the design phase the thermal behavior of the module and/or pack is usually simulated in a virtual environment using advanced computational tools. This step involves detailed reconstruction of the pack geometry, determining the steady-state and transient thermal response of the battery in order to identify nonuniform heat accumulation in any part of the pack and to verify compliance with the design requirements identified (e.g., the maximum temperature or pressure gradients within the pack must be below the specified values). The load estimates for the cooling fan or pump are calculated using standard correlations. Another advantage of performing the flow simulations is the capability to perform sensitivity analyses. The design engineer can identify factors that the thermal response of the battery is most sensitive to, and as such prioritize these design requirements in terms of determining the right control strategy (such as modulating the flow rates and inlet fluid temperatures).

3.5.1.5 Designing, Building, and Testing the Preliminary BTMS

Based on the computational analysis, the auxiliary components such as fans, pumps, and heat exchangers together with active elements such as the heaters and evaporator coils are suitably sized. The constraints for the control unit, as identified in the previous steps, are implemented. A comparison of alternative systems with respect to performance, energy requirement, complexity, and maintenance is often carried out at this stage. Factors such as ease of operation and reliability of the different components are considered when estimating the cost of the BTMS. Once the preliminary design is finalized, a bench-top battery pack is integrated with the thermal management unit to validate the results from the simulation phase. Practical limitations on the hardware, such as lead times for the different cooling components and maximum offset in the temperature are calibrated. The designer revisits the previous steps as often as necessary until the design specifications are accomplished. The prototype unit is then subjected to evaluation of the thermal control strategies based on the vehicle's operating environment (e.g., various climactic conditions).

The prototype is installed in a vehicle and the testing is carried out using a vehicle dynamometer to simulate the drive conditions.

3.5.2 Optimization

Optimization of the final unit usually requires a few iterations of the steps outlined above, considering design factors such as the safely allowable pressure drop and temperature range. Other constraints such as impact of the operating temperature on battery performance and life, vehicle performance, cost, and ease of maintenance are considered at the system level. The following is a design example to optimize the flow channel based on the equations shown in the previous section.

Consider the air-cooled channel described above. The range for the hydraulic radius and the flow rate shown in Figure 3.22 were obtained from the design objectives. Other constraints usually include a specification of the maximum permissible change in temperature across the flow channel (i.e., $\Delta T_{1,max}$) and a preference for laminar flow conditions (a specified value for the Reynolds number, e.g., 2,400). The heat generation rate—which can be measured using a calorimeter—is set to 2W, for illustrative purposes. Figure 3.23 shows the bounds imposed by different constraints (1) a preferred value for ΔT_{net} is between 3.5 and 4.5°C, (2) the minimum pressure drop desirable to maintain a ceiling on the parasitic load is 40 kPa, (3) the maximum pressure drop that the flow channel can withstand safely is set at 110 kPa, (4) the nonuniformity in the cooling channel (ΔT_1) should not exceed

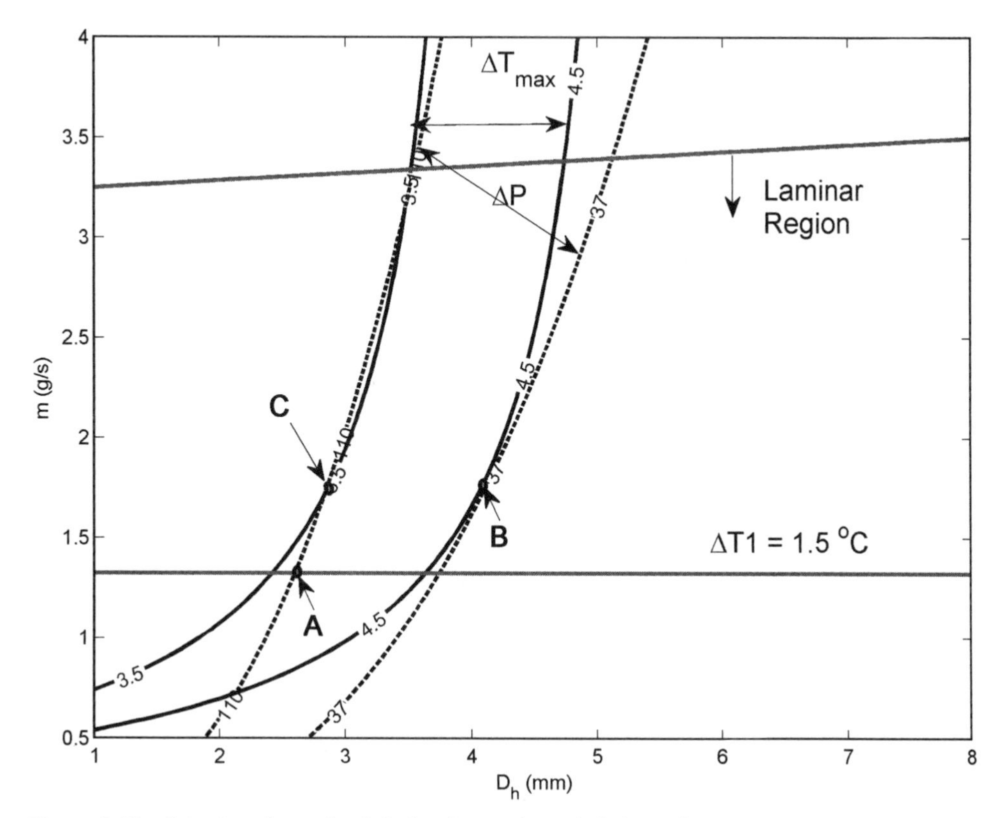

Figure 3.23 Selecting the optimal design for an air-cooled channel.

1.5°C to ensure good performance of the module, and finally, laminar flow conditions are preferred.

From (3.13)–(3.18), contours corresponding to the pressure and temperature bounds can be made similar to those shown in Figure 3.23. In addition, the contours corresponding to a fixed value for ΔT_1 (which is independent of the channel diameter) and a fixed Reynolds number are superimposed to identify the operating window. The smallest channel diameter corresponds to the maximum heat transfer coefficient; this point is shown as *A* in the figure. Point B corresponds to the design with the minimum value for the pressure drop, and point C to the optimum design that maximizes the heat transfer coefficient within the maximum allowable pressure drop across the channel while keeping the ΔT_{net} value at its minimum specification. A similar analysis for the oil-cooled system is shown in Figure 3.24, where the optimum hydraulic diameter is about 3.9 mm and the flow rate is about 1.18 g/s—much lower than that for the air-cooled channels due to the properties of the coolant.

If on the other hand, one were to opt for a phase change material (e.g., wax) to remove the heat from the skin of the cells, paraffin wax has a heat of fusion of about 200 J/g and a density of 0.82 g/cc in the molten state. About 30% of the module volume is accessible for the phase-change material packaging, when the hydraulic diameter is 3.9 mm. At the 2W per cell generation rate considered above, for a 2C discharge, the average heat generated per cell will be 3.6 kJ over a 30-minute period. Using the properties for the wax provided above, the minimum

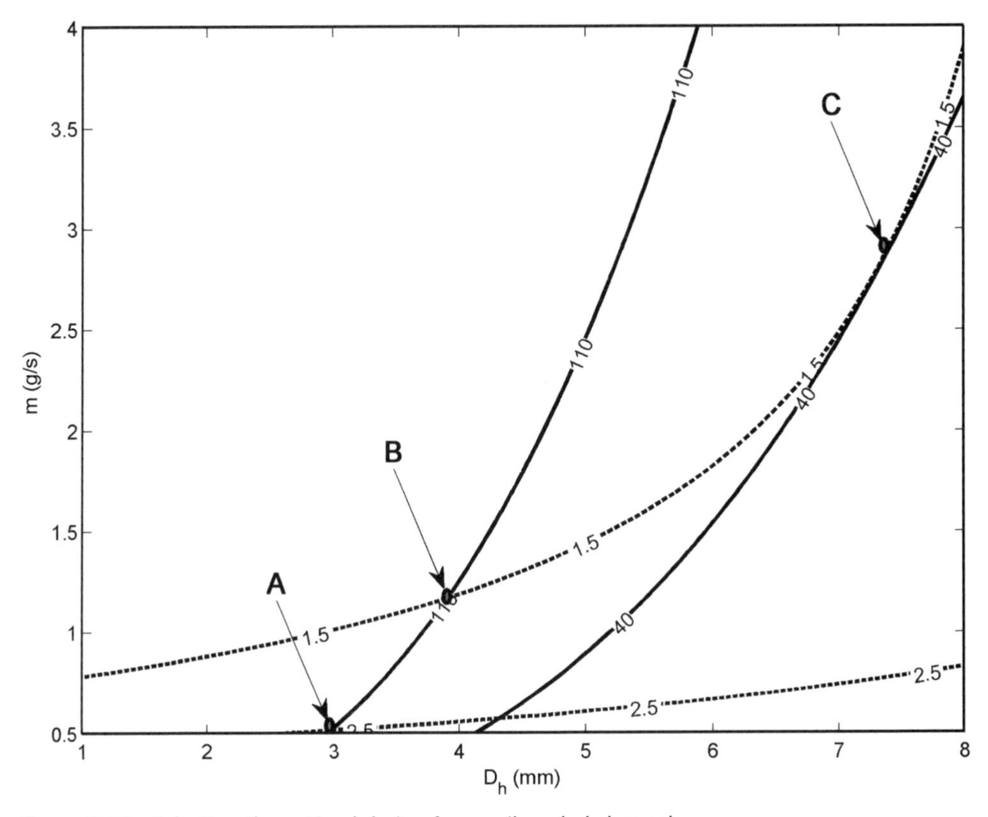

Figure 3.24 Selecting the optimal design for an oil-cooled channel.

thickness for the phase-change material is 1.44 mm, which leaves 2.5 mm for the convective coolant to carry the heat away from the module. From Figures 3.23 and 3.24, one can see that in order to maintain the pressure drop and hence the mass flow rate within acceptable limits, the only viable coolant in this scenario is air.

In the next example, let us consider a staggered arrangement of cells within a module, as shown in Figure 3.25, which have a jacket of oil-cooled channels surrounding each cell with the optimal values we just identified, and air-circulated through the module.[2]

For this case, the heat transfer coefficient is obtained from the Nusselt number (which is defined as the ratio between the convective to conductive heat transfer across the boundary) using the following expression:

$$h = \frac{k\mathrm{Nu}}{D_h} \tag{3.19}$$

The Nusselt number can be obtained from standard tables or by using standard correlations such as the one shown below [4–6]:

$$\mathrm{Nu} = 0.3 + \frac{0.62\,\mathrm{Re}^{0.5}\,\mathrm{Pr}^{0.3333}}{\left[1+\left(\frac{0.4}{\mathrm{Pr}}\right)^{(2/3)}\right]^{0.25}\left[1+\left(\frac{\mathrm{Re}}{282000}\right)^{(5/8)}\right]^{0.8}} \tag{3.20}$$

Once the Nusselt heat transfer coefficient is known, the skin temperature of the cell is related to the inlet and exit stream temperatures using the following expression for a staggered arrangement of 12 or more cylindrical cells per row:

$$\left(\frac{T_s - T_e}{T_s - T_i}\right) = \exp\left(-\frac{\pi \mathrm{D_h} N h}{\rho C_p V N_T S_T}\right) \tag{3.21}$$

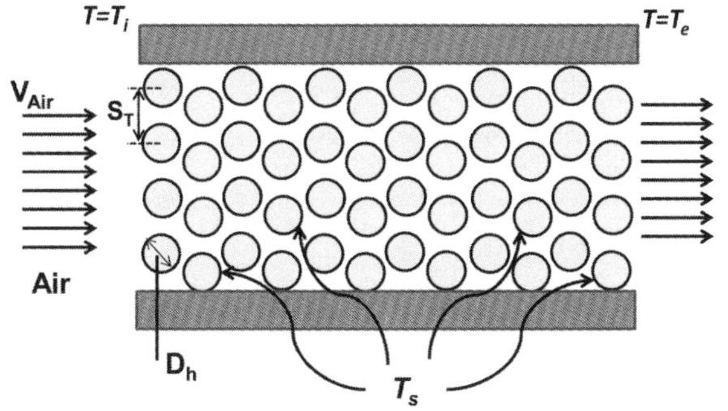

Figure 3.25 A module comprised of 18650-type cylindrical cells arranged in a staggered orientation with air flowing in between the cells to carry the heat away from the cooling jackets.

2. Data for this example was obtained from Yuksel, T., and J. Michalek. "Development of a Simulation Model to Analyze the Effect of Thermal Management on Battery Life," SAE Technical Paper 2012-01-0671, 2012.

In this example we use 12 cells per row and four rows of cells (i.e., N_T = 4); V is the volumetric flow rate of air across the module and S_T is the transverse pitch as shown in Figure 3.26. The constraints to incorporate include a specification of the maximum allowable pressure drop, temperature bounds, and minimizing the cost of the parasitic load. The temperature of the exit stream (T_e) should be as high as possible in order to maximize the amount of heat removed from the cell surface but at the same time at least 2.5°C cooler than the skin temperature (T_s) for the heat transfer to be effective. If the inlet stream temperature (T_i) is assumed to be 20°C and the skin temperature on the cells are assumed to be at 30°C, and the maximum allowable pressure drop is set at a tenth of an atmosphere (~ 10 kPa); the results are shown in Figure 3.26. The cost of the parasitic load is shown on a relative scale—it is assumed to vary linearly as the mass flow rate corresponding to the velocity of the inlet stream of air for the chosen size of the flow channel, and the values are normalized to be set equal to unity at a pressure drop corresponding to 10 kPa and the exit stream temperature of 27.5°C, in line with the constraints outlined above. A few trade-offs can be read from the chart: if the maximum pressure drop is maintained at 10 kPa but the temperature of the exit stream is allowed to be 3°C lower than the skin temperature of the cells, the cost of the parasitic load can be reduced by 15%, whereas setting this difference to be 2°C increases the cost by 30%. For a

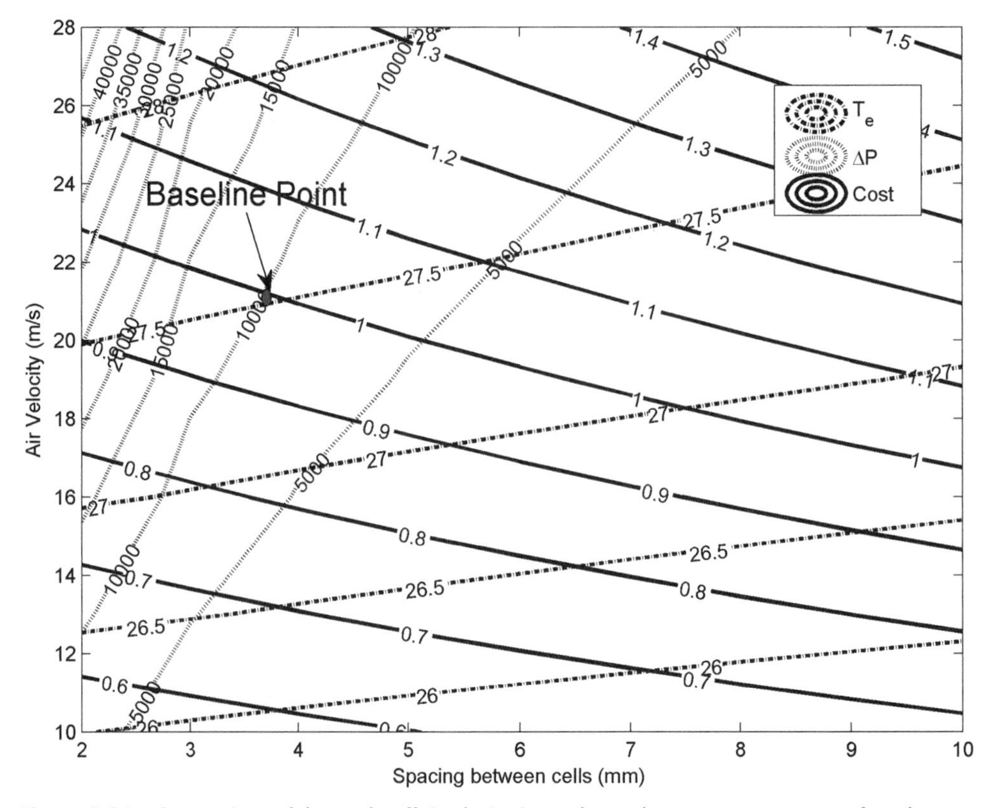

Figure 3.26 Comparison of the trade-offs in designing a thermal management system for a battery module comprised of 18650-type cells as shown in Figure 3.25: reducing the exit temperature value by 0.5°C reduces the normalized cost by 15% compared to the baseline chosen, whereas increasing this value by 0.5°C increases the cost by 30%.

set value of $T_s - T_e$ (in our case, 2.5°C), if a higher pressure drop (e.g., 5 kPa) can be tolerated, there is a corresponding reduction in the cost (by 5% in this example).

3.6 Conclusions

In summary, the battery pack performance, and thus the performance of an EV or HEV, is affected by its operating temperature and the degree of temperature gradient in the pack. Thermal issues are often of concern in a battery pack designed for vehicles because of higher power and more aggressive charge/discharge profile. Thermal analysis must be performed to properly design thermal management systems for all electric vehicles. This chapter used fundamental heat transfer principles to design thermal management units for individual cells and modules. The temperature distribution depends strongly on the properties of the coolant and the design of the flow channels. A pack with no convection can reach unacceptably high temperature levels. Even with reasonable air flow rates, the temperature in the pack can vary significantly. Attention must be given to proper thermal design of the modules and packs. The last example indicates that adding ventilation holes improves the thermal performance of a battery module installed on a vehicle.

References

[1] Gu, W. B., and C. Y. Wang. "Thermal-Electrochemical Coupled Modeling of a Lithium-Ion Cell," presented at the Fall Meeting of the Electrochemical Society, Honolulu, HI, 1999.

[2] Botte, G. G, R. E. White, and Z. Zhang. "Thermal Stability of $LiPF_6$–EC:EMC Electrolyte for Lithium Ion Batteries, *J. Power Sources,* Vol. 97–98, 2001, pp. 570–575.

[3] Pesaran, A. A. "Battery Thermal Management in EVs and HEVs: Issues and Solutions," presented at the Advanced Automotive Battery Conference, Las Vegas, NV, 2001.

[4] Incropera, F. P., D. P. Dewitt, T. L. Bergman, and A. S. Lavine. *Introduction to Heat Transfer,* Fifth Edition, Hoboken, NJ: John Wiley & Sons, 2006.

[5] Bird, R. B., W. E. Stewart, and E. N. Lightfoot. *Transport Phenomena*, Second Edition, New York: John Wiley & Sons, Inc., 2006.

[6] Churchill, S. W., and M. Bernstein. "A Correlating Equation for Forced Convection from Gases and Liquids to a Circular Cylinder in Crossflow," *J. Heat Transfer, Trans. ASME* Vol. 99, 1977, pp. 300–306, doi:10.1115/1.3450685.

CHAPTER 4

Battery Life

Energy and power of electrochemical storage systems fade due to multiple degradation mechanisms, some of which are related to cycling, but others which are related to time. To accurately predict life across a range of scenarios, sometimes five to 10 degradation mechanisms must be characterized and modeled, but often just a few mechanisms dominate.

Chapter 2 used a simple example to conceptualize a battery electrode as a container of water. Water height in the container represented electrochemical potential, while water volume represented stored charge. Taking this example a bit further to include degradation, the water "battery" can degrade either due to loss of water from the system (i.e., loss of the charge shuttle in the battery) or due to shrinking or deforming the water storage container, representing electrical isolation or chemical/phase change of active material in the electrode.

4.1 Overview

For well-designed, mature Li-ion systems, dominant degradation mechanisms include solid electrolyte interface (SEI) film growth on the negative electrode that consumes cycleable Li—the charge shuttle—from the system. Growth of SEI causes both battery resistance growth and capacity loss. The rate of SEI growth with time accelerates with high temperature and high state of charge (SOC). For Li-ion applications with frequent, deep cycling, electrode active material loss/isolation may dominate. The rate of material loss with each cycle depends on depth of discharge (DOD) and C-rate. Material loss is a common fade mechanism for emerging high energy density battery chemistries whose electrode active materials experience large expansion and contraction during cycling.

4.1.1 Physics

Characterized by their physics, failure modes may be:

1. Mechanically induced (e.g., gas buildup or vibration leading to packaging failure);
2. Chemically induced (e.g., chemical reactions that proceed with time, with rate dependent on temperature and chemical state);

3. Electrochemically induced (e.g., side reactions driven or accelerated by electrochemical (dis)charge processes);
4. Electrochemomechanically induced (e.g., material failure associated with volumetric changes and mechanical stresses caused by electrochemical (dis) charge process reactions);
5. Thermal coupling of all of the above.

Regarding purely mechanical-induced failures not related to the electrochemical state, these can be characterized with traditional accelerated vibration testing and structural analysis.

Regarding chemical and electrochemical fade mechanisms—whether they involve electron transfer or not—both proceed due to reaction processes. The distinction between chemical and electrochemical degradation is purely conceptual, as an aid to determine whether the degradation is more related to calendar time or number of cycles.

Across all mechanisms, thermal behavior plays an important role, with chemical, electrochemical, and mechanical degradation rates tightly coupled to temperature. Chemical and electrochemical reaction rates and transport properties are highly temperature-dependent, as are mechanical strength and elasticity properties. Differential expansion and contraction of cell components with temperature also causes thermally induced mechanical stress. Methods to predict battery thermal response, discussed in Chapter 3, are a necessary tool for life prediction.

4.1.2 Calendar Life Versus Cycle Life

It is common to think of battery useful lifetime in terms of number of charge/discharge cycles. For example, one might expect a battery to last for 1,000 cycles before reaching end of life (EOL). Commonly EOL is defined using some measure of faded performance relative to beginning of life (BOL). However it is important to recognize that battery chemistries may also degrade even when not in use due to unwanted chemical reactions. For batteries spending much of their lifetime in storage, such as with standby batteries for uninterruptible power supplies, time spent at a given temperature and SOC will often dictate application lifetime. Different battery chemistries each have different sweet spots for achieving long calendar life. Li-ion chemistries last longest when stored at partial SOC. Lead-acid chemistries last longest when stored at full SOC. Both chemistries suffer from short calendar life at hot temperatures, > 40°C.

The goal of aging tests is to characterize, as quickly as possible, degradation mechanisms across the anticipated range of duty-cycles and temperatures. Tests are designed to accelerate the rate of these mechanisms, compressing the time needed for testing. Given aging data, lifetime predictive models are then used to extrapolate performance fade forward in time to some EOL condition. It is not straightforward how to extrapolate cycling results that were compressed into, say, 6 months, to a relevant performance fade prediction at, say year 5 or year 10 for the actual device application. In order to extrapolate forward in time, a lifetime model must properly separate and/or capture interdependencies of cycling- and calendar-de-

Table 4.1 Failure Modes Categorized by Physics

Mechanical	Structural failure of packaging and structures
	Characterized with vibration tests, readily accelerated with vibration magnitude and rapid accumulation of vibration cycles
Chemical	Side reactions occurring during rest
	Rate dependent on temperature and chemical state
	Characterized by storage tests, accelerated with elevated temperature at various chemical states, including full charge/discharge extremes
Electrochemical	Side reactions driven by charge rate (sometimes discharge rate)
	Influenced by electrical cycling on the cell, with rate influenced by temperature-dependent reaction and transport properties
	Characterized by accelerated cycling in the temperature and potential window exciting the reaction
Electrochemomechanical	Degradation caused by material expansion/contraction during electrochemical-thermal cycling
	Influenced by material properties of system, occurrence of phase changes with cycling, packaging, external body forces, charge/discharge rate, temperature, chemical state, and mechanical damage state
	Characterized by accelerated cycling across a matrix of relevant duty cycles

pendent degradation mechanisms. Further attention to this subject is given later in this chapter.

Calendar life expresses the theoretical lifetime of a battery when sitting at rest at a given temperature and SOC. Cycle life is a term mainly associated with the aging of a battery during repeated charge/discharge. Cycle life is always related to the specific cycling protocol used during the test, particularly the severity of the (dis) charge cycle. Less intuitive but still important, cycle life also depends on the state of the battery during rest periods interspersed between cycles.

In the test environment, calendar-dependent degradation is commonly measured using storage tests accelerated with temperature and mapped versus SOC, as shown in Figure 4.1, produced using the graphite/NCA life model. This model is further discussed in Section 7.7. Cycle-dependent degradation is characterized by subjecting the battery to back-to-back cycles with little rest. Conducting additional aging experiments that vary the amount of rest between cycles helps to separate calendar from cycling degradation and establish possible couplings between the two that need to be accounted for in a lifetime model. Figure 4.2 shows the performance fade with cycling at several values of DOD, and number of cycles per year. The average SOC of the cell across its lifetime is held constant in these simulations.

It is critical when extrapolating cycle-life predictions from test data to also include appropriate calendar-life limiting terms in the extrapolation model.

4.1.3 Regions of Performance Fade

Mature electrochemical storage technologies will experience gradual capacity and resistance changes throughout life, with fade rate dependent on duty-cycle, temperature, and age. Figure 4.3 shows an example of nonlinear performance fade com-

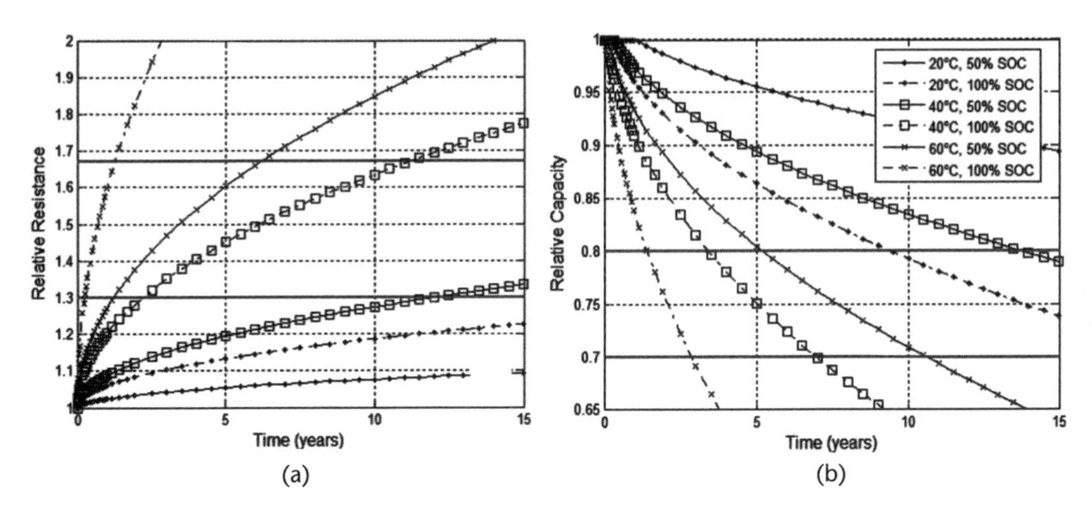

Figure 4.1 Performance fade with storage: (a) change in resistance, and (b) change in capacity.

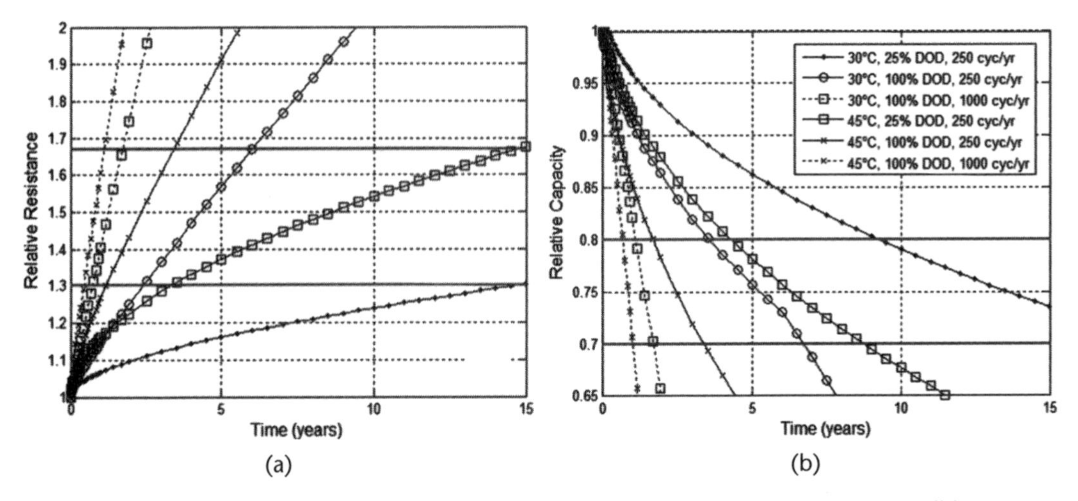

Figure 4.2 Cycle-life predictions from two different test protocols of a Li-ion graphite/NCA cell for geosynchronous orbit application. (Reproduced from [2].)

mon for Li-ion cells. In this example, performance fade processes may be separated into several regions:

Region I: *Break-in* region, dependent on initial cycles. Figure 4.3 shows a slight initial performance increase, though a slight performance decrease is also possible.

Region II: *Decelerating fade* region, strongly dependent on calendar aging processes and moderately dependent on cycling.

Region III: *Accelerating fade* region, strongly dependent on cycling.

Actual lifetime performance data always contains noise, and compared to Figure 4.3, the transitions between regions will be less apparent. For example, the transition between regions II and III may be more gradual and appear as a linear

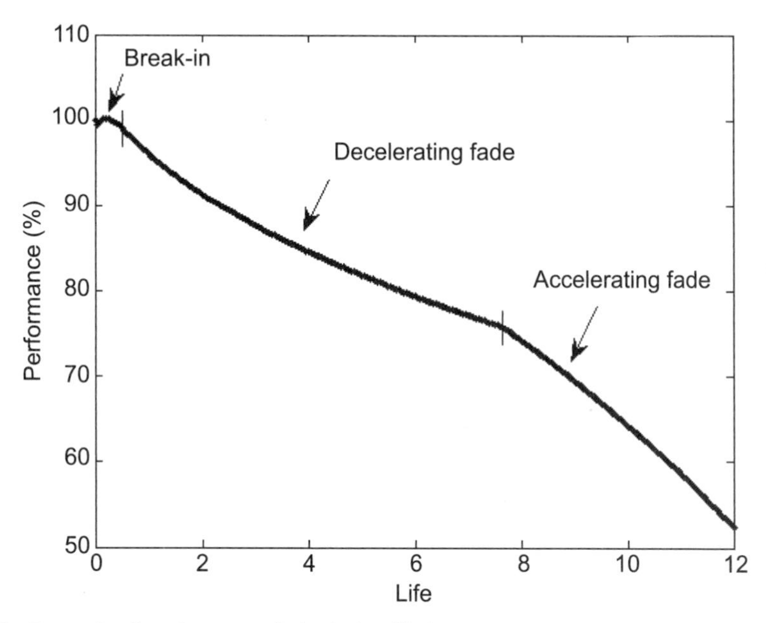

Figure 4.3 Example of performance fade during lifetime.

region of fade processes. Depending on the aging condition, some regions in Figure 4.3 may not exist at all. Under severe cycling for example, regions I and II may be absent, leaving only the accelerating fade, region III.

Prior to the application of a device, as part of the manufacturing process, Li-ion cells undergo formation cycles at the factory. These cycles are typically conducted at low (dis)charge rates and elevated temperature. Formation cycles establish an initial SEI layer or surface film at the negative electrode surface, consuming some cycleable Li from the system and causing a loss of up to 10% capacity during the first few cycles.

During the break-in region, cells can experience performance changes ranging from rapid decrease to slight increase in initial performance, as shown in Figure 4.3. Rapid decrease could result from incomplete formation cycles at the manufacturing plant, or under high temperatures could simply be rapid SEI growth due. Slight increase in capacity is possible in systems containing excess lithium that may be released during initial cycles. Slight decreases in resistance may result from small fractures opening in electrode active materials, introducing new surface area to the system and relieving some of the residual mechanical stresses of the manufacturing process. Generally, increases in initial performance are only evident at moderate to low temperatures. The existence of break-in processes complicates the identification of fade rates from the first weeks or months of cell aging experiments.

The decelerating fade region II is the ideal place for the battery to spend its lifetime. In Li-ion batteries, SEI growth is an example mechanism whose growth rate slows with age. As SEI grows in thickness, the diffusion rate of chemical species from the electrolyte through the film to the negative surface slows down. Diffusion limited reactions fade track with ~ $t^{1/2}$, giving a decelerating performance fade trajectory. Kinetic-limited chemical reactions may initially track with ~ t, but also slow down or stabilize as reservoirs of chemical species are consumed in the reactions effect.

Eventually though, some accelerating process will likely overtake the graceful fade. Such processes can be chemical, although for a mature, stable system are generally electrochemical or electrochemomechanical-dependent. An example chemical process causing accelerating fade is Mn dissolution/SEI growth in the graphite/$Li_yMn_2O_4$ Li-ion system. Once Mn dissolves, it can migrate through the electrolyte to the graphite-negative electrode, where the Mn serves as a catalyst for faster SEI growth.

Due to electrode material loss or cycleable Li loss, the electrochemical window of positive and negative electrodes can shift relative to one another, introducing some new side reaction that was previously not a problem. For example if negative electrode loss outpaces cycleable Li loss, Li plating may begin to occur late in the battery's life during low temperature charging.

An example of mechanical-coupled processes that cause accelerating fade in Li-ion systems is mechanical damage to the SEI/electrode accelerating SEI growth/electrolyte decomposition. Another is active material loss/isolation that is accelerated as fewer remaining sites are stressed at greater levels for the same cycling profile.

4.1.4 End of Life

For a given application, the designer must size the battery such that power/energy requirements are met all throughout lifetime. In an actual application, failure will be first noticed at the limits of system design points, such as for the combination of discharge at high rates, cold temperatures, and/or low SOC extremes. This means power/energy requirements must be met at EOL under worst-case cold temperature. It is sometimes said that automotive batteries die during the summer, but their funeral is held in the winter.

EOL criteria differ for energy versus power applications and for individual cells versus complete battery pack systems. EOL for energy applications is commonly taken as when the device's useable energy fades to 70% to 80% of original energy. Useable energy is primarily influenced primarily by capacity loss but is also influenced by power fade. For example, consider an EV application where sufficient discharge power for vehicle acceleration is available at 5% SOC at BOL. At EOL, that same discharge power might only be available above 15% SOC. Capacity loss of, say 20%, plus this extra 10% of useable energy loss due to power fade, would bring the device to EOL with remaining useable energy at 70% of BOL.

EOL for power applications is commonly taken as when the device's useable power fades to 70% to 80% of original power. Again, remaining energy plays a role. The power needs of the application must be met across some useable energy range to meet the longest duration power draw requirement of the application, with some margin. For an HEV, this might be dictated by a worst-case drive cycle, with several repeated acceleration events lasting 5 to 10 seconds each.

Periodically during battery aging tests, the aging (dis)charge cycle is interrupted to run reference performance tests (RPTs). The RPT objective is to map useable power and energy changes over the duration of the aging test. Figure 4.3 shows an example of power-mapped versus SOC using the hybrid pulse power characterization (HPPC) test protocol discussed in Chapter 2. At BOL, the power needs are met over a wide energy range, 52% ΔSOC. At EOL, pulse power capability and total energy have shrunk, reducing the useable energy range capable of meeting the application's power requirement to a narrow window, 12% ΔSOC in Figure 4.4.

The designer must trade off system design with device lifetime. Overdesigning a system with excess energy and power or with a powerful thermal management system will equate to longer life, albeit with penalties of extra cost, mass, and volume. Design trade-offs are commonly conducted to meet both years of life and number of cycles requirements for some reasonably aggressive (i.e., 95th percentile) duty cycle and environment.

Designing a battery with excessive energy usually leads to greater incremental cell cost than sizing with excessive power. As a rule of thumb, size battery pack excess energy as small as possible within the margin of error of life-predictive models and oversize on power to maximize the amount of energy that is useable, particularly late in life.

4.1.5 Extending Cell Life Prediction to Pack Level

Differential aging of cells in a pack or large system may occur for several reasons. Cell manufacturing variability causes cell BOL performance differences and slight aging process differences. Differential aging is also driven by temperature gradients across cells in a pack. Depending on pack electrical configuration, cell balancing system topologies and possible active switching of individual series strings, cells, or strings may experience different electrical (dis)charge cycles that can impact aging. Cell balancing is further discussed in Chapter 7.

Cell aging models introduced in this chapter may be extended to pack-level by capturing cell BOL and aging process differences using Monte Carlo techniques overlaid with models that predict cell electrical and thermal imbalances during cycling (see Chapters 2 and 3, respectively).

Once each individual cells' performance fade characteristics are estimated, their individual resistances and capacities can be combined to determine impact on

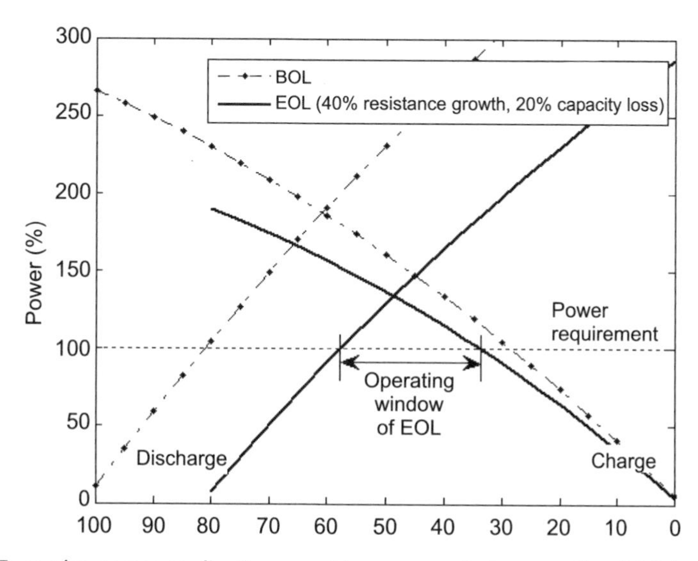

Figure 4.4 For pulse-power application, useable energy window meeting (dis)charge power requirements will shrink over lifetime due to resistance growth and capacity loss.

pack performance. For n cells in series (or n groups of parallel-connected supercells in series), the resistance of the overall pack will be the sum of the resistances of the individual cells,

$$R_{pack} = \sum_{n} R_i \tag{4.1}$$

Pack capacity will vary depending on whether the pack utilizes a passive or active balancing system. For series-connected strings of cells, a system utilizing passive cell balancing can only dissipate extra energy from strong cells. In this case, the pack capacity will be limited to the weakest cell in the pack,

$$Q_{pack} = \min_{n}(Q_i) \tag{4.2}$$

For a battery pack with active cell balancing, the system can transfer energy amongs cells and the pack capacity will be nearly

$$Q_{pack} = \operatorname{mean}_{n}(Q_i) \tag{4.3}$$

Equation (4.3) assumes 100% efficiency of the active cell-balancing system. Losses will occur dictated by circuit design and the amount of cell mismatch being processed.

The remainder of this chapter discusses gradual cell-level performance fade due to electrochemical/thermal/mechanical degradation processes leading to EOL. However, infant mortality due to "bad" cells or sudden system failures are also possible. This is particularly true in a complex system with hundreds or thousands of cells, tens of sensors, wiring harnesses and connectors, battery management system circuitry, cooling systems fans, pumps, and cooling fluids. Failure modes of each of these system components must be comprehensively addressed through failure modes, effects, and criticality analysis (FMECA).

4.1.6 Fade Mechanisms in Electrochemical Cells

Table 4.2 categorizes fade mechanisms for electrochemical cells. Degradation originates due to various physical mechanisms that play out at length-scales ranging from micron-sized electrode active particles to the length-scale of cell packaging.

Chemical, electrochemical, and mechanical-coupled degradation rates are each differently addressed while a cell is being designed. Chemical degradation processes may be slowed by modifying the electrochemical recipe, introducing electrolyte additives to stabilize reactions, or coating electrode particles to provide a barrier to inhibit reactivity at the particle surface. Electrochemical degradation processes may be modified by shifting or limiting the electrochemical window over which the material is cycled. Chemical and electrochemical reaction rates may also be controlled by electrode surface area. Larger particles with less surface area are good for slowing the rate of chemical reactions. Small particles with large surface area are best for promoting facile electrochemical reactions in high-power systems. Hence, a trade-off between calendar life and performance exists when selecting electrode particle size. Mechanical-coupled degradation may be influenced by tuning particle morphology, optimizing cell packaging and manufacturing processes.

Table 4.2 General Degradation Mechanisms in Electrochemical Cells

Length-Scale	*Mechanism*	*Physics*
Particle	Impedance film on electrode surface due to deposition of side reaction byproducts	(E)CT
	Chemical instability with electrolyte causing impedance film on electrode surface, dissolution of byproducts into electrolyte	(E)CT
	Lattice instability impacting particle bulk	C(T)
	Electrochemical grinding due to material expansion/contraction during cycling, particle fracture, fade of transport properties	ECTM
Composite electrode	Binder decomposition, creep	(C)TM
	Electrical isolation of regions of electrode due to mechanical stress and fracture	TM
	Electrical isolation of regions of electrode due to impedance films on electrode surface	(E)CT
	Ionic isolation of regions of electrode due to pores clogged with side reaction byproducts	(E)CT
	Current collector corrosion	CT
Electrolyte	Decomposition, fade of ionic transport properties, gas generation	(E)CT
	Precipitation of side reaction byproducts into pores impeding ionic transport	(E)CT
Separator	Mechanical stress, viscoelastic creep causing pore closure	TM
	Loss of electrical isolation between two electrodes (internal short) due to mechanical failure of separator (or solid electrolyte), metallic dendrite or foreign object penetrating separator	M
Cell	Delamination of electrode sandwich due to mechanical forces such as gas pressure build-up and inadequate compression	(E)CTM
	External forces and vibrations damaging cell packaging, electrodes, separator, electrical pathways	M
	Loss of electrolyte through cell walls or impurity ingress such as water in humid environments	CT

E = electrochemical: cycling-dependent mechanism
C = chemical: time-dependent mechanism (though generally coupled to electrochemical state)
T = thermal: temperature-dependent mechanism
M = mechanical: mechanical stress-dependent mechanism leading to strain and fracture
Parenthesis indicate weak coupling

Once the cell is fielded, the specific cycling, temperature, and pressure history of the cell will dictate its lifetime.

4.1.7 Common Degradation Mechanisms in Li-ion Cells

Lithium-ion cells can experience all of the fade mechanisms noted in Table 4.2. Before discussing specific mechanisms in detail, Figure 4.5 shows an overview of the typical electrochemical operating window for a Li-ion battery. For negative electrodes that operate at a working potential below ~1.1V (Li/Li+), such as graphite depicted in the figure, the electrolyte will continuously reduce throughout life forming and continuously growing the SEI film layer on the negative. Operating any negative electrode material at potentials below 0V vs Li can cause lithium plating on the surface of the negative. All well-known Li-ion electrolytes suffer from instability and cannot operate at cell potentials above 4.3 to 4.4V without oxidizing at the positive electrode. This inhibits the introduction of several promising candidate

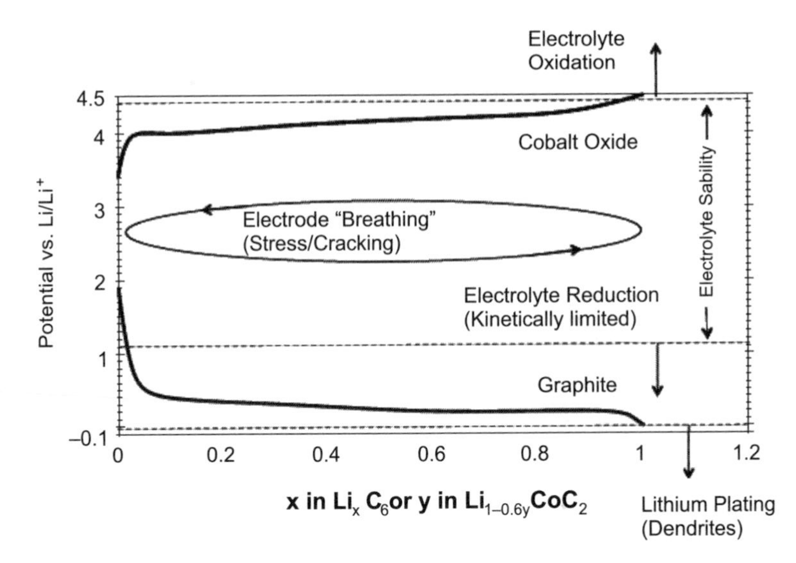

Figure 4.5 Typical electrochemical operating window of Li-ion battery. (Adapted from [3] and [4].)

high voltage Li-ion positive electrodes. Energy density of Li-ion systems could be increased if some new electrolyte or surface coating could be identified that allows higher operating potentials. Finally, in Figure 4.5, charge/discharge cycles are associated with stresses of "electrode breathing" as materials shrink and expand.

High temperature and high end-of-charge voltage (EoCV) accelerate many of the reactions below. As rules of thumb, a 15°C increase in temperature or a 0.1V increase in EoCV will cut life in half. Low temperature may also accelerate degradation. Slow transport of Li-ions at low temperature leads to excessive concentration gradients in active particles that cause Li plating and particle fracture.

The next sections will describe the following degradation mechanisms in further detail:

Electrochemical reactions

1. SEI formation and growth
2. Lithium plating
3. Binder decomposition
4. Current collector corrosion
5. Electrolyte decomposition
6. Metal oxide positive electrode decomposition

Electrochemomechanical processes

7. SEI fracture and reformation
8. Particle fracture
9. Electrode fracture
10. Separator viscoelastic creep
11. Gas build-up and mechanical consequences

4.1.7.1 SEI Formation and Growth

The solid electrolyte interphase layer is critical to the operation of Li-ion batteries utilizing graphite, silicon, and other low-voltage negative electrodes. Graphite electrodes for example operate in a potential window 0 to 300 mV above a lithium reference electrode (Li+/Li), whereas typical organic-based electrolytes are unstable below 1.3V (Li+/Li).

Due to this incompatibility, electrolyte products are reduced at the surface of the negative electrode. A few negative electrode active materials operate in a higher-voltage window and largely avoid SEI formation. An example is lithium titanate, operating at ~1.5V (Li/Li+). Such high-voltage negative electrodes provide longer potential calendar life, but at the cost of energy density due to lower overall cell operating voltage.

System-level consequences of SEI growth are (1) sharp initial drop in capacity during formation cycles (irreversible), followed thereafter by (2) gradual loss of cycleable Li causing capacity loss (irreversible), and (3) growth of impedance film layer on negative surface causing power loss (irreversible, although film layer can be damaged leading to temporary recovery of power), as well as (4) self discharge (partly reversible).

The typical model of SEI growth assumes that an electrolyte-solvent species (e.g., ethylene carbonate ($C_3H_4O_3$)) diffuses through the SEI and is reduced at the graphite/SEI interface to form lithium carbonate (Li_2CO_3) [5] and/or ethylene dicarbonate ($(CH_2OCO_2Li)_2$) [6]. Of these two reactions, the former consumes cycleable Li from the negative electrode; the latter consumes Li+ from the electrolyte salt.

In these models, SEI growth rate is limited by diffusion of the solvent and/or kinetics of the one-electron reduction of EC molecules. The diffusion rate-limitation is well supported by data for multiple Li-ion chemistries, where performance fade tracks closely with $\sim t^{0.5}$. A mixed diffusion-kinetic rate limitation is also possible, with fade $\sim t^z$, $0.5 < z < 1$. Over the long term, availability of Li-ions, electrons, and other participating species may also play a limiting role.

SEI formation is an important step in Li-ion manufacturing. Formation cycles are conducted with initial charging at very low C-rate to avoid cointercalation of solvent into the active material along with Li-ions. As shown in Figure 4.6, solvent cointercalation can cause exfoliation or cracking of graphene layers, as well as excessive gas generation.

The initial formed SEI layer may have a dense compact structure influenced by electrolyte additives such as vinelyne carbonate. In all, chemistry, surface morphology, treatments, coatings, C-rate, and temperature are used to tune SEI initial formation.

Once the cell is placed into service, SEI growth will proceed in a stable manner throughout lifetime controlled by transport or kinetic processes. If limited by a diffusion transport process, SEI growth proceeds with square-root-of-time dependence. A mixed diffusion/kinetic-limited process may proceed slightly faster. A species-limited growth or SEI-densification process may cause SEI growth to proceed slightly slower. During this stable phase, SEI growth rate is mainly influenced and accelerated by high temperature and high state of charge in the negative electrode. Charge rate plays a role.

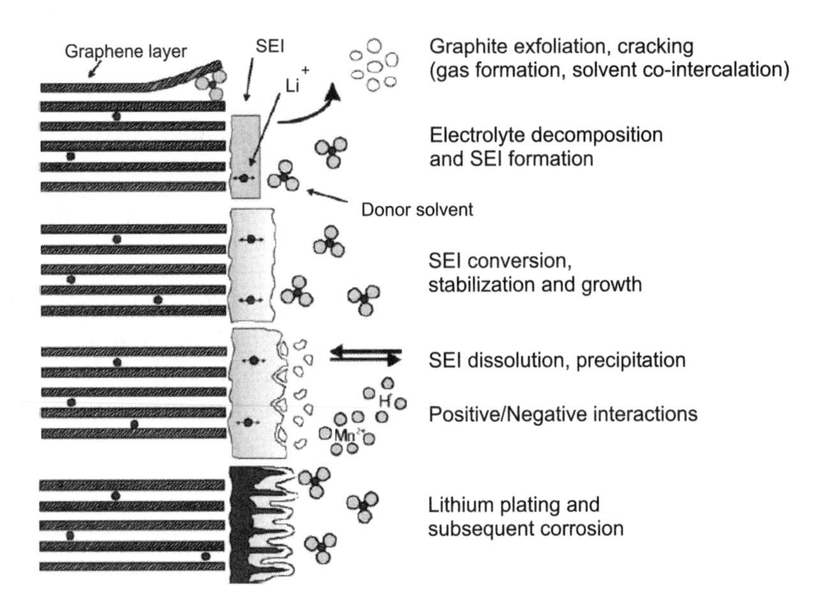

Figure 4.6 SEI formation and growth processes. (Reproduced from [7].)

Beyond the stable phase, SEI growth rate can accelerate causing more sudden performance fade due to

- Catalysis of SEI reaction(s) by impurities such as manganese or copper deposited on the negative electrode surface.
- Mechanical-coupled processes such as surface exfoliation and SEI microcracking.
- SEI dissolution into electrolyte if exposed to high temperatures (> 60°C). At low temperatures, dissolved SEI can re-precipitate in pores, inhibiting ionic transport.

In addition to dissolution, brief high-temperature exposure can cause rapid damage. Exothermic reactions, possible precursors to thermal runaway, may occur above 80°C. High temperature may also initiate chemical conversion of organic SEI to inorganic products that are more stable but greatly impede Li+ ionic transport.

4.1.7.2 Lithium Plating

Lithium plating can occur on the negative electrode due to a preferential electrochemical reaction that deposits metallic Li on the electrode surface as opposed to the normal intercalation or conversion reaction process. Plating occurs during charging at a potential window < 0V vs Li+/Li. Similar to SEI formation and growth, low-voltage negative electrodes will experience Li-plating more readily than high-voltage negative electrodes.

Li-plating may be purely kinetically controlled, competing with the normal charge transfer process at high charge rates. But given that the kinetic overpotential of graphite intercalation reaction is small, it is most common for Li plating to occur when graphite surface concentrations are saturated with Li at full charge where equilibrium potential of graphite approaches zero. If both effects are combined, Li plating is most likely to occur at cold temperatures where diffusion from particle surface to bulk is sluggish and at high charge rates. Kinetic and diffusion-transport processed are depicted in Figure 4.7.

Of the Li that is plated on the negative surface, a portion may be discharged reversibly and an open-circuit voltage of 0V vs Li^+/Li, while some portion will be irreversibly lost. Plated Li that becomes disconnected from the electrode will result in irreversible capacity fade and reaction with the electrolyte. Reaction with the electrolyte can be an exothermic precursor to thermal runaway. This is most likely for cold-temperature charging followed by sudden warming.

Plating will preferentially occur at sharp corners of electrode particles, near the separator, or in inhomogeneous regions of the electrode coating. These are all regions of the cell where local electronic conduction is much faster than ionic conduction relative to neighboring regions region. Once Li begins to plate, it grows in the form of a dendrite, continually plating at the tip of the dendrite due to the excellent electronic conductivity of the dendrite.

Another preferential location for Li plating is at the edge of a negative electrode sheet. Mismatch between positive and negative electrode sheets, for example, can result in excess positive at some location in the cell. During charging that region of positive will provide a source of Li without sufficient negative host material to accept the Li. Li will then plate at the edge of the negative electrode sheet. Li plating at electrode edges is easily mitigated by making the negative electrode sheets slightly larger than the positive sheets so they overhang a millimeter or so to accommodate manufacturing intolerances during electrode stacking/winding processes.

As shown in Figure 4.8, lithium dendrites take the form of a thin whisker structure that branches. The thin structure can penetrate the separator, growing within the pores. Should a dendrite fully bridge across the separator from the negative to the positive electrode, the cell will experience an internal short. If the dendrite does not burn itself out, the internal short may generate enough heating to lead to thermal runaway.

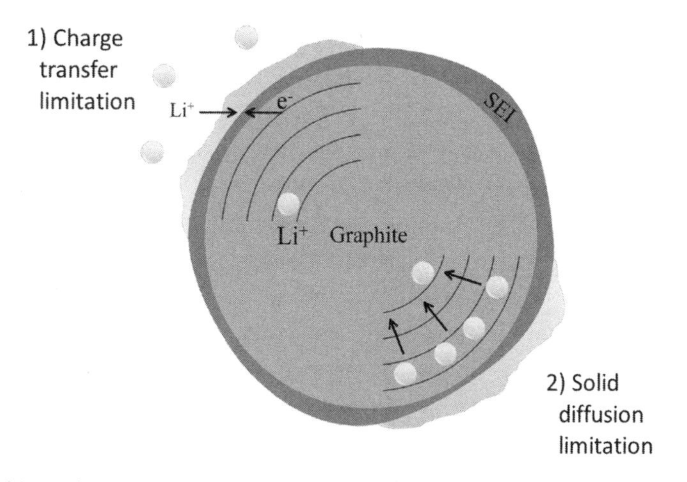

Figure 4.7 Lithium plating at the surface of graphite. (Reproduced from [8].)

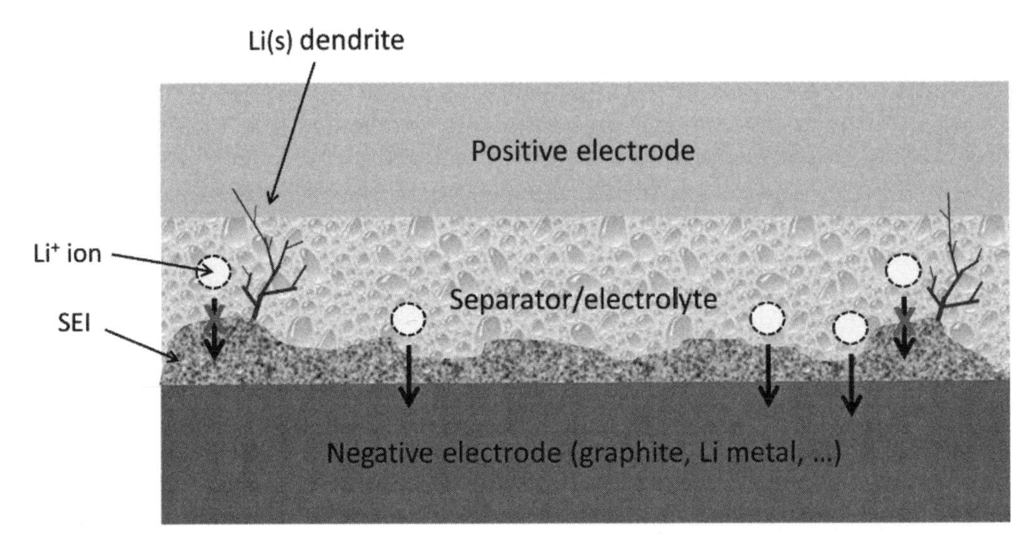

Figure 4.8 Dendrite formed from Li plating/deposition reaction. (Reproduced from [9].)

4.1.7.3 Binder Decomposition

Polyvinylidene-difluoride (PVdF) and other fluorine-containing polymer binders may react with the negative electrode to form LiF. High temperatures can damage binder, degrading its mechanical properties. Viscoelastic creep or slight displacement of electrode particles may occur under cycling-induced stress and can occur at temperatures well below the glass transition temperature of the binder.

Binder decomposition damages particle-to-particle and particle-to-collector adhesive strength. Coupled with cycling, portions of the electrode may detach and become electrically isolated.

4.1.7.4 Current Collector Corrosion

Current collector materials are selected to be compatible with the electrochemical voltage window of each electrode. For Li-ion cells, materials are typically copper for the negative and aluminum for the positive.

If copper is exposed to high positive voltage vs Li+/Li, for example if the cell is left in an overdischarged state, >3V vs Li^+/Li, copper may dissolve into the electrolyte [10], contaminate the negative electrode surface, and serve as a catalyst for accelerated SEI growth. Corrosion can be accelerated by presence of impurities such as H_2O and HF in the electrolyte.

Corrosion will also damage the mechanical adhesive strength of the electrode to the collector. Such damage may lead to delamination and isolation of the electrode from the collector. Aluminum and other collectors may be precoated with graphite to improve stability and adhesion.

4.1.7.5 Electrolyte Decomposition

Electrolyte can be reduced at the negative electrode during SEI formation and growth as Li+ is consumed. For LiPF6 salt systems, PF6 may be converted to HF in presence of moisture. Electrolyte additives serving as HF scavengers reduce consequences of HF acid attack on electrodes that can decompose surface film layers and dissolve metals from metal-oxide positive electrodes.

Electrolyte can oxidize at the positive electrode when accelerated at high voltages. The oxide layer acts as a resistive surface film on the positive. Both reduction and oxidation reactions cause gas generation. The reactions are accelerated with high SOC, temperature, and charge rate.

4.1.7.6 Metal-Oxide Positive Electrode Decomposition

Metal-oxide positive electrodes can experience chemical decomposition. Typical metal-oxide electrodes in Li-ion batteries include $LiMn_2O_4$, $LiCoO_2$, Li(NiaMnbCo$_1$-a-b)O_2, and Li(NiaCobAl$_1$-a-b)O_2, respectively referred to as LMO, LCO, NMC, and NCA. This is in contrast to graphitic negative electrodes and (FEP) positive electrodes whose chemical composition is quite stable.

Decomposition occurs due to disordering, lattice instability, dissolution, and other surface effects. Disordering and lattice distortion chemical changes may be coupled with electrochemical and mechanical processes.

Disordering. Metal oxides can experiencing a structural disordering reaction in which metal ions exchange sites with lithium ions, degrading the material's ability to intercalate Li. Pure $LiNiO_2$ is unstable in this manner. Doping with aluminum (Al) or cobalt (Co) largely stabilizes the material. Site exchange of manganese (Mn) and Li has also been reported (see Figure 4.9).

Lattice instability. For all metal-oxide positive electrodes, the low stoichiometry/high voltage range is thermodynamically unstable, rendering some 30% to 50% of the theoretical capacity unusable. Approaching full delithiation, Li(metal) O_2 materials transition from monoclinic to hexagonal phase with large volume change, causing the lattice to collapse. For LMO, the high stoichiometry range, corresponding to a discharged cell, is also unstable. By discharging the cell too low, fully intercalated LMO can accept additional Li and a Mn^{3+} Jahn-Teller distorted

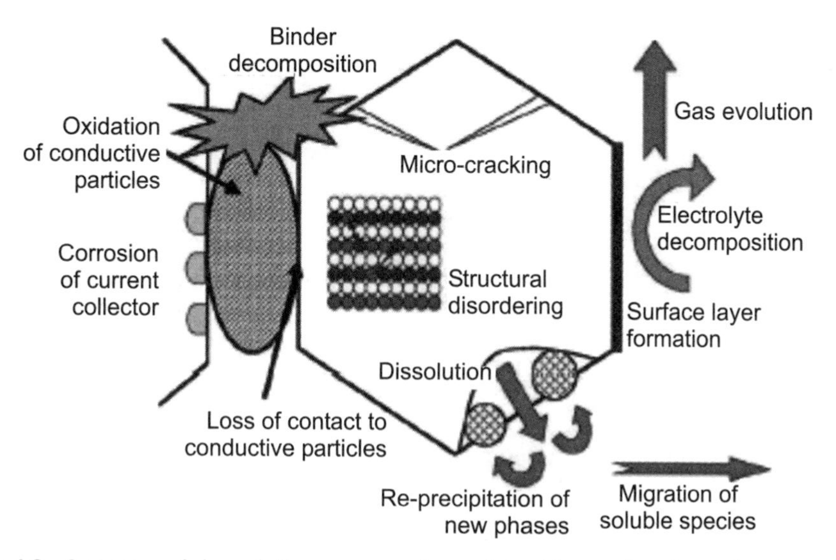

Figure 4.9 Summary of degradation processes for metal-oxide positive electrodes. (Reproduced from [7].)

tetragonal phase is formed without the ability to reverse back to the nominal spinel structure when the cell is recharged.

Dissolution. Chemical dissolution of transition metals may occur, particularly at extremes of charge or discharge. While Co dissolution does not cause great consequences for performance fade, Mn is a known catalyst for accelerated SEI growth once it inevitably migrates/deposits on the negative electrode, as shown in Figure 4.10. Co may dissolve from $LiCoO_2$ if charged above 4.2V vs Li+/Li. Manganese can dissolve under several scenarios. In the discharged state, trivalent Mn ions disproportionate into tetravalent and divalent ions:

$$2Mn^{3+} \rightarrow Mn^{4+} + Mn^{2+}$$

Divalent Mn ions are soluble and dissolve into the electrolyte. Missing Mn sites are replaced with Li leading to an impedance film on the positive electrode surface. Mn may also dissolve in a chemical delithiation reaction with HF, a byproduct of LiPF6 salt reaction with H_2O impurity [11].

Overall, the rate of dissolution processes can be mitigated by reducing surface area, doping, or mixing materials (such as LMO mixed NCA or LCO), and by including electrolyte additives that act as HF scavengers.

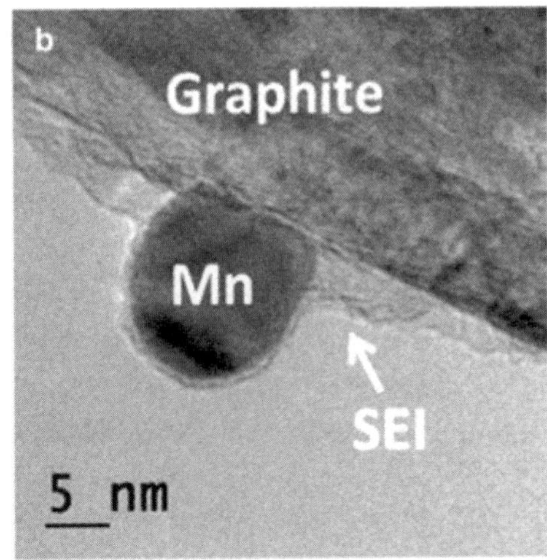

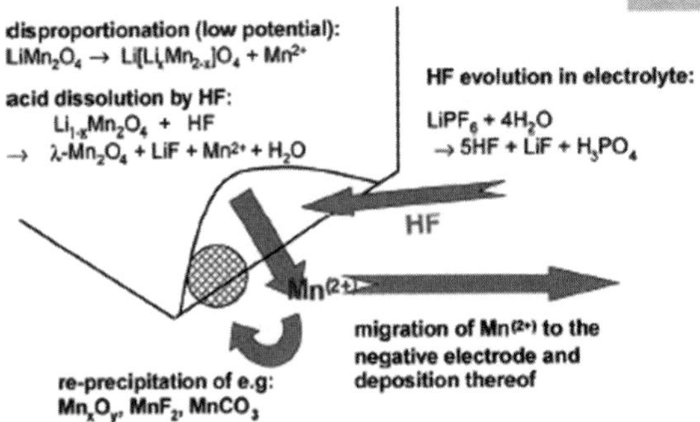

Figure 4.10 (a) Processes for manganese chemical dissolution. (Reproduced from [7]). (b) Manganese migrated to negative electrode, catalyzing accelerated SEI growth rate at negative. (Reproduced from [11].)

Surface effects. Surface impedance layer has been shown for NCA and other nickelate materials, particularly when charged to voltages above 4.2V vs Li^+/Li. These effects may be attributed to electrolyte oxidation at the surface, $LiPF_6$ decomposition and/or oxygen loss leading to a deficient oxide phase at the surface, possibly a rock-salt structure with low Li+ ionic conductance. Gases such as oxygen (O), carbon dioxide (Co_2), carbon monoxide (CO), carbon (C), and hydrogen (H) can evolve due to the various side reactions at the positive surface.

4.1.7.7 SEI Fracture and Reformation

Active material particles expand and contract during (de)intercalation and/or electrochemical conversion processes. With expansion and contraction of the active particle lattice, chemically stable surface films can experience microcracking. Cracks in surface films provide new sites for additional reaction. While electrochemical growth of SEI is predominantly calendar-dependent, SEI fracture and reformation is a typical cause for acceleration of calendar fade processes with mild to moderate cycling. Aside from SEI fracture due to particle strain, particle-to-particle displacements due to binder viscoelastic creep may also open up new sites for SEI growth, causing cycling-dependent acceleration of nominally calendar-dependent side reactions.

4.1.7.8 Particle Fracture

Intercalation- or conversion-induced lattice expansion/contraction can cause crack development in negative and positive electrode active materials. Particle strain, stress, and fracture is driven by bulk changes in SOC as well as concentration gradients. Bulk strains due to changes in SOC are a consequence of ΔSOC swings experienced during (dis)charge processes, largely unavoidable. Phase change regions of the ΔSOC—with large strains and lattice distortions—might be avoided during cycling. Concentration gradient or diffusion-induced strains are exacerbated by high C-rates and low temperatures.

This degradation mode is a major cause for faster degradation of cell performance with deep (dis)charge cycling. Partial (dis)charge cycles cause less mechanical stress than full cycles.

The order of magnitude of volume change during full discharge and charge is around 5% for metal-oxide positive electrodes, 10% for graphitic negative electrodes, and 300% for silicon negative electrodes. High-strain materials obviously experience greater stress and fracture. This continuous stress and fracture process during electrochemical cycling is commonly referred to as electrochemical grinding.

Silicon and many other high-energy density electrodes under investigation unfortunately rely on electrochemical grinding to operate. Their high energy density comes at the price of limited cycle life due to large volume expansion. SEI is constantly damaged and regrown, continuously consuming cycleable Li from the system, degrading capacity. Nonetheless, composites of silicon and graphite, Si nanoparticles, and various other nanostructures may improve mechanical resilience to cycling.

Particle fracture failures will occur when particle regions are in tension, not compression. Tension will occur at the outer edge of a particle when deintercalated at a high rate. Tension will occur in the center region of a particle when intercalated at a high rate. Fracture zones will have poor solid-state diffusion transport properties due to breach of the particle lattice. Alternately, fractured regions may also act as new pores, absorb electrolyte, and provide favorable ionic transport paths.

4.1.7.9 Electrode Displacement and Fracture

Stress and fracture can also play out at the electrode-level and higher length-scales. Stresses may be intercalation- or conversion-induced, thermally induced by variable thermal expansion rates of cell components, influenced by gas generation stressing the cell electrode stack or wound jellyroll. Such failures often are apparent at edges of electrodes or at center core regions of cylindrical wound cells where high temperatures and high stresses are superimposed.

Electrode displacement and fracture is exacerbated by binder failure, current collector corrosion, and surface film growth mechanisms mentioned earlier. Mechanical degradation relieves some of the residual stresses of manufacturing during jellyroll winding/stacking and electrolyte filling. Electrolyte filling causes separators and binders to swell some 10%.

As with binder failure, current collector corrosion, and surface film growth, regions of fractured electrodes will have locally poor electronic conductivity. Poor electronic conductivity causes remaining healthy regions of the cell to be cycled at progressively higher and higher rates. Inhomogenous (dis)charge elevates the stresses on remaining active sites and promotes the occurrence of detrimental side reactions, both accelerating fade.

4.1.7.10 Separator Viscoelastic Creep

Polymer and polymer-composite separators have viscoelastic mechanical properties. In the short term, behavior is elastic, meaning that deformation will recover when stress is removed. Under constant stress, however, such as in a tightly wound jellyroll or compressed electrode stack, separator material can flow over time. Flow of separator material can close pores and impede transport of charged species in the electrolyte such as Li+.

4.1.7.11 Gas Buildup, Electrode Isolation

Gas generation is a consequence of electrolyte reduction/oxidation side reactions at the electrode surfaces. Gas generation causes pressure buildup that imparts mechanical stresses on electrodes and cell components. During life, gas generation will cause expansion of cell packaging and cell internal components. Porous electrodes and separators may hold some gas bubbles. The remaining gases are stored in empty spaces in the cell packaging. Both effects can cause cells to noticeably expand over lifetime. Delamination of local regions of the electrode will accelerate stress of the remaining healthy regions of the cell, causing rapid performance fade.

In the extreme, excessive gas generation will trigger the cell to vent. Hard metal can packages are designed to release gases through purpose-designed vents. The failure point of soft pouch cells can be somewhat tuned through pouch or module design. Clearly, cell venting will lead to electrolyte dryout and rapid performance decline. A danger with any cell venting is that $LiPF_6$ salt will react with moisture in air to form HF, a carcinogen. Li-ion electrolytes are flammable.

4.2 Modeling

This section discusses models of cycle and calendar fade based on chemical, electrochemical, and electrochemical-mechanical mechanisms. It is rarely known a priori what degradation mechanisms must be included in a life model for a specific cell technology. Often several alternative degradation mechanisms must be hypothesized, formulated in a model, and confirmed/refuted based on statistical accuracy of the model compared with available data.

Section 4.2.1 discusses physics-based degradation models, with examples focusing on Li-ion systems. Physics degradation models complement electrochemistry models of Chapter 2. Section 4.2.2 discusses semiempirical degradation models that complement equivalent circuit and systems models.

Physics-based degradation models offer predictive capability and extensibility across chemistries, as well as insight on how to improve cell design. Physics models, however, are complex to develop and characterize from aging experiments.

For the system designer, semiempirical models provide a more straightforward approach to predict life from accelerated aging experiments. Given sufficient aging test data to populate semiempirical models, such models are useful for trade-off studies between battery excess energy/power sizing, lifetime/warranty, temperature, and allowable (dis)charge protocol. Included in discussion of systems engineering, Chapter 7 presents example aging test data and describes regression of semiempirical life models to data.

4.2.1 Physics-Based

A physics-based life model builds on the electrochemical performance model introduced in Chapter 2 by introducing rate laws that describe property changes of the performance model over the course of simulated lifetime. In rough order of importance, Table 4.3 lists common electrochemical model parameters that change versus lifetime. These parameters of the performance model serve as states of the life model. Life-model states effectively contain the minimum set of information needed to define battery health and performance at any point during lifetime.

Example degradation rate laws for the SEI film thickness on the surface of the negative electrode, volume fractions on the negative and positive volume fractions, respectively ($\delta_{film,n}$, ε_n, and ε_p) are given in the next sections.

4.2.1.1 Reaction/Transport Models

Reactions at the negative electrode are presented as an example of mixed kinetic/transport-limited electrochemical thermal reactions. Figure 4.11 summarizes reactions occurring at the negative electrode. Whereas the electrochemical performance model of Chapter 2 included only an intercalation reaction at the negative, $j^{\mathrm{Li}} = j^{\mathrm{Li}}_{\mathrm{intercalation}}$, Figure 4.11 includes six additional side reactions:

Table 4.3 Electrochemical Model Properties that Change with Lifetime

Property (Units)	*Description*	*Related properties (i = n,p)*	*Physics*
$\delta_{film,n}$, $\delta_{film,p}$ (m)	Impedance film thickness on negative and positive surface such as SEI	$R_{film,i} = \delta_{film,i}/\kappa_{film,i}$ $\kappa_{film,i} = f(T)$	(E)CT
ε_n, ε_p	Electrode active material volume fraction (changes as electrode damage accumulates)	$q_i = \varepsilon_i/F\ (A\ \delta_i\ c_{s\ max,i}\ \Delta\theta_i)/F$	ECTM
$D_{s,n}$, $D_{s,p}$ (m^2/s)	Solid diffusion in negative and positive (changes as electrode damage accumulates)	$D_{s,i} = f(\varepsilon_i)$	(ECT)M
$a_{s,n}$,$a_{s,p}$ (m^3)	Interfacial surface area (changes as electrode damage accumulates)	$a_{s,i} = f(\varepsilon_i)$	(ECT)M
σ_n, σ_p (S/m)	Electronic conductivity of electrode (changes as electrode damage accumulates)	$\sigma_i = f(\varepsilon_i)$	(ECT)M
$c_{e,0}$ (mol/m^3)	Average concentration of electrolyte salt (consumed by side reactions)	$D_e = f(c_e,T)$, $\kappa_e = f(c_e,T)$	ECT
ε_e	Electrode porosity (gradually reduced by impedance films and byproducts of other side reactions in the electrolyte)	$D_e^{eff} = D_e\ \varepsilon_e^{1.5}$ $\kappa_e^{eff} = \kappa_e\ \varepsilon_e^{1.5}$	ECT

E = electrochemical: cycling-dependent mechanism

C = chemical: time-dependent mechanism (though generally coupled to electrochemical state)

T = thermal: temperature-dependent mechanism

M = mechanical: mechanical stress-dependent mechanism leading to strain and fracture

- 1,2: Li+ is lost from the electrolyte forming SEI product $(CH_2OCO_2Li)^2$ [6];
- 3: Cycleable Li is lost from the negative forming SEI product Li_2CO_3 [5];
- 6: Li is plated during charging at the negative surface forming Li(s), partially reversible [5];
- 4: Plated Li forms SEI product Li_2CO_3, irreversible [5];
- 5: Plated Li is stripped during subsequent discharge [5].

The overall total reaction at the negative is [A/m^3].

$$j^{\text{Li}} = j^{\text{Li}}_{\text{intercalation}} + \sum_{i=3}^{6} j_i^{Li} \tag{4.4}$$

Film Growth

Assuming the negative film consists of a homogeneous mixture of reaction products, a rate law for growth of film thickness [m/s] is

$$\frac{\partial \delta_{film}}{\partial t} = -\frac{a_n}{F}\left[\frac{1}{2}\left(\frac{M}{\rho}\right)_{(CH_2OCH_2Li)_2}\left(j_2^{Li}\right) + \frac{1}{2}\left(\frac{M}{\rho}\right)_{Li_2CO_3}\left(j_3^{Li} + j_4^{Li}\right) + \left(\frac{M}{\rho}\right)_{Li(s)}\left(j_5^{Li} + j_6^{Li}\right)\right] \tag{4.5}$$

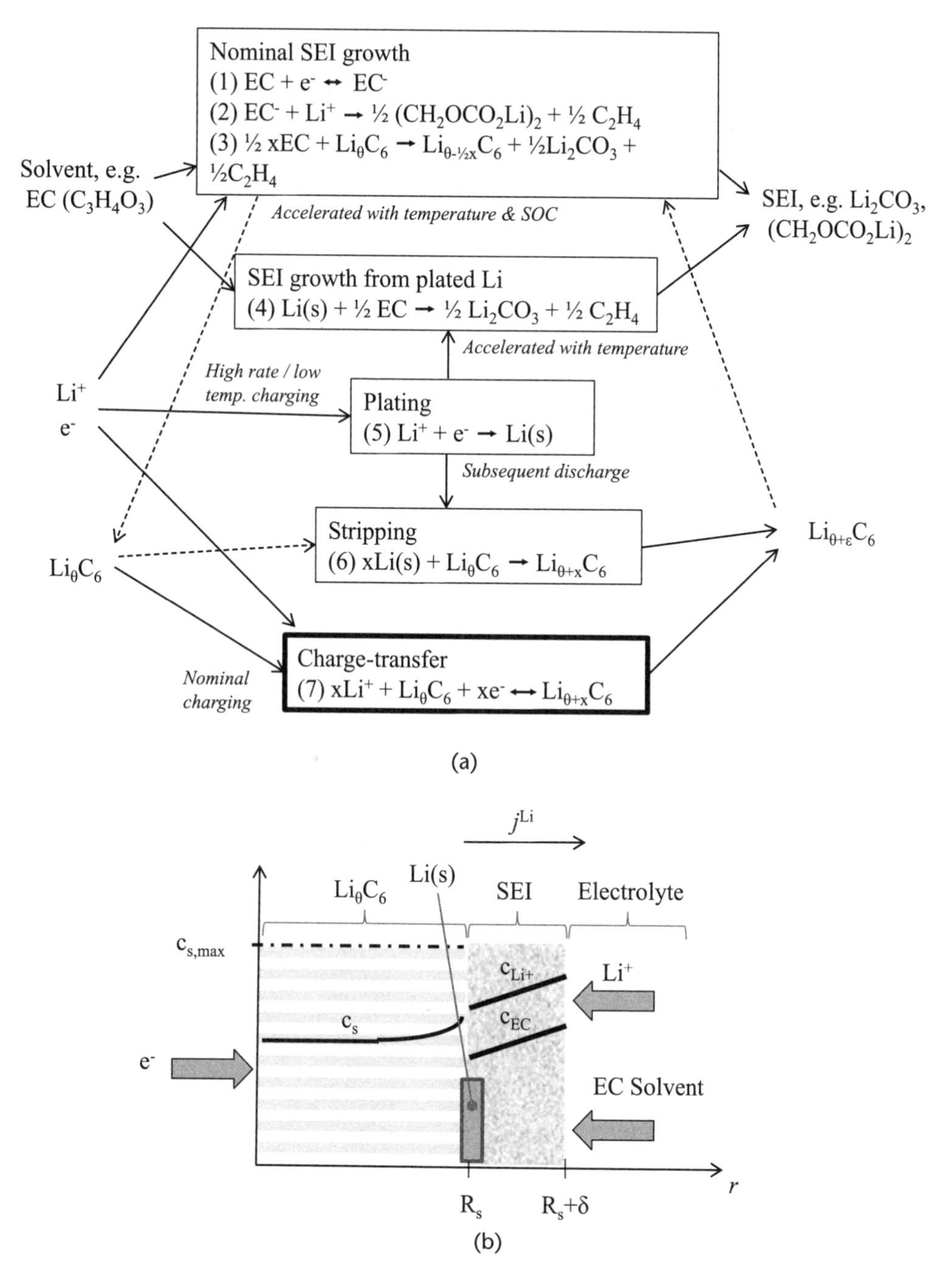

Figure 4.11 Multireaction transport model at the Li-ion graphite negative electrode (a) reactions, and (b) transport.

where a_n is electrode specific surface area [m²/m³], M is molecular weight [g/mol] and ρ is density [g/m³]. Film resistance is dependent on constituent species and ionic conductivity of species

$$R_{film} = R_{sei}^{0} + \delta_{film}\left[\left(\frac{M}{\rho\kappa}\right)_{(CH_2OCO_2Li)_2} c_{(CH_2OCO_2Li)_2(SEI)} + \left(\frac{M}{\rho\kappa}\right)_{Li_2CO_3} c_{Li_2CO_3(SEI)} + \left(\frac{M}{\rho\kappa}\right)_{Li(s)} c_{Li(s)}\right] \quad (4.6)$$

where c_i are concentrations and κ_i ionic conductivity of individual species.

Mass Conservation

Following the typical SEI model [5, 11], solvent diffusion is a transport-limitating step,

$$\frac{\partial c_{EC}}{\partial t} = D_{EC}\frac{\partial^2 c_{EC}}{\partial r^2} - \frac{\partial \delta}{\partial t}\frac{\partial c_{EC}}{\partial r} \tag{4.7}$$

with boundary conditions at the graphite/SEI interface, $r = R$,

$$-D_{EC}\frac{\partial^2 c_{EC}}{\partial r^2} + \frac{\partial \delta}{\partial t}c_{EC} = \frac{j_2}{a_n F} \tag{4.8}$$

and at the SEI/electrolyte interface, $r = R_s + \delta$,

$$c_{EC} = \varepsilon_{film} c_{EC}^{0} \tag{4.9}$$

where SEI volume fraction is $\varepsilon_{flim} = a_n \delta_{film}$. The remaining species equations are assumed to not be transport-limited,

$$\varepsilon_{film}\frac{\partial \dot{c}_{(CH_2OCO_2Li)_2}}{\partial t} = -\frac{j_2}{2F} \tag{4.10}$$

$$\varepsilon_{film}\frac{\partial \dot{c}_{Li_2CO_3}}{\partial t} = -\frac{j_3 + j_4}{2F} \tag{4.11}$$

$$\varepsilon_{film}\frac{\partial \dot{c}_{Li(s)}}{\partial t} = -\frac{j_5 + j_6}{2F} \tag{4.12}$$

Kinetics

Common practice is to model the intercalation reaction using Butler-Volmer kinetics, with irreversible side reactions described with Tafel kinetics.[1] The intercalation reaction follows

1. The Butler-Volmer formulism is most commonly used to capture electrochemical charge transfer. Strictly speaking, the Butler-Volmer approach cannot consider multiple reactions in a thermodynamically consistent manner. Full treatment requires solving a system of microscopically reversible [13, 14], including both charge-neutral heterogeneous reactions and charge-transfer reactions. Rate equations are written for heterogeneous phase change reactions, such as adsorption of a liquid electrolyte species onto an electrode surface. Additional species conservation equations must be solved for surface sites and vacancies as site concentrations are required to solve reaction rates.

$$j_{\text{intercalation}}^{Li} = a_n i_{0,n} \left[\exp\left(\frac{\alpha_{a,n} F}{R_{ug} T} \eta_n \right) - \exp\left(-\frac{\alpha_{c,n} F}{R_{ug} T} \eta_n \right) \right] \tag{4.13}$$

$$\eta_n = \phi_s - \phi_e - U_n^{ref} - \frac{j_{total}^{Li}}{a_n} R_{film} \tag{4.14}$$

Kinetics of EC reduction, reaction 1, are described by the Tafel equation,

$$j_{sei} = -a_n i_{0,sei} \left[-\exp\left(-\frac{\alpha_{c,n} F}{R_{ug} T} \eta_{sei} \right) \right] \tag{4.15}$$

$$\eta_1 = \phi_s - \phi_e - U_1^{ref} - \frac{j_{Li}^{total}}{a_n} R_{film} \tag{4.16}$$

Similar Tafel kinetic equations are written for reactions 3–6.

Following this example of negative electrode film growth, similar models may be developed for electrolyte oxidation and metal dissolution (e.g., Cu, Mn) and lattice dislocation.

4.2.1.2 Mechanical Stress

Coulombic-throughput or energy-throughput are sometimes used as proxies to describe mechanical stress-induced fade and are regressed to experimental capacity data [15]. These models are difficult to extend to a wide range of cycling conditions [16].

Mechanical stress effects have been modeled at length scales ranging from particle level to electrode sandwich level [9, 17], cell level [18, 24], and pack level [19]. Particle stress investigations have shown, for example, possible failure during fast-rate charging where intercalation-rate of Li into a negative electrode active particle drives a faster rate of expansion at the outer radii of the particle, generating high-tensile stress in the inner core that may lead to fracture, as shown in Figure 4.12(a). During deintercalation, high-tensile stress occurs at the outside of the particle; see Figure 4.12(b).

Overall, stresses may be concentration, temperature, or pressure/force induced. Summarized in Figure 4.13, sources of stress that influence cell electrochemical performance and lifetime include

- Residual stress of manufacturing (e.g., jellyroll winding tension, cell packaging, binder and separator swelling during electrolyte filling processes);
- External loads and forces (e.g., packaging of cells within modules and packs, possible vibration or impact loads on battery packs in the application environment);

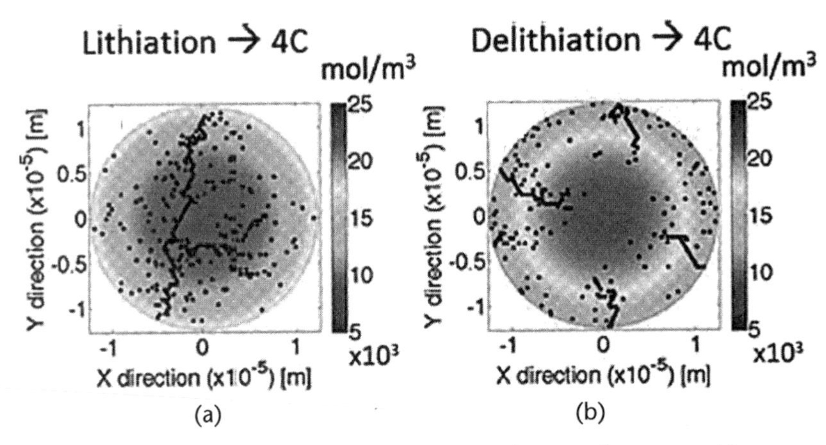

Figure 4.12 (a) High rate charge of negative particles leads to tensile stress and fracture evolution from the center. (b) High rate discharge leads to tensile stress and fracture at the surface. Darkened lines within the particle represent regions of fracture. (Reproduced with permission from the Electrochemical Society [20].)

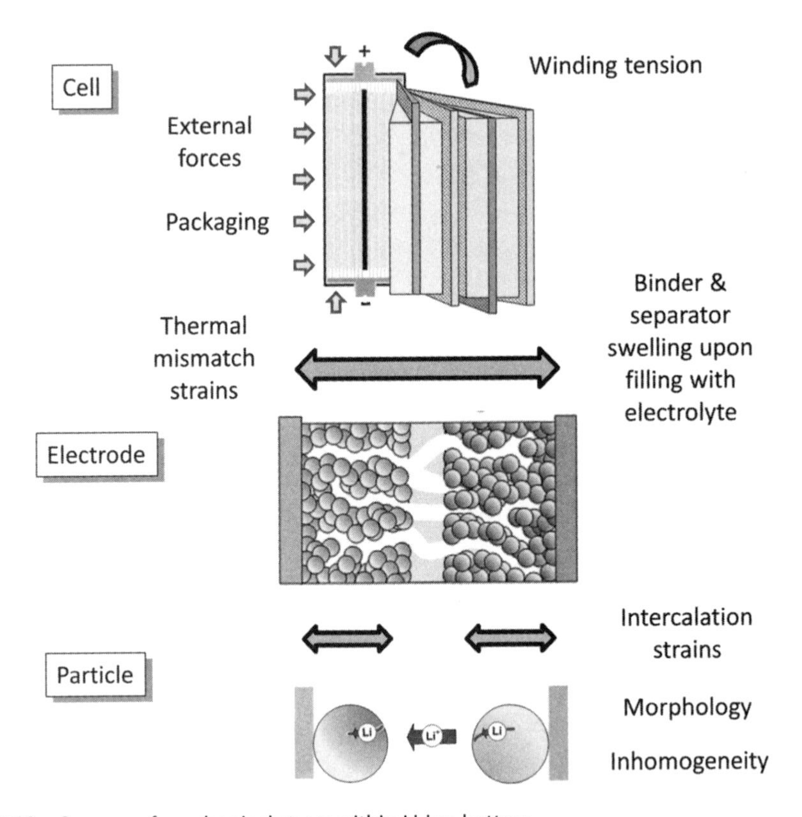

Figure 4.13 Sources of mechanical stress within Li-ion battery.

- Stresses during (dis)charge processes (e.g., material expansion/contraction with concentration and temperature);
- Changes during lifetime (e.g., gas pressure buildup, damage of composite material with accompanying changes in mechanical properties).

Some of the above analyses may be treated using finite element models to analyze the solid mechanics behavior of the cell and its components, neglecting details of electrochemical cycling. For a lifetime predictive model, however, it is desirable to predict accumulated damage and performance fade associated with micromechanical stresses. The discrete element method (DEM) is one practical method to study micromechanical stress and fracture as it evolves with charge/discharge cycles. Such a model has been implemented at the active material particle length-scale [11] although the model may need to include higher length scales to capture all relevant mechanical degradation mechanisms influencing cell lifetime. It is more typical in the literature to introduce empirical equations that relate cycling current or C-rate with loss of negative and positive active material, $d\varepsilon_n/dt$ and $d\varepsilon_p/dt$, respectively [21].

$$d\varepsilon_i/dt = k0 * \exp(-Ea/Rt) * |I(t)| \tag{4.17}$$

4.2.2 Semiempirical Models

Surrogate models of physical degradation mechanisms may be written as simple formulas to match aging trends observed in cell-level life test data. This follows common practice in the accelerated stress testing and life verification literature. Compared to physics models, these surrogate models are relatively easy to formulate, regress to aging data, and make predictions that include statistical confidence intervals.

As an example, a model for resistance growth may assume that resistance growth versus calendar time, t, and number of electrochemical-thermal-mechanical cycles, N, are additive:

$$R = a_0 + a_1 t^z + a_2 N \tag{4.18}$$

A surrogate model for capacity fade may assume capacity, q, is limited either by cycleable Li in the cell, q_{Li}, or electrode active sites, q_{sites}

$$q = \min(q_{Li}, q_{sites}) \tag{4.19}$$

$$q_{Li} = b_0 + b_1 t^z + b_2 N \tag{4.20}$$

$$q_{sites} = c_0 + c_2 N \tag{4.21}$$

This model [22] reproduces several common features observed in capacity fade data:

1. Graceful fade regime observed for cells aged under pure storage conditions ($b_1 +^z$) or in low-to-moderate cycling conditions (b_2N)
2. Linear fade regime (c_2N) observed either

a. Immediately starting at BOL for a cell repeatedly cycled at moderate-to-high stress levels such as during an accelerated cycling test
b. Following an initial graceful fade controlled by SEI growth, where capacity degradation rate suddenly accelerates

The above semiempirical life model may be further generalized:

Let $y(t,N)$ be a measured performance metric (e.g., R or q), described as a function of multiple degradation states, x_i, (e.g., $y = f(x_1,x_2,x_3...)$). Table 4.4 lists suggested equations for tracking degradation state variables, x_i.

Examples of diffusion, kinetic, and mixed diffusion-kinetic processes include SEI film formation, film formation on metal oxides, and electrolyte reduction/oxidation. Examples of break-in processes include release of reserve Li into the system that was initially unavailable, reordering of disordered lattice during initial cycles, or microcracking of particles opening up new electrochemical active surface area.

Table 4.4 Performance Fade Trajectory Equations for Generic Degradation Mechanisms

Mechanism	*Trajectory Equation*	*State Equation*	*Fitted Parameter*	*Physics*
Diffusion-controlled reaction	$x(t) = kt^{1/2}$	$\dot{x}(t) = \frac{k}{2}\left(\frac{k}{x(t)}\right)$	k, rate (p = 1/2)	(E)CT(M)
Kinetic-controlled reaction	$x(t) = kt$	$\dot{x}(t) = k$	k, rate (p = 1)	(E)CT
Mixed diffusion/kinetic	$x(t) = kt^p$	$\dot{x}(t) = kp\left(\frac{k}{x(t)}\right)^{\left(\frac{1-p}{p}\right)}$	k, rate p, order, 0.4<z<1	(E)CT(M)
Cyclic fade, linear	$x(N) = kN$	$\dot{x}(N) = k$	k, rate (p = 0)	E(T)M
Cyclic fade, accelerating	$x(N) = \left[x_0^{1+p} + kx_0^p(1+p)N\right]^{\frac{1}{1+p}}$	$\dot{x}(N) = k\left(\frac{x_0}{x(N)}\right)^p$	k, rate p, order, $0 \geq p$ > 1.5	E(T)M
Break-in process	$x(t) = M(1 - \exp(-kt))$ or $x(N)$	$\dot{x}(t) = k(M - x(t))$	M, limit of fade k, rate	ECTM
Sigmoidal reaction formula	$x(t) = M\left[1 - \frac{2}{1 + \exp(kt^p)}\right]$ or $x(N) =$	$\dot{x}(t) = \frac{2Mkp\ X(t)\exp(kX(t))}{\left[1 + \exp(kX(t))\right]^2}$ $X(t) = \left\{\frac{1}{k}\ln\left(\frac{2}{1 - x(t)/M} - 1\right)\right\}$	M, limit of fade k, rate p, order	ECTM

E = electrochemical: cycling-dependent mechanism

C = chemical: time-dependent mechanism (though generally coupled to electrochemical state)

T = thermal: temperature-dependent mechanism

M = mechanical: mechanical stress-dependent mechanism leading to strain and fracture

Examples of cyclic fade include side reactions driven by electrochemical cycling such as Li plating or stress and fracture driven by intercalation- and thermal-induced strains. Regarding intercalation-induced stress/strain and fracture, voltage-limited cycles that decline in intensity as the battery degrades may exhibit nearly linear fade behavior, $p\sim0$. Cycles with constant coulombic (Ah) throughput may exhibit mild accelerating trend, $p\sim1$, as remaining electrode sites are stressed at progressively higher levels to maintain the same cycle. Cycles with constant energy throughput (Wh) may exhibit even faster accelerating trends, $p>1$.

The model, once regressed to cell-life test data, may be extrapolated versus time, t, and cycles, N, to some EOL condition. Examples of regression to data are given in Section 7.7. But briefly, the model regression steps are:

1. Hypothesize degradation mechanisms. Create a trial function combining trajectory equations from Table 4.4, $y = f(x_1,x_2\ldots)$, to describe performance fade for hypothesized fade mechanis.
2. Local model regression. Separately regress model $y(t,N)$ to each aging test condition to find degradation rates, k_i. (Orders z_i and p_i may either be regressed or fixed at constant values that best represent mechanism hypotheses).
3. Rate model regression. Plot rates, ki, versus duty-cycle aging stressors such as temperature (T), SOC, DOD, and C-rate. Hypothesize rate equations that functionally describe fade rate versus stressors (e.g., $k = k_{ref} x\ f$ (T,SOC,DOD,C-rate,...)). A suggested set of rate model equations is given in Table 4.5. Regress rate model coefficients. Polynomials or lookup tables may be used.
4. Global model regression. Substitute rate equations into performance fade model to create a model with functional dependence on duty-cycle, $y(t,N;$

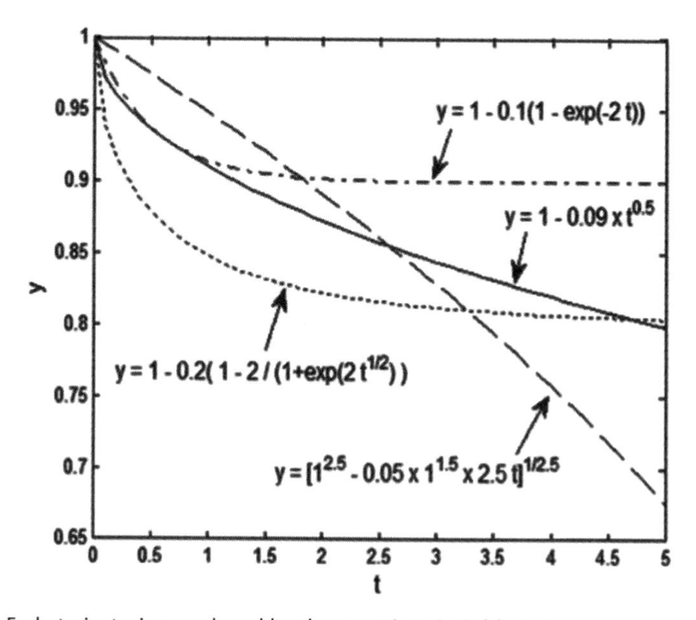

Figure 4.14 Fade trajectories produced by the equations in Table 4.4.

Table 4.5 Example Rate Model Equations to Describe Acceleration of Fade Rate Versus Storage or Cycling Condition Stress Factors

	Acceleration Term	*Fitted Parameter*	*Physics*
Arrhenius acceleration with temperature	$\theta_T = \exp\left[\frac{-E_a}{R_{ug}}\left(\frac{1}{T(t)} - \frac{1}{T_{ref}}\right)\right]$	E_a	ECTM
Tafel acceleration with voltage	$\theta_V = \exp\left[\frac{\alpha F}{R_{ug}}\left(\frac{V_{oc}(t)}{T(t)} - \frac{V_{oc,ref}}{T_{ref}}\right)\right]$	α	ECT(M)
Bulk intercalation stress	$\theta_{\Delta DoD} = \left(\frac{\Delta DOD_i}{\Delta DOD_{ref}}\right)^\beta$	β	EM
Intercalation diffusion gradient stress	$\theta_{C-rate} = \left(\frac{C_{rate,i}}{C_{rate,ref}}\right)\left(\sqrt{\frac{t_{pulse,i}}{t_{pulse,ref}}}\right)$	—	EM
Thermal stress	$\theta_{\Delta T} = (\Delta T_i)^\gamma$	γ	TM

E = electrochemical: cycling-dependent mechanism
C = chemical: time-dependent mechanism (though generally coupled to electrochemical state)
T = thermal: temperature-dependent mechanism
M = mechanical: mechanical stress-dependent mechanism leading to strain and fracture

T,SOC,DOD,C-rate,...). Repeat regression of model parameters to best represent global (entire) dataset.

5. Assess statistical validity of model. Iterate on mechanism hypotheses and trial functions. Compare regression statistics (e.g., mean-square error, R^2, adjusted R^2) across trial models, select the most statistically relevant model.

4.3 Testing

For the cell designer, objectives of aging/life tests are to identify main failure modes. Knowledge of failure modes is fed back into the design process so that they may be reduced through design improvements. For the system designer, an important objective of aging tests is to measure performance fade rates to inform a lifetime predictive model. The life model then guides proper sizing of system BOL excess energy and power. Life-model uncertainty must be accommodated by oversizing energy and power, which introduces extra cost. Clearly a goal in life modeling is to minimize uncertainty in lifetime predictability.

As discussed previously, cell performance degrades due to

- Calendar age, *t*, spent at a given temperature and SOC;
- Number of (dis)charge cycles, *N*, together with the rate, temperature, and SOC window of the cycles;
- Coupled calendar and cyclic aging mechanisms, introducing path dependency of the aging process.

To aid discussion of life testing of Li-ion batteries, Table 4.6 provides a summary of dominant aging mechanisms and typical dependence on operating condition.

The life test and verification process involves multiple steps:

1. Quantify expected duty cycle and environment. Collect statistics from field, or use system models (Chapters 2 and 3) to simulate behavior of a cell within the energy storage system. Statistically categorize scenarios to identify reference worst-case duty cycles for use in aging tests (e.g., 95th percentile (dis)charge power profile, temperature, and pressure).
2. Define EOL condition. Using data from the cell supplier or rules of thumb (e.g., EOL at 20% to 30% fade), roughly size battery. Determine level of capacity and/or resistance fade that represents EOL. Note that EOL may be under some specific operating condition (e.g., pulse power requirement at cold temperature and low SOC).
3. Design of experiments. Create matrix of aging test (dis)charge power profiles, temperatures, pressures to be tested.
4. Design RPTs. RPTs provide repeatable benchmark measurements of resistance and capacity throughout life. RPTs are typically conducted monthly under moderately accelerated aging conditions or weekly under highly accelerated aging conditions.
5. Conduct aging experiments. Run aging test campaign for several months to years, depending upon life predictive confidence level needed to field design.
6. Conduct diagnostic experiments. Noninvasive experiments such as electrochemical impedance spectroscopy (EIS) may be conducted periodically as part of RPTs. Invasive experiments may be conducted at the end of the aging experiments (e.g., cell teardown and construction of half cells using extracted electrode samples to separately measure remaining performance of individual electrodes).

Table 4.6 Typical Li-ion Battery Degradation Processes and Dependence on Operating Conditions

Electrochemical Reactions	*Dependence*
SEI formation and growth	High T, SOC_{max}, high charge rate
Lithium plating	Low T, high charge rate
Binder decomposition, failure	High T, pressure
Current collector corrosion	Overdischarge, overcharge, storage at high T
Electrolyte decomposition	High T, Overcharge, SOC_{max}

Electrochemomechanical Processes	*Dependence*
SEI fracture and reformation	High T, high C-rates
Particle fracture	Low T, high C-rate, large DOD
Electrode fracture	Low T, high C-rate, large DOD
Separator viscoelastic creep	High T, high stack pressure
Gas buildup and mechanical consequences	High T, high SOC, High charge C-rates

7. Fit life models to data. Regress semiempirical models or physics-based life models to resistance growth and capacity fade measured during the aging experiments.
8. Refine system design and control strategies. Determine amount of excess power/energy at BOL needed to achieve desired life. Design thermal management system and determine temperature control set points. Set current and power limits versus SOC and T to avoid regions of operation that cause excessive damage.

4.3.1 Screening/Benchmarking Tests

Aging tests may be carried out in several rounds. Prior to designing a full set of experiments, it is useful to run a small set of screening tests to identify dominant mechanisms to be excited and appropriate stress levels for the full aging test campaign.

Benchmarking tests provide useful reference data for comparisons across technologies. The standard data helps to track technologies from different vendors, variations on manufacturing batches, and design heritage. Suggested screening/benchmark tests are given in Table 4.7.

When designing a new cell chemistry, geometry, or package, it is always less expensive to perform aging studies on small-scale samples, where possible. For example, new electrochemical material couples are commonly tested in mAh-size coin cell formats. But full validation of lifetime for a large cell or system design is difficult to scale from small-scale results, particularly for mechanisms coupled to thermomechanicalelectrical effects. Physics-based models offer potential to provide a route to scale-up small cell aging test results to large cell virtual designs.

4.3.2 Design of Experiments

Design of experiments—an entire discipline in its own right—deals with setting up a test matrix of what levels and combinations of stress factors to explore the degradation space. Objectives are to maximize statistical significance and minimize the number of experiments needed to map the degradation space.

Experiments should outline a set of stress factors that are orthogonal; that is, are uncorrelated and efficiently contrast one another. This minimizes duplicate data collected by tests run at different aging conditions. For example, SOC and

Table 4.7 Typical Screening/Benchmarking Tests

Cycling/Storage Condition	*Temperature*	*Purpose*
Calendar storage	55°C to 60°C	Assessment of calendar life. Cells that last ~1 year in this harsh condition might last ~10 years at 20°C to 30°C.
100% depth of discharge cycling	20°C to 30°C	Best-case assessment of cycle life, conducted at favorable temperature similar to application.
100% depth of discharge cycling	45°C to 60°C	Worst-case assessment of cycle life, conducted at worst-case temperature. Useful for identifying calendar/cycling-coupled failure modes likely to occur in the actual application.

open-circuit voltage are highly correlated variables. Only one of these two should be chosen as a stress factor.

The degradation rate versus stress factor multidimensional space is often too expensive to explore with a full factorial tests (i.e., an experimental design that varies on one stress factor at a time). Techniques to map the space with partial factorial experiments include Fisher, Box, and Taguchi methods.

They should be selected based on expected accelerating factors influencing degradation rate. Application duty-cycle scenarios should be statistically analyzed to identify statistical levels of stressors (e.g., 95th percentile T, C-rate, OCV or SOC, DOD). For calendar fade, time-average and maximum values of T and SOC (or OCV) are significant stressors. For power applications, or when thermal behavior is a concern, root mean square (RMS) current is an important metric to match.

For cycle-life limitations, DOD plays a significant role, with acceleration at low temperatures attributable to brittle material properties and at high temperatures attributable to binder and separator creep. Average C-rate and pulse time of (dis) charge (or square root of pulse time) are also important metric that describe severity of concentration gradients. Rather than time-averaging statistics to represent a duty cycle, it may be preferable to analyze duty cycles by breaking them down into microcycles, for example, by taking moving averages of C-rate, or using the so-called rainflow algorithm from the mechanical fatigue literature. Together with C-rate, other cyclic stress factors can be binned and plotted. An example histogram of C-rate, temperature, and SOC versus pulse time as shown in Figure 4.15.

Calendar and cyclic degradation are highly correlated stress factors, not orthogonal to one another. For life extrapolation, calendar degradation should be separated from cycling degradation to the maximum extent possible. It is therefore useful to include cycles-per-day, N/t, as a factor in the test matrix. One extreme of the test matrix may be run at 100% duty cycle; that is, with little or no rest between each charge/discharge cycle. Other portions of the test matrix should include cycling experiments run with at a partial duty cycle, where rest is included in between each cycle where cycles-per-day, N/t, approaches a rate closer to what is expected in the application. Table 4.8 shows an example life test matrix. Due to manufacturing variation, replicates should be included at several test conditions so that the experiments also quantify cell-to-cell aging process variation.

4.3.3 RPTs

RPTs serve to benchmark performance changes throughout lifetime. They are typically conducted monthly; however, it is desirable to conduct RPTs more frequently in situations where fade rates are highly accelerated, such as at high temperature/high SOC/high C-rate/high DOD cycling. RPTs most commonly consist of low rate (≤1C) capacity measurements and HPPC tests to map resistance versus SOC. In order to decouple the effect of resistance growth from the loss in measured capacity, it is useful to add a constant voltage hold at not just the end of the charge, but at the end of the discharge as well. Alternately, the capacity at a constant current can be measured at a very low C-rate (C/20 to C/50); these measurements are also useful for observing changes in the open circuit potential with aging. Often for a test matrix such as that shown in Table 4.8, RPT tests will be run at a common temperature, in the range of 23°C to 30°C, to facilitate comparison of aging results across conditions.

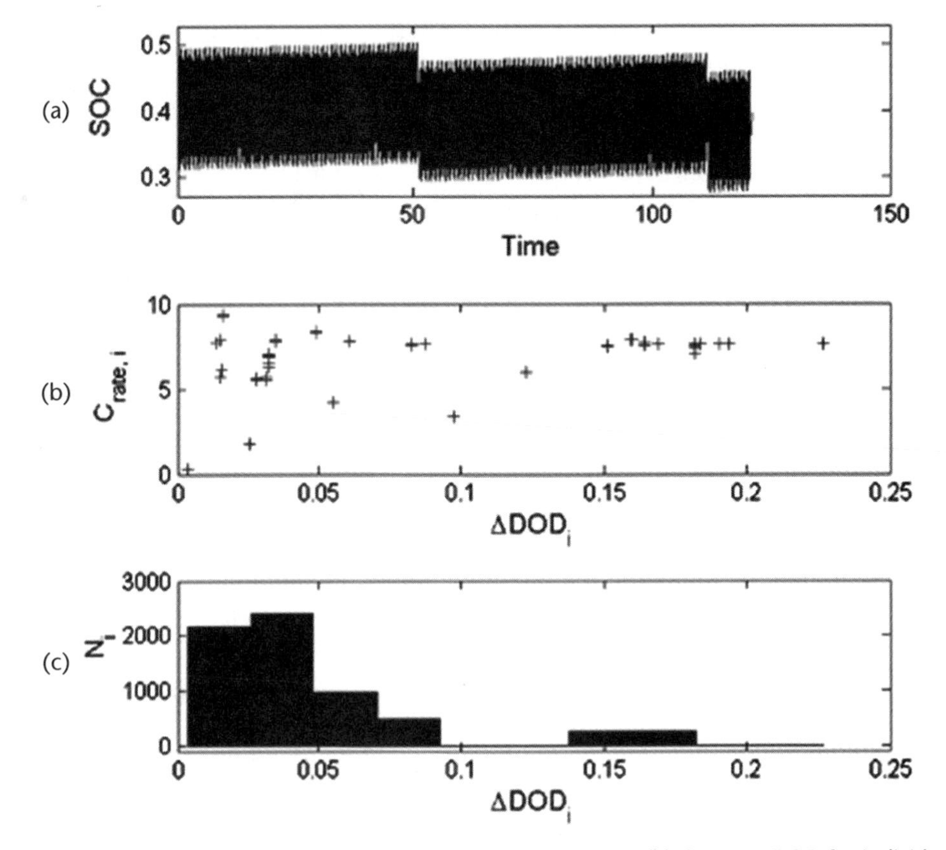

Figure 4.15 Pulse power application duty cycle (a) SOC vs time, (b) C-rate vs DOD for individual microcycles discretized using rainflow algorithm, and (c) histogram of microcycles.

Ideally, the RPTs influence the degradation rate of the given aging condition as little as possible. For this reason, a partial HPPC test might be run using only pulses at a few SOCs rather than the full HPPC. Instead of running the RPT at a common reference temperature, RPTs may be run at the aging-test temperature to eliminate the influence of changing temperature. EIS is a minimally invasive technique and is thus well suited for RPT testing. Medium- to high-frequency sweeps (>0.01 Hz) are relatively quick to run and provide useful information on interfacial processes impacting fast transient power capability. It takes more time to obtain information on transport-limiting processes that govern available energy.

4.3.4 Other Diagnostic Tests

In addition to EIS, other minimally invasive tests include:

- Electrochemical dilatometry measurement of cell package strain during cycling. Large strains may be correlated with phase changes (possibly with poor reversibility), plastic, or viscoelastic deformation of electrodes and separator and gas generation.
- Entropic measurements may similarly indicate phase changes.
- With Li loss, the electrochemical windows of the two electrodes will shift with respect to one another. This shift may be quantified by

Table 4.8 Example Cell Life Test Matrix, with Numbers Indicating Number of Cell Replicates

Storage Aging Test Matrix

	Temperature		
SOC	30°C	45°C	55°C
100%	3	3	1
50%		3	1
20%			1

Storage + Cycle Aging Test Matrix

			Temperature			
			0°C	25°C	45°C	45°C
C-rate	*ΔDOD*	*Maximum SOC*	100% duty cycle (no rest between cycles)			50% duty cycle
Medium	Low	95%	1	1	1	
	Medium	90%	1	2		
		100%	1	2	1	
High	Medium	90%	1	2	1	
		100%		2	1	1
	High	95%		1	1	1

- Plotting dQ/dV from each cycle and observing shifts in peaks that are a characteristic signature of electrode phase changes. Figure 4.16 provides an example.
- Using a single particle model that captures OCV behavior versus stoichiometry of negative and positive electrodes, fitting beginning and end of discharge stoichiometries for both electrodes to low rate (dis)charge data.
- Implanting a reference electrode in a cell or in the vicinity of a cell inside a shared electrolyte bath such as in Figure 4.17.

Tearing the cell apart can provide further diagnoses of failure mechanisms, using microscopy, surface, or bulk measurement techniques to observe chemical and morphology changes. A good example of teardown and analysis is given by Cannarella [24]. Half cells, symmetric cells, or three-electrode cells may be constructed from samples punched from extracted electrodes. Beware of deducing degradation rates from half-cell data where one electrode is the working electrode and the other is both reference and working electrode. Half cells using Li metal essentially have an infinite reservoir of Li. They can therefore operate for hundreds of cycles at low coulombic efficiency and still show reasonable performance despite their unstable rate of Li consumption (e.g., silicon).

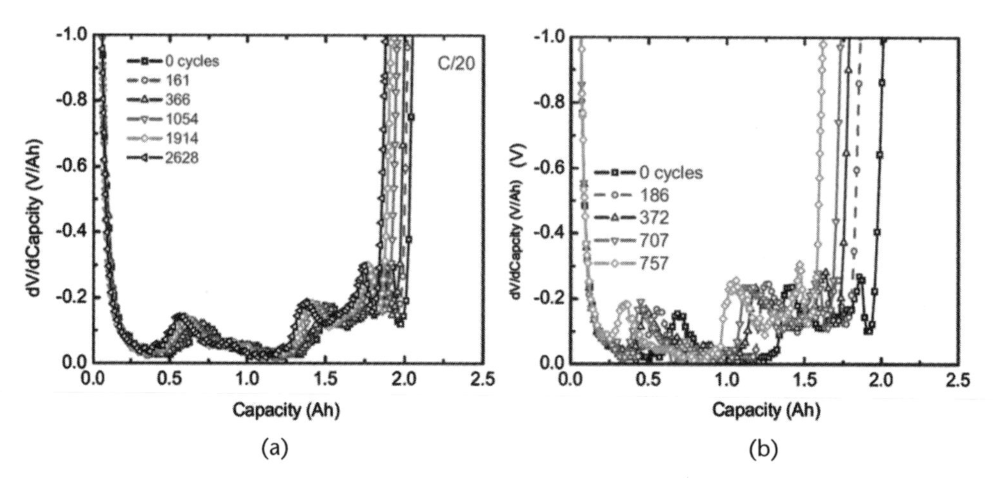

Figure 4.16 Test results from graphite/FeP Li-ion cell aged with 90% DOD cycling at C/2 rate and 60°C. (a) Capacity loss as measured at the C/20 rate during RPTs, and (b) same data, expressed as differential voltage versus capacity. This plot more clearly shows shifts in OCP curves. For this aging condition, the first peak near 0.0 Ah corresponds to the deintercalated FeP electrode. As discharge proceeds from left to right, the remaining peaks correlate with phase changes in the graphite electrode. The distance between each of these peaks stays constant, indicating graphite active material loss is negligible. Rather, the peaks shift to the left, indicating cycleable Li is being lost from the system with consequence that the graphite electrochemical window shifts to progressively lower SOCs as the cell ages, limiting capacity. (Reproduced with permission from the Electrochemical Society [23].)

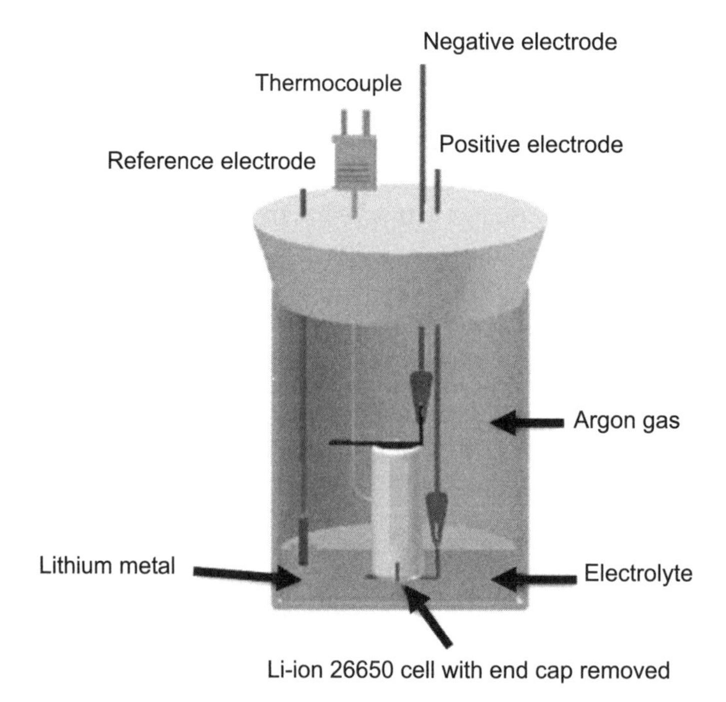

Figure 4.17 Cylindrical cell with bottom cap removed and sealed in a special fixture with reserve electrolyte bath and Li metal reference electrode. (Reproduced with permission from the Electrochemical Society [23].)

References

[1] Christophersen, J. P., E. Thomas, I. Bloom, and V. Battaglia, "Life Validation Testing Protocol Development," DOE Vehicle Technologies Annual Merit Review, February 26, 2008, http://energy.gov/sites/prod/files/2014/03/f12/merit08_christophersen_1.pdf.

[2] Hall, J., T. Lin, G. Brown, P. Biensan, and F. Bonhomme. "Decay Processes and Life Predictions for Lithium Ion Satellite Cells," 4th Int. Energy Conversion Engineering Conf., San Diego, CA, June 2006.

[3] Gur, I. "Advanced Management and Protection of Energy Storage Devices," ARPA-E Program Meeting, January 9, 2013.

[4] Srinivasan, V. "Batteries 101" of How Can I Make a Li-Ion Battery Work Better? ARPA-E Program Meeting, January 8, 2013.

[5] Perkins, R. D., A. V. Randall, X. Zhang, and G. L. Plett. "Controls Oriented Reduced Order Modeling Of Lithium Deposition on Overcharge." *J. Power Sources*, Vol. 209, No. 1, 2012, pp. 318–325.

[6] Safari, M., and C. Delacourt. "Simulation-Based Analysis of Aging Phenomena in a Commercial Graphite/LiFePO4 Cell," *J. Echem. Soc.,* Vol. 158, No. 12, 2011, pp. A1436–A1447.

[7] Vetter, J., P. Novák, M. R. Wagner, C. Veit, K. -C. Möller, J. O. Besenhard, M. Winter, M. Wohlfahrt-Mehrens, C. Vogler, and A. Hammouche. "Ageing Mechanisms in Lithium-Ion Batteries." *J. Power Sources,* Vol. 147, 2005, p. 269.

[8] Legrand, N., B. Knosp, P. Desprez, F. Lapicque, and S. Raël. "Physical Characterization of the Charging Process of a Li-Ion Battery and Prediction of Li Plating by Electrochemical Modelling," *J. Power Sources,* Vol. 245, 2014, pp. 208–216.

[9] Shi, D., X. Xiao, X. Huang, and H. Kia, "Modeling Stresses in the Separator of a Pouch Lithium-Ion Cell." *J. Power Sources,* Vol. 196, 2011, pp. 8129–8139.

[10] Braithwaite, J. W., A. Gonzales, G. Nagasubramanian, S. I. Lucero, D. E. Peebles, J. A. Ohlhausen, and W. R. Cieslak. "Corrosion of Lithium-Ion Battery Current Collectors," *J. Electrochem. Soc.* Vol. 146, No. 2, 1999, pp.448–456.

[11] Xiao, X., D. Ahn, Z. Liu, J. -H. Kim, and L. Lu. "Atomic Layer Coating to Mitigate Capacity Fading Associated with Manganese Dissolution in lithium Ion Batteries," *Echem. Comm.,* Vol. 32, 2013, pp. 31–34.

[12] Phloehn, H. J., P. Ramadass, and R. E. White. "Solvent Diffusion Model for Aging of Lithium-Ion Battery Cells," *J. Electrochem. Soc.,* Vol. 151, No. 3, 2004, pp. A456–A462.

[13] Colclasure, A. M., and R. J. Kee. "Thermodynamically Consistent Modeling of Elementary Electrochemistry in Lithium-Ion Batteries." *Electrochim. Acta,* Vol. 55, 2010, pp. 8960–8973.

[14] Colclasure, A. M., K. A. Smith, and R. J. Kee. "Modeling Detailed Chemistry and Transport for solid-Electrolyte-Interface (SEI) Films in Li-Ion Batteries." *Electrochim. Acta*, Vol. 58, 2011, pp. 33–43.

[15] Peterson, S. B., J. Apt, and J. F. Whitacre. "Lithium-Ion Battery Cell Degradation Resulting from Realistic Vehicle and Vehicle-to-Grid Utilization." *J. Power Sources*, Vol. 195, 2010, pp. 2385–2392.

[16] Wang, J., P. Liu, J. Hicks-Garner, E. Sherman, S. Soukiazian, M. Verbrugge, H. Tataria, J. Musser, and P. Finamore. "Cycle-Life Model for Graphite-$LiFePO_4$ Cells." *J. Power Sources,* Vol. 196, 2011, pp. 3942–3948.

[17] Renganathan, S., G. Sikha, S. Santhanagopalan, and R. E. White. "Theoretical Analysis of Stresses in a Lithium Ion Cell." *J. Echem. Soc.,* Vol. 157, No. 2, 2010, pp. A155–163.

[18] Sahraei, E., R. Hill, and T. Wierzbicki. "Calibration and Finite Element Simulation of Pouch Lithium-Ion Batteries for Mechanical Integrity." *J. Power Sources,* Vol. 201, 2012, pp. 307–321.

[19] Sahraei, E., J. Campbell, and T. Wierzbicki. "Modeling and Short Circuit Detection of 18650 Li-Ion Cells Under Mechanical Abuse Conditions." *J. Power Sources,* Vol. 220, 2012, pp. 360–372.

[20] Barai, P., and P. P. Mukherjee. "Stochastic Analysis of Diffusion Induced Damage in Lithium-Ion Battery Electrodes." *J. Electrochem. Soc.,* Vol. 160, No. 6, 2013, pp. A955–A967.

[21] Safari, M., and C. Delacourt. "Simulation-Based Analysis of Aging Phenomena in a commercial Graphite/LiFePO4 Cell." *J. Electrochem. Soc.,* Vol. 158, No. 12, 2011, pp. A1436–A1447.

[22] Smith, K., M. Earleywine, E. Wood, and A. Pesaran. "Battery Wear from disparate Duty-Cycles: Opportunities for Electric-Drive Vehicle Battery Health Management." American Control Conference, Montreal, Canada, June 27–29, 2012.

[23] Liu, P., J. Wang, J. Hicks-Garner, E. Sherman, S. Soukiazian, M. Verbrugge, H. Tataria, J. Musser, and P. Finamore. "Aging Mechanisms of $LiFePO_4$ Batteries Deduced by Electrochemical and Structural Analyses." *J. Electrochem. Soc.,* Vol. 157, No. 4, 2010, pp. A499–A507.

[24] Cannarella, J., and C. B. Arnold. "Stress Evolution and Capacity Fade in Constrained Lithium-Ion Pouch Cells." *J. Power Sources,* Vol. 245, 2014, pp. 745–751.

CHAPTER 5

Battery Safety

Safety, in the context of Li-ion batteries, has received a lot of attention over the decades, and more so as the size of batteries used in the different target applications has drastically increased. One of the key technological barriers in scaling up the size of batteries, together with uniformity and catering to newer use patterns, is the safe handling of the battery. The amount of energy stored in a vehicle battery pack is over three orders of magnitude in comparison with that in a laptop battery module. In turn, the requirement to manage such high energy content in a confined space under a wide variety of operating conditions over 10 or more years of usage poses challenges very different from the handling of small modules or cells individually. This chapter elaborates on the chemical, electrochemical, and thermal events that result in safety issues often encountered with Li-ion batteries, the differences in the nature of problems inherent to large-format cells and modules utilizing a large number of cells, and test methods commonly used in the community to evaluate the safety of Li-ion batteries. In keeping with the format of this text, the first section discusses the background reactions, followed by a mathematical description of the different abuse events. An experimental section outlining the test methods followed by best practices in the industry towards addressing the aforementioned challenges conclude this chapter.

5.1 Safety Concerns in Li-Ion Batteries

As described in earlier chapters, a Li-ion cell is comprised of several components with inherently different material properties. For example, the cathode material has thermal properties akin to ceramics whereas the binder is made of polymers. The combination of the mechanical rigidity of the cathode particles across a wide range of temperatures and the flexibility of the binder material are crucial for efficient performance and packaging of the cell, but these factors in turn introduce constraints on the operating regime for the cell. In this example where the mechanical properties change with temperature, the binder material becomes brittle at very low temperatures, resulting in dislodging of the cathode particles from the electrode during prolonged exposure to cold temperatures. Another example, is the different coefficients of thermal expansion for the current collectors (copper or aluminum foils) and the separator (typically, polyethylene, or polypropylene), resulting in uneven stress buildup across the different layers of a wound cell. Consequently, the lower operating temperature of the cell should then be restricted to that temperature above which the wear introduced to the components is acceptable. Storing or operating the cell below a certain temperature often results in performance issues and at the extreme limits, the failure of one or more components triggers a series of events that dissipate a large amount of the energy stored in the cells in an uncontrollable fashion. Similar constraints exist on the electrochemical, chemical, thermal, and mechanical stability of the different constituents of the cell. The issue of battery safety is further complicated by the change in these constraints with the size and age of the cells, not to mention the assembly of modules and packs with a multitude of cells—each with a different set of initial parameters and aging pattern. Factors compromising safety of large-format Li-ion cells can be classified into categories detailed in the following sections.

5.1.1 Electrical Failure

Several of the commonly used battery materials have a tight tolerance on the voltage they can be safely exposed to before undesired changes to the chemical structure are realized. For example, the vast majority of the cathode materials and solvents used in the electrolytes that are described in Chapter 2 disintegrate when exposed to voltages higher than 4.5V, releasing large amounts of heat, together with molecular oxygen—often resulting in venting of the cell casing together with copious amounts of smoke and/or fire. Usually, the electrical circuitry around the cell is designed to prevent supply of energy to the cell after a set voltage limit is triggered, resulting in prevention of unintentional overcharge; however, an occasional failure of the protective electronics is often attributed to this failure mode. Another commonly encountered electrical failure is the short-circuit of the cell(s) or occasionally the battery pack. These events are attributed to improper shielding, poor choice of voltage isolation, or failure to properly ground the battery pack. Such electrical failure modes are often combated using a sleuth of redundant circuitry in place—on some occasions in excess of three layers. Conventional battery packs rely on the presence of a positive temperature coefficient (PTC) device or a current interrupt device (CID) to combat short circuits. However, in large battery packs, localized failure introduces additional challenges due to propagation of the heat generated from isolated cell failure. Insulation of the pack from the vehicle grounding, having separate control units for the pack from the modules and in some cases the individual cells, segregation of modules limited to a maximum voltage (typically 50V), and thermal insulation of the different modules within a battery pack are often developed as part of the design strategy.

5.1.2 Thermal Failure

Excessive temperature leads to evaporation of the organic solvents in the electrolyte, in turn causing swelling of the pouch material in which the cells are encapsulated. If the cell is exposed to temperatures higher than 80°C, the thermal stability of the components begins to deteriorate, resulting in other reactions. The solid/electrolyte interface (see Chapter 4) disintegrates at 85°C to 105°C: the electrolyte is no longer protected from the excessive reduction potential at the anode/electrolyte interface and forms highly resistive layers that block the pores of the anode.

These extra resistive layers only result in further increase in the cell temperature. At 128°C, the polymer constituents of the separator begin to melt and lose their physical integrity—the pores across which the electrolyte carries the ions between the two electrodes begin to collapse and the cell resistance increases further due to limited ion transport between the two electrodes. In turn, when the polymers in the separator reach their melting points, there is no effective mechanical separation between the two electrodes, resulting in an electrical short circuit. Subsequent rise in cell temperature triggers thermal decomposition of the electrolyte and in turn the components of the electrodes themselves. These reactions are very rapid and drive one another in a vicious cycle, resulting in vast amounts of energies (of the order of several hundred kilojoules per Ah of the cell capacity) commonly referred to as thermal runaway reactions once the cell reaches the point of no return. Whereas lower capacity cells (e.g., 18650s) have traditionally employed safety measures such as shutdown separators to delay or prevent thermal runaway, localized failure in large-format cells often results in inadequate response times for such mechanisms to engage in a protective action. Large-format cell manufacturers are increasingly relying on single-cell controllers in order to enable quicker isolation of faulty cells. However, as mentioned before, due to the concentration of large amounts of energy in constrained volumes, safety in large battery packs remains a challenge best overcome by preventive measures such as a strong casing or heat-absorbent packaging material.

5.1.3 Electrochemical Failure

A combination of extreme voltages and temperature often hampers the electrochemical stability of the battery materials. The disintegration of the cathode material during the thermal runaway reactions is mainly attributed to the inability of the transition metal oxide lattice to effectively bond the oxygen atoms at very low lithium concentrations. As a result, when the cell is overcharged, oxygen is dislodged from this material, which then falls apart quickly like a deck of cards and releases a large amount of heat at very short periods of time. In order to combat the issue of overcharge, large-format battery packs used in different applications today limit the operating window to about 65% of the available energy. Operation at the safe window provides better abuse tolerance should the lithium concentrations drop at select spots within the large battery pack. Another major limitation of Li-ion batteries is the inability to operate safely at low temperatures: the transport properties of the cell components at temperatures below –10°C are typically two or more orders of magnitude slower than those at 25°C. As a result, high rate discharge or a quick charge on a battery stored at cold temperatures for prolonged periods of time result in the cell releasing the energy as waste heat.

Another problem that stems from the high resistance at lower temperatures is the shift in the local electrode potentials; as described in Chapter 4, the anode voltage is lowered farther during charge if the cell resistance is higher. Around 0V, the lithium content of the anode is stable in its metallic form. Thus, at low temperatures, the anode tends to plate lithium in the form of dendrites owing to the high resistance. This effect is more pronounced at higher rates of charge (since the anode potential drops faster). *Lithium plating* is thus a serious safety concern that can result in puncturing of the separator and a short circuit across the electrodes.

The fully charged cells are particularly vulnerable to the plating problem owing to a combination of a local potential favorable for plating and the availability of large amounts of lithium at the anode toward the end of charge. At the design phase, a slightly larger form factor is introduced for the anode in order to accommodate any excess lithium without plating. However, in large-format cells, a second failure mechanism that can contribute to plating is the uneven distribution and growth of local resistances within the cell— often caused by defective material selection or design flaws—which results in some parts of the cell being overcharged while others are being underutilized. A third possibility is the wide spread in the internal resistance and/or cell capacity of multiple cells used to build a large battery pack. To mitigate such failure, a conservative operating window (both for temperature and voltage) is specified for Li-ion cells. The C-rates at which the cells can be operated at low temperatures is also restricted. Makers of large-format cells often require much tighter tolerances on uniformity of the cell components and the distribution in the initial cell properties among the different batches compared to the cells used in consumer electronics and single-cell applications. Emphasis on abuse-tolerant chemistries, including cathode compositions stable across wider voltage windows and electrolyte additives to mitigate overcharge or flammability has in particular been the focus of developmental efforts in battery research and development (R&D) since the advent of large-format cell applications.

5.1.4 Mechanical Failure

Mechanical failure in large-format Li-ion cells includes rupture of the current collector owing to excessive tension as the electrodes expand and contract repeatedly during cycling, the peeling of electrode active material from the current collector often due to poor choice of slurry composition during the coating of active material on to the current collectors, binder failure at extreme temperatures leading to dislodging of particles from the electrode leading to electrical short circuit, mechanical compliance of the separator material including a mismatch between the elastic properties in the transverse and the winding directions, large pore sizes and/or lower tortuosity resulting in higher probability for short circuits, swelling or shrinkage of the polymer at extreme temperatures, and mechanical defects that appear on the electrodes during the slitting or calendaring operations (e.g., edge effects on coatings, sharp burrs on current collectors). Many of the design factors used in small-format cells do not readily translate to the large cells. For example, the length of the electrode increases several folds in large-format cells, and the inner winds of prismatic cells often encounter mechanical stresses far higher than the ultimate stress of the current collectors, resulting in splitting of the electrode. Such issues are overcome in advanced cell designs by introducing multiple jelly rolls within the same cell container or by employing the stacked cell design.

5.1.5 Chemical Failure

Generation of hydrofluoric acid from the decomposition of the fluoride-based electrolyte interacting with moisture content that originates from contamination leads to corrosion of the cell casing and the cell components. This issue is heightened in large-format cells due to nonuniformity in the sealing/welding strength across the larger dimensions of the container, the availability of excess electrolyte to combat dry-up of the cell within the cell can, as well as the generation of larger volumes of gaseous side products in the large-format cell designs, resulting in higher pressure threshold for the failure of the cell container.

5.2 Modeling Insights on Li-Ion Battery Safety

Utilizing the mathematical framework provided in earlier chapters of this book, we now discuss some case studies to shed light on aspects of battery safety pertaining to large-format cells, which will help the reader appreciate the nuances in cell and battery design across different applications. The heat generated from abuse reactions that are electrochemical in nature is computed using (3.3) with particular attention devoted to the change in kinetic and transport properties with temperature. Individual material balance for each chemical species is introduced to capture chemical reaction rates. Heats of reaction for the different chemical reactions can be measured as described in the next section of this chapter.

5.2.1 Challenges with Localized Failure

Conventional cell design utilizes the closing of the pores (~200 nm in size) in the separator due to melting of the polymer material during a thermal event to prevent ions from being rapidly transferred from one electrode to the other during a short circuit. This protective feature, often referred to as the shutdown separator, depends on uniform melting of the separator across the cross section of the cell to prevent the short-circuit flux from further heating up the cell.

However, as shown in Figure 5.1, the heat generation pattern during a short circuit is very different for a large-format cell from that in a conventional cell. The heat generation is localized (i.e., the region where the short circuit happens can be far away from the other parts of the cell). As a result, even though larger-current densities resulting in higher local temperatures flow across the short circuit, there are parts of the cell sufficiently distant from the short where the temperature does not reach the melting point of the separator material, effectively rendering the shutdown mechanism ineffective. In addition, a secondary heating pattern emerges close to the positive and negative terminals as the result of a lower-resistance pathway near the tabs for the charges to cross the voltage barrier. An alternate design toward improving safety should consider better heat distribution within large cells and additional protection features—for example, a mechanism to isolate the cell based on its voltage.

5.2.2 Effectiveness of Protective Devices in Multicell Packs

The next example considers the effectiveness of the PTC devices used in 18650 cells when the cells are connected together to form a high-capacity battery module. The electrical resistance of the PTC device increases as a function of its temperature. When a cell heats up due to a safety issue, the PTC prevents any current from the adjacent cells connected in parallel contributing to the heat generation across the short circuit. Figure 5.2 shows the distribution in maximum temperature across the

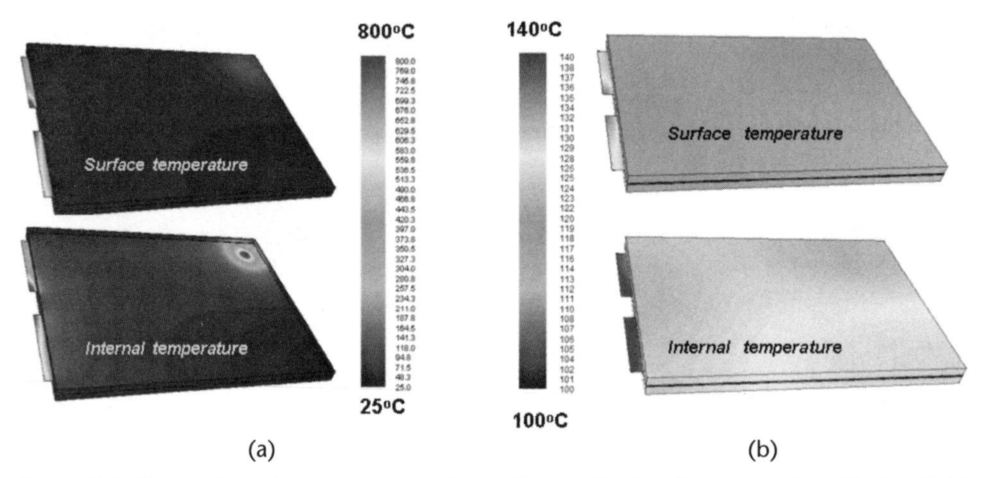

Figure 5.1 Comparison of temperature evolution 8 seconds after short circuit in a 20-Ah cell (a) versus a 400-MAh cell (b). Both cells are of identical chemistry and design; however, even though it results in a higher maximum temperature at the location of the short, the short-circuit current for large-format cells does not trigger surface temperatures to rise sufficiently faster as to enable detection of the problem and safeguard the cell against propagation. For the smaller cell, the maximum temperature is far below that of the larger cell; however, the average cell temperature exceeds the melting point of the separator, thus providing the shutdown functionality.

different cells in a module. The module is comprised of 80 cylindrical cells of the 18650 type, each of 2.2-Ah capacity connected in the 16P-5S configuration. As seen in Figure 5.2(a), the temperature difference between the cells adjacent to the faulty cell and the cells at the periphery of the module reaches as high as 100°C. The difference in the individual cell temperatures is indicative of the lag in propagation of the heat generated from one cell to the others. Depending on the spacing between the cells, the heat generated due to thermal runaway in one cell may propagate within seconds to adjacent cells or can be arrested to within a few millimeters near the cell.

Figure 5.2(b) shows the current across the PTC device within each cell as a function of time for an intracell distance of 100 μm and an assumed short-circuit resistance of 20 mΩ. As observed, the currents across the PTC devices of some cells within the modules do not drop even after several minutes, indicating an inherent limitation in the safety feature of the battery module. In such cases, the nonuniform heating across the module limits the functionality of the PTC device, and the design of a safe cell does not translate into a safe battery pack. Additional monitoring of bank voltages or cell temperatures is necessary to isolate the faulty cell/module in such instances.

5.2.3 Mechanical Considerations

The third example deals with scaling up an existing design in order to build larger wound format cells. Current distribution across the bigger jelly roll is often an issue and is quite commonly identified by cell manufacturers who in turn provide for additional tabs across the length of the electrode to provide a balanced utilization of the electrode material. As a result, an 18650-type cell with 2.4 Ah or less capacity usually has one tab for each electrode across which the current to (or from) the entire electrode (which is ~ 70 cm long) flows. When the cell capacity is increased

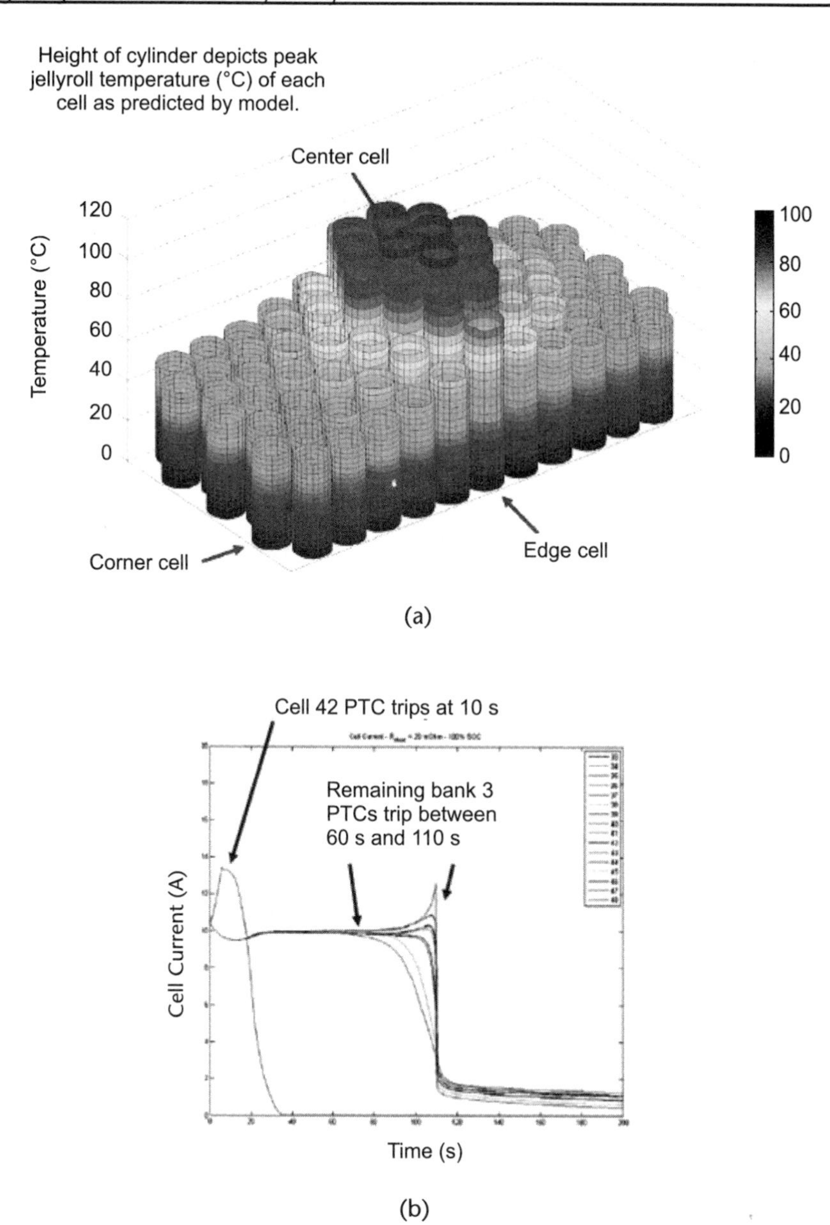

Figure 5.2 (a) Maximum temperature distribution on a 16P-5S battery module comprised of 18650 cells and (b) corresponding delay in PTC activation among the different cells within the module. These simulation results indicate that even when using conventional cell designs with established safety mechanisms, high-capacity battery packs trigger failure modes not encountered in conventional modules.

to greater than 2.6 Ah, a second tab is introduced. Larger-format cells have multiple tabs, and some cylindrical cells designed for automotive applications have a continuous tab that is crimped after the jelly roll is made to be attached to the bus bars. However, the often overlooked factor that heavily influences the yield as well as the life of the cells is the mechanical stress the increased number of winds introduces to the jelly roll.

As shown in Figure 5.3, the temperature rise within the cylindrical cells can rise by as much as 50°C for 3C operating currents, and if the cooling mechanism in

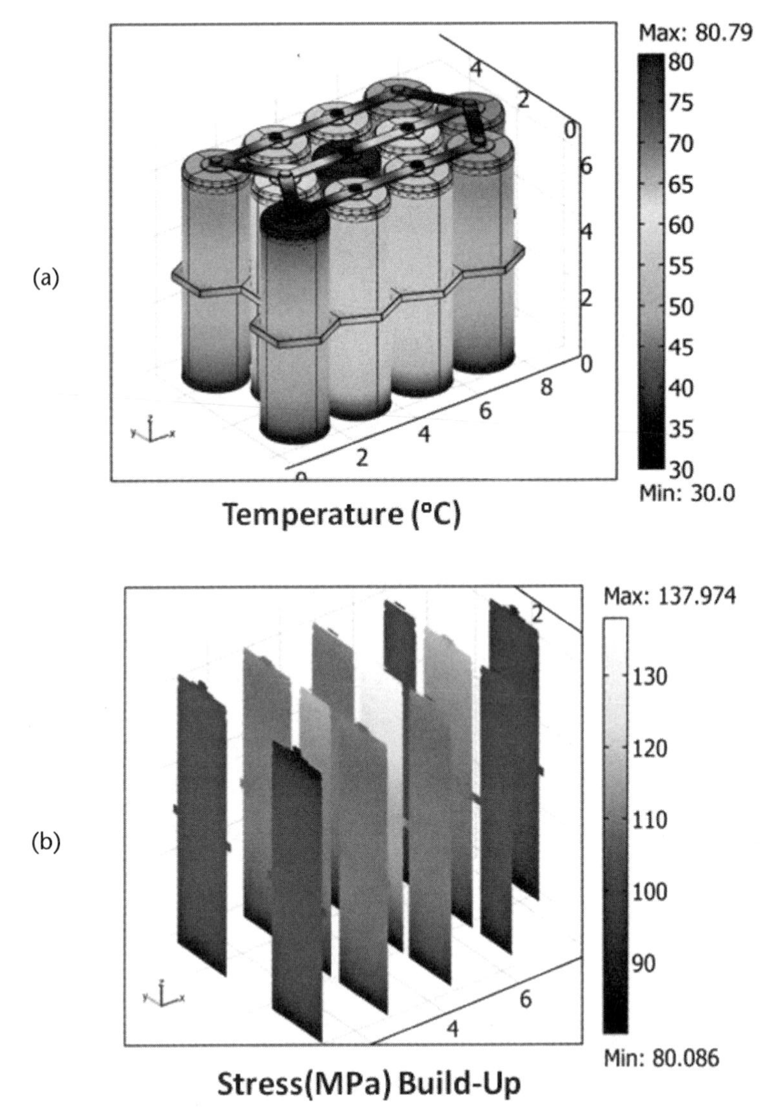

Figure 5.3 Stress buildup in a module comprised of cylindrical cells due to uneven distribution of cell temperature within the module. The cooling plate is located at the bottom of the module in this example, and the heat transfer from the cells located in the interior of the module to the ambient is not adequate, which translates to poor mechanical behavior of the components within the jelly roll. The manufacturing process should take into account such disparities when determining the winding tension for the jelly roll.

place is not uniform, the mechanical stresses that develop within the cells will differ widely—often resulting in rupture of the current collectors.

5.2.4 Pressure Buildup

One of the most frequently debated queries in the industry is about the amount of pressure the packaging should be able to withstand before the cell vents in the event of a thermal runaway. Whereas the heat generation mechanisms and the temperature evolution across the different stages of failure are relatively well characterized, little information is available on the amount of pressure the cell generates and how to develop a rational approach toward designing a safety vent to contain the failure within a battery pack. Figure 5.4 shows the calculated values of pressure per unit value of cell thickness for a 24-Ah cell with a given set of parameters for the chemistry and the thermal runaway reaction kinetics. Whereas the actual numbers for the different chemistries and reactions vary with the design and manufacturer, the trends are worth investigating: the chemical reactions resulting in gaseous species are the least contributing factors for the generation of pressure within the cell. These reactions are triggered earlier on in the runaway process, but the volume of gaseous species generated is not sufficient to trigger venting of the pouch. The elasticity of the pouch material in turn keeps the pressure generation due to mechanical deformation under check—until the cell temperature reaches sufficiently high values to cause evaporation of the volatile components. At this point, the pressure within the cell rises abruptly and crosses the yield strength of the seal, resulting in a vent. Such a systematic analysis serves as a means to determine the moduli of the pouch material and the temperature range across which the material is expected to retain its elasticity.

The determination of weld strengths in metallic cans as well as the seals in pouch material is often determined arbitrarily by the cell manufacturer based on prior experience with 18650 or prismatic cells <1 Ah. However, as mentioned earlier, the ratio of active to passive components in large-format cells is significantly higher than that for the traditional single-cell applications. For example, the amount of electrolyte used in the 24-Ah pouch cell is about 10 times that used in a 2.6-Ah 18650 type cell, and the volume of the larger-format cell does not scale linearly with the electrolyte content to accommodate for pressure buildup. Adopting a

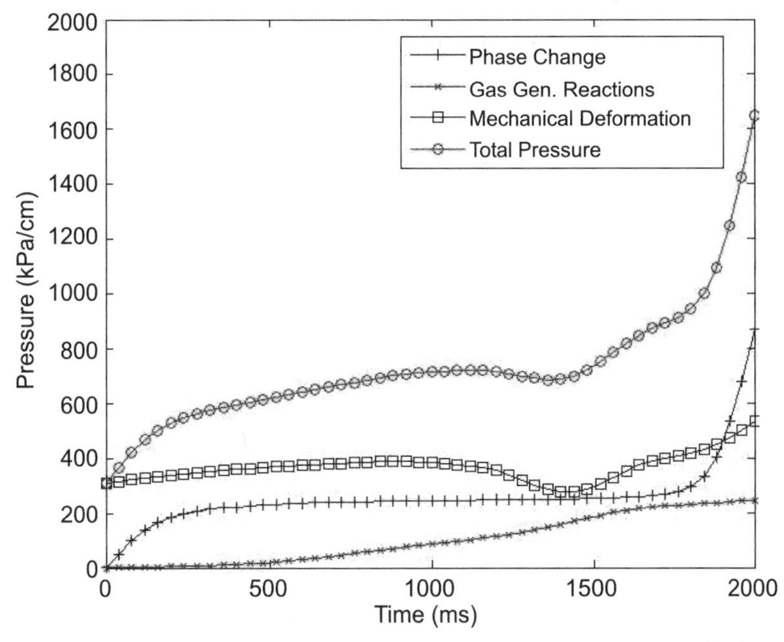

Figure 5.4 Different factors contribute to the buildup of pressure within the cell when a thermal runaway reaction happens: volatile components in the electrolyte undergo a phase change, several chemical reactions generate gaseous products, and the container undergoes mechanical deformation to accommodate the rise in pressure. The choice of pouch material and the determination of weld/seal strengths should account for these factors.

more systematic approach as outlined above will result in improving the uniformity in abuse response of the cells.

5.2.5 Designing Protective Circuitry to Combat Short Circuit

Short circuit of cells in a battery pack is a major concern among the different safety issues commonly encountered in the field. The short circuit may result from an external trigger (example.g., a flood leading to electrical contact across the leads of the cells in a battery pack that has been mechanically compromised) or due to an internal defect that originates during the manufacturing process (e.g., due to the presence of some metallic impurity within the jelly roll). Regardless of the origin, a short circuit in a Li-ion cell leads to field incidents that range from failure of individual cells or modules followed by a rapid heat up of parts of the pack to uncontrolled venting and smoke from the battery pack. However, the rate of incidence of such failure is so rare that a repeatable laboratory test to capture the response of the cell to a short circuit is difficult. The different test procedures under development to characterize short-circuit response of a cell are outlined in Section 5.3.3; but this example focuses on the electrical nature of the different types of short circuit that happen in a cell and its implication toward improving safety of the battery.

Basically, there are four components within a battery that can lead to an electronic short circuit: the cathode active material, the aluminum current collector on which it is coated, the anode active material (typically carbon), and the copper current collector on which it is coated. Any short-circuit results in components of the anode compartment (i.e., either carbon or the copper foil) coming into contact with the components of the cathode compartment (i.e., the cathode active material or the aluminum foil). Thus there are four permutations of short circuit that are theoretically possible within a cell:

Type I short, where the cathode and anode active materials come into direct contact;

Type II short, where the aluminum foil comes into contact with the carbon on the anode;

Type III short, where the copper foil comes into contact with the cathode active material;

Type IV short, where the copper and aluminum foils come into contact with each other.

The external short circuit would be classified as a type IV short since the current collector tabs come into electrical contact with each other for this case. All these types of short circuits can possibly originate internal to the cell. However, given the cell construction for most of the existing cell designs, the type II short is least likely, merely based on the area of each component that has a reasonable probability of coming into contact with the other. For a given cell design and contact area during a short circuit, when the heat generation rates are simulated for each type of short circuit, type II shorts accumulate the maximum amount of heat due to the right range of resistances across the short-circuit pathway. A type IV short is very conductive since the contact is between two metallic foils; as a result of the low resistances, the heat generation for this type of short circuit is lower, whereas a short involving the cathode active material is so resistive that the currents passing across the type I and type III shorts are lower compared to the others, and as a result, the heat generation rate for these scenarios are usually the lowest.

The temperature evolution during the different types of short circuits in a 400 MAh cell is shown in Figure 5.5. There are several implications to improve battery safety: first, these results do not directly translate to large-format cells, which are designed either with several-fold increments in energy that eventually the cells will enter into thermal runaway. However, the response time and the thermal load that the cooling system should be able to accommodate can be computed using the methods outlined in Chapter 3 for different cell sizes and form factors. Second, the design of high-power cells intuitively considers lower impedances for the cathode. Whereas it is not surprising that low impedance cathodes result in lower amounts of heat generation, there is a trade-off between the amount of heat generated during normal operation and the potential for generating excessive heat owing to good conductivities during a short-circuit. Cell manufactures should then consider designing the cathode slurry with minimum values for electronic conductivity that will meet the performance requirements. The last and most significant implication is around the design of safety features that will protect the cell in the event of a short circuit. There are several possible solutions: for instance, since the short circuit between the current collectors is deemed safe for the cells discussed in the example, it may be reasonable to provide an extra winding of the copper current collector on the outside of the jelly roll so that any unintentional short circuit during a mechanical crush will result in a type III or IV short rather than a type I short,

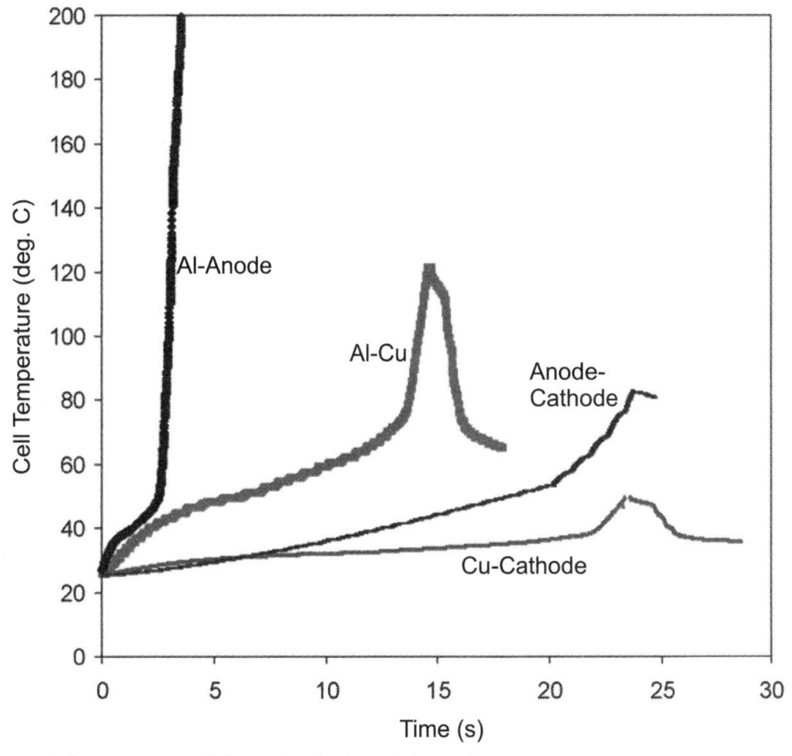

Figure 5.5 Different types of short circuits in a Li-ion cell.

or a thermally activated fuse can be incorporated as part of the cell design to ensure contact between the desired layers of the jelly roll, thus minimizing the damage external to the cell.

5.3 Evaluating Battery Safety

Evaluation of Li-ion batteries for safety is a highly contested area in the industry. Despite attempts by several committees, the experimental procedures for abuse testing of batteries vary considerably among the different sections of the industry. This section elaborates on the tools and practices commonly used by the battery community in this context. The subsequent sections address progress towards building a consistent set of test procedures.

The test procedures related to the component level are mainly developed and advocated by the cell manufacturers. It is in the interest of the pack designer to understand the material level limitations in order to be able to distinguish issues that can be addressed at the pack-design phase versus those that need to be addressed at the cell level. As elucidated in the previous section, the safety of individual cells does not automatically translate to the safety of the battery pack built using those cells. However, as discussed in the examples in Section 5.2, each part of the cell plays a crucial role individually and in conjunction with the other components of the cell. Understanding safety at the component level is inevitable in building a safe battery pack. Determination of properties that determine safety at the component level is largely carried out for the active ingredients using an accelerated rate calorimeter that provides information on how fast the abuse reactions take places and the associated heats of reaction. The passive components such as the separator film are evaluated for mechanical and thermal stability.

5.3.1 Measurement of Reaction Heats: Accelerating Rate Calorimeters

The accelerating rate calorimeter (ARC) or adiabatic calorimeter was devised by Dow Chemical Company in the 1970s and was patented in 1984. It is a great tool for evaluating exothermic chemical reactions. The instrument was developed to understand the thermal hazards associated with reactive chemicals, including peroxides, explosives, batteries, and other reactive materials. A general schematic of an accelerating rate calorimeter is shown in Figure 5.6. The basic instrument involves a sample holder, typically referred to as a bomb, which is placed inside a reaction chamber. The sample holder and the reaction chamber are kept at the same temperature through the control and use of heaters. The heaters ensure that an adiabatic condition exists and there is no heat transfer between the sample holder and the cell. The sample holder is typically made from a high-strength material such as titanium or Hastelloy® so as to withstand the pressure buildup during an exothermic reaction. The amount of reactive material placed in the sample holder is dependent on the size of the holder and the expected exothermic reaction.

Once a sample material is placed in the holder, the container is sealed in the reaction chamber and the instrument goes through a heat/wait/seek cycle, as shown in Figure 5.7. Initially, the sample and chamber are heated to a user selectable

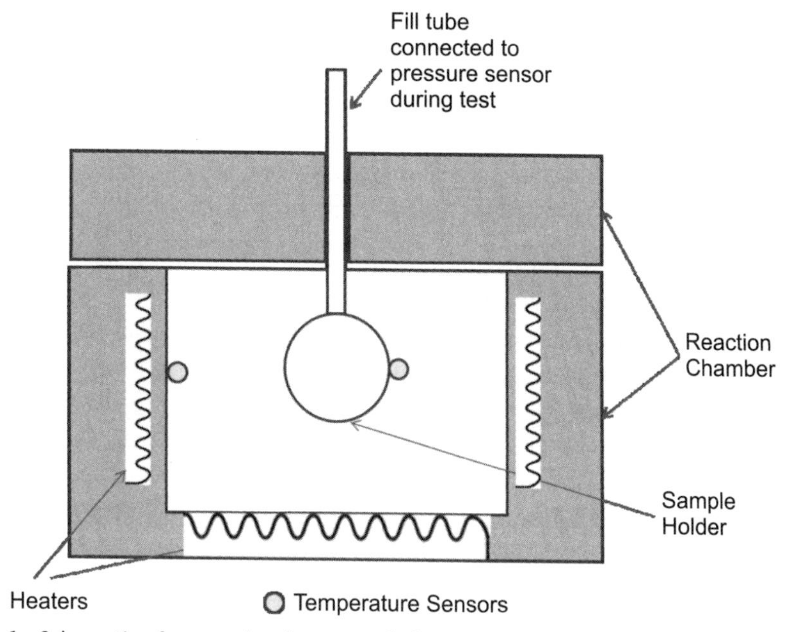

Figure 5.6 Schematic of an accelerating rate calorimeter.

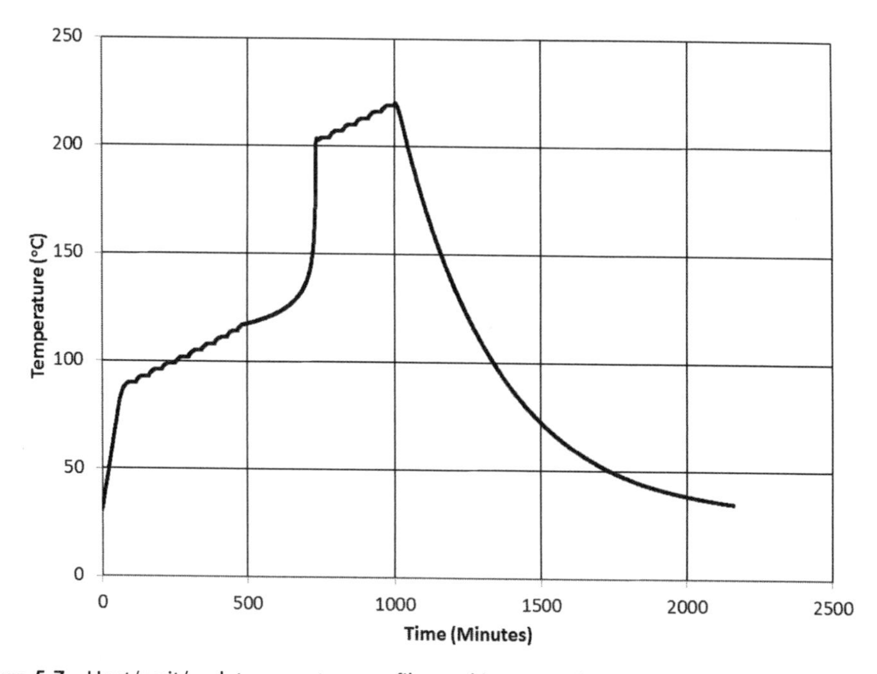

Figure 5.7 Heat/wait/seek temperature profile used in an accelerating rate calorimeter.

temperature. The temperature is then held for a set amount of time (wait) in order to detect (seek) an exothermic reaction. If no reaction is detected, then the proportional-integral-derivative (PID) heater controllers will increment the temperature of both the chamber and sample holder. The heat/wait/seek process will be repeated until an exothermic reaction is sensed. Once an exothermic reaction occurs, the temperature of the reaction chamber will be controlled to the temperature of the sample holder; the rise in temperature at times may exceed several 100°C/min. If an adiabatic condition is maintained during the chemical reaction, then the system provides an accurate measurement of the heat data along with the onset temperature, the maximum heating rate, and the enthalpy of reaction. Typically, tests will be performed over several days so as to ensure that the onset temperature is accurately characterized, and the temperature is incremented only a fraction of a degree between heat/wait/seek periods.

Limitations associated with ARCs. It should be noted that no accelerating rate calorimeter is truly adiabatic—heat will be lost from the sample to its holder. A phi factor, the ratio of the thermal mass of the sample and holder to the thermal mass of the sample alone, is used to correct the calorimetric result for this heat loss. Another limitation associated with this type of calorimeter is that the sample holder typically only has a single highly accurate resistive temperature detector (RTD), thermistor, or thermocouple placed on its exterior. It is assumed that this single temperature sensor on the sample holder represents the temperature of the entire sample vessel. Essentially, the assumption is that the sample and its container have an infinite thermal conductivity or no mass—an obviously inaccurate assumption. Furthermore, it is extremely difficult to accurately match the temperature between the reaction chamber and the sample holder during the exothermic reaction because some reactions occur too quickly.

5.3.1.1 ARCs and Testing Batteries.

ARCs are used to assess the thermal stability of electrolytes, cathodes, anodes, and cells. It is well known that the cathode material has the strongest influence on a cell's thermal stability. Table 5.1 shows the properties for some of the commercially available cathodes on the market today. $LiCoO_2$ cells are the most common chemistry used in consumer electronics due to the high energy density of the cell.

Table 5.1 Sample Characteristics of Cathode Electrode Materials

Material	*Specific Capacity mAh/g*	*Midpoint V vs. Li at C/20*	*Comments*
$LiCoO_2$	155	3.9	Still the most common. Co is expensive.
$LiNi_{1-x-y}Mn_xCo_yO_2$ (NMC)	140-180	~3.8	Capacity depends on upper voltage cut off. Safer and less expensive than $LiCoO_2$
$LiNi_{0.8}Co_{0.15}Al_{x0.05}O_2$ (NCA)	200	3.73	High capacity. About as safe as $LiCoO_2$
$LiMn_2O_4$ (Spinel)	100-120	4.05	Poor hight temperature stability (but improving with R&D). Safer and less expensive than $LiCoO_2$
$LiFePO_4$ (LFP)	160	3.45	Synthesis in inert gas leads to process cost. Very safe. Low Volumetric energy
$Li[Li_{1/9}Ni_{1/3}Mn_{5/9}]O_2$	275	3.8	High specific capacity, R&D scale, low rate capability
$LiNi_{05}Mn_{1.5}O_2$	130	4.6	Requires an electrolyte that is stable at high voltage

From: [1].

However, these cells are thermally unstable due to the breakdown of the cathode and its subsequent evolution of oxygen at higher temperatures [2]. Figure 5.8 shows the thermal response from ARC experiments on several different cathode materials incorporated into 18650 cells. The $LiCoO_2$ 1.2 Ah cell has the lowest onset temperature and also the highest heating rate. In contrast, the $LiFePO_4$ cell has the highest onset temperature and the lowest heating rate—the $LiFePO_4$ cathode does not evolve oxygen during its decomposition and thus its relatively benign thermal response as compared to $LiCoO_2$ during this calorimetric experiment.

ACRs can also be used to assess the thermal response of the component materials within a battery. Figure 5.9 shows the thermal response of the individual cell components within a cell. The experiment was performed by removing the battery component materials from a fully charged NMC cell. The component materials were resealed into a 18650 container with additional electrolyte. As can be seen from the figure, each cell component has different onset temperatures as well as maximum heating rates. When these components are combined into the full cell, the thermal runaway onset temperature is approximately 220°C, which is different from the individual components within the cell. The difference is primarily due to the interactions of the components incorporated into the cell as well as the component differences in specific heat.

Adiabatic calorimeters are very useful tools when addressing the safety aspects of cell design. They can be used to assess how:

- An electrolyte additive affects the flammability temperature of an electrolyte;
- An artificial SEI layer affects the evolution of oxygen during cathode decomposition at higher temperatures;
- The individual battery components affect the safety of the battery system.

In the end, the ARC is a vital tool for assessing the abuse tolerance of all a batteries components and features.

5.3.2 Thermomechanical Characterization of Passive Components

The safety of a Li-ion cell depends equally on the response of the passive components as much as it does on the ingredients that react during an abuse scenario. The

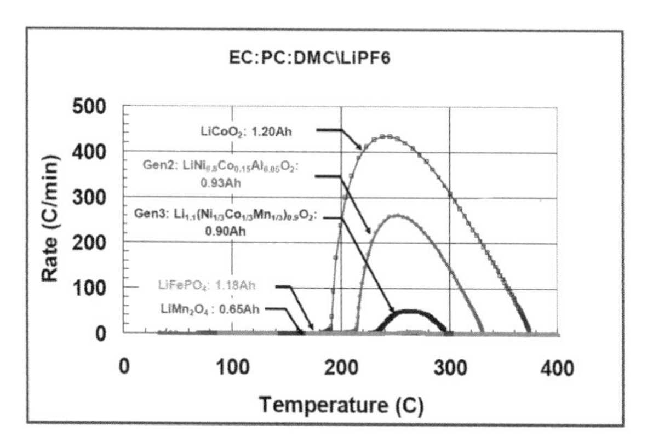

Figure 5.8 ARC experiments performed on different cathode materials. (Source: Doughty [3].)

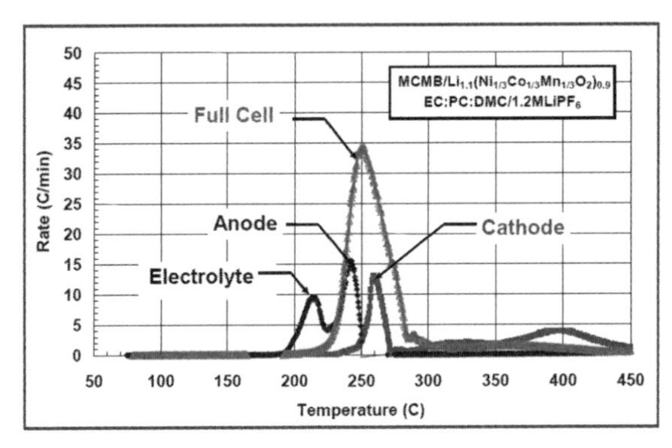

Figure 5.9 ARC experiments performed on the individual components in a NMC cell. (Source: Doughty [3].)

mechanical characterization of the passive components such as the separator, pouch material, and current collectors is even more important in the context of large-format cells than it has traditionally been in the battery manufacturing process.

Testing of the thermal-mechanical properties of the cell components is carried out using a tensile-strength tester using standard American Society for Testing and Materials (ASTM) methods.[1] A standard vice to hold the sample of prespecified dimensions is subjected to tensile stresses as a function of temperature. Detailed test procedure for mechanical characterization of polymer films used in Li-ion batteries is described in other standard references on Li-ion batteries [4]. This section is included here only to highlight the differences in test methodology for large-format cells. For instance, the width of a separator film in a jelly roll for an 18650-type cell typically exceeds that of the electrodes by 3% to 4% whereas such tight clearance is not expected for a large-format cell for two reasons: the temperature distribution across the larger cell varies quite a bit for wound cells, and the tolerance of the jelly roll within the can is not nearly as close as that for the 18650 format. With the advent of continuous tabs that are crimped on either end of the jelly roll, the separator width is maintained just below the point where it does not interfere with the crimping of the tabs. Similarly for stacked format cells, the cell is more forgiving of the mechanical properties of the separator than it is of the shrinkage when the cell is exposed to high temperatures. These factors allow for the design of separators with larger pores, for example, to allow for the higher power rating of the automotive cells.

Also with lower winding tensions, the mechanical strength along the thickness direction of the different layers is not as crucial as that for tightly wound low-capacity cells. The large-format pouch cells are encased in hard casing for single-cell applications; for the more widely targeted application within larger battery packs, the packaging of the battery pack implies that the design of the membranes should

1. See for example, ASTM D5947-96, Standard Test Methods for Physical Dimensions of Solid Plastics Specimens; ASTM D2103, Standard Specification for Polyethylene Film and Sheeting; ASTM D3763, Standard Test Method for High-Speed Puncture Properties of Plastics Using Load and Displacement Sensors; ASTM D1204, Standard Test methods for Linear Dimensional Changes of Non-rigid Thermoplastic Sheeting or Film at Elevated Temperatures; and ASTM D882, Standard Test Method for Tensile Properties of Thin Plastic Sheeting, ASTM International.

focus on optimizing thermal properties and tolerance to a wider temperature window that the large-format cells are designed for use under, rather than the traditional emphasis on the mechanical strength. For example, the trends shown in Figure 5.10 indicate that films approach the yield strength at much lower strain values when the test is performed at lower temperatures. These results when considered on their own indicate the need for a stronger membrane. However, from Figure 5.4 it is clear that mechanical deformation is not the limiting factor that triggers the violent response when a cell undergoes an abuse reaction. In this context, a lower resistance membrane will result in less heat generation as well as exhibit a lower propensity for plating of lithium at the lower temperature extremes for which the automotive batteries are designed, and thus has a better chance of averting a catastrophic response from the cell.

5.3.3 Cell-Level Testing

Laboratory testing of the abuse response of Li-ion cells comprises of subjecting the cell to extremes of temperature and mechanical stress together with electrical short circuits. The metric used for evaluating the cells varies significantly based on the cell design, target application, and the end-user requirements. This section outlines some of the standard test practices and highlights the differences between testing of small cells typically used in consumer electronics applications versus larger-format cells used in larger-capacity modules/packs. In each of the following tests, a pass/fail designation is assigned based on prespecified criteria. Examples of standards governing the test-procedure are included in Table 5.2. However, this is an evolving discipline, and there are several participants actively shaping the discussion as the understanding on abuse testing of large-format lithium ion batteries matures.

5.3.3.1 Short-Circuit Tests

These tests involve subjecting the cells to an internal or external short circuit under a known set of conditions (state of charge, heat transfer rate, and ambient

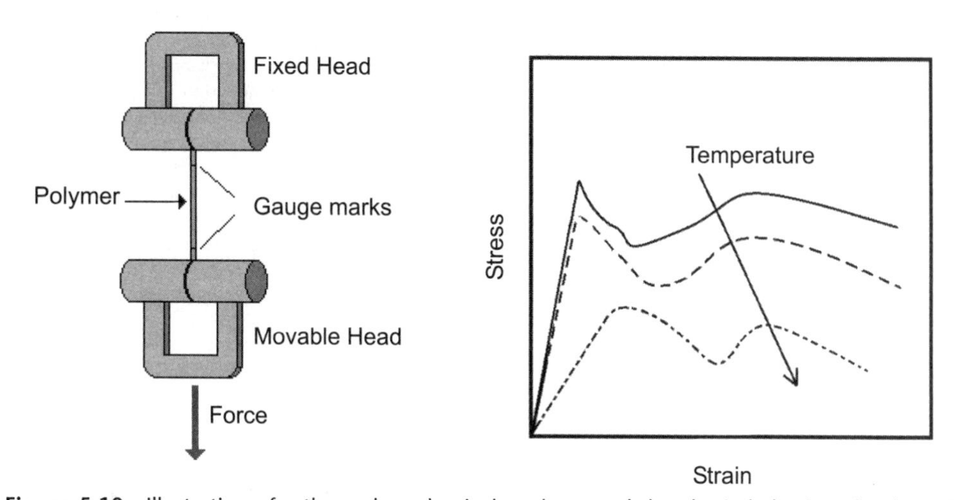

Figure 5.10 Illustration of a thermal-mechanical analyzer and the elastic behavior of polymeric films as a function of temperature.

Table 5.2 Example of Standards Governing Testing of Large-Format Li-Ion Cells

Test	*SAE J2464*	*FreedomCAR*	*IEC 62660*	*UL2580*	*Korea MVSS 18-3*
Thermal shock	5 cycles between 70 and –40°C; hold cells for 1 hour; modules/packs for 6 hours	5 cycles between 80 and –40°C; hold cells for 1 hour; modules/packs for 6 hours	30 cycles between 65°C and –20°C; repeat test at various SOC	Same as J2464, except that the temperature extremes are from 85 ± 2°C to –40 ± 2°C	
Mechanical shock	18 each 25g shocks (XYZ negative and positive directions × three times)	RMS acceleration of 27.8 m/s2	500 m/s² (50g) 6 msec 60 each (XYZ negative and positive directions × 10 times) HEV cells at 80% SOC and EV cells at 100% SOC	Test according to SAE J2464 using half sine wave load profile with 25g amplitude	The battery should be subjected to 10 shocks in each axis in half sine wave, 30g amplitude and 15-ms duration
External short circuit	One hard short (≤ 5 mΩ) and one moderate short at a resistance similar to that of battery at 25 ± 5°C	Apply a hard short of ≤5 mΩ in less than 1 second at 20°C; hold for 10 minutes	≤5 mΩ for 10 minutes at 20°C; the sample rate for voltage and current recording should be ≤ 10 ms	Total circuit resistance ≤ 20 mΩ; repeat test at a load that draws a maximum current ≥15% of the short-circuit protection current	Resistance = 50 mΩ for 1 hour or until no detectable current flow for 5 minutes; initial SOC = 80%
Overcharge	Charge cells at two rates: (a) 1C rate and (b) at the max. usable current to 200% SOC	32 A to 200% SOC	Charge until the cell voltage reaches 2× maximum voltage or 200 % SOC	Charge the pack per manufacturer's recommendations, with one module precharged to 50% SOC and others at 0%	Charge to 1.5 times the nominal voltage with 32 A constant current; final SOC = 150%
Crush	Crush to 85% of initial dimension; hold for 5 minutes; continue to 50%	Crush to 85% of initial dimension; hold for 5 minutes; continue to 50%	Crush to 85% of initial dimension; or until force ≥1000 × weight of cell; or until cell voltage drops by one third of initial value; use a 150 mm diameter sphere for prismatic cells or a 150 mm diameter bar for cylindrical cells	Use ribbed test platen Crush all 3 axes per SAE J2464; Max force = 100 kN; test article may be installed in a protective framework representative of what is provided in the vehicle	
Open flame test	10 minutes at 890°C to simulate a fuel fire (see also J2929, J2579)	10 minutesat 890°C to simulate a fuel fire		Subject fully charged battery to uniform fire along the length at its bottom until at least one thermocouple shows >590 °C for 20 minutes	Apply heat between 890 and 900 °C for 2 minutes to bottom of the battery at 80% SOC
Rollover test	Complete revolution in 1 minute, then rotate the battery in 90° increments for one full revolution	One complete revolution for 1 minute in a continuous, slow-roll fashion, then rotate the battery in 90° increments for one full revolution		Rotate sample at 100% SOC, at a continuous rate of 90°/15 s; testing should subject the sample to a 360° rotation in three mutually perpendicular different directions	Same as FreedomCAR; observe leakage; rotate the battery in 90° increments; hold the battery for 1 hour at each position

temperature). A broad range of tests fall into this category. An external short-circuit test conducted by the Japanese Industrial Safety Standard, JIS C8714, which caters to portable Li-ion cells and batteries for use in portable electronics is carried out by connecting the terminals of the cell at 100% SOC across a resistor 80 ± 20 mΩ and monitoring the cell temperature for 24 hours or until it returns to the chamber temperature (set to 55°C), whereas Underwriters Laboratories (UL) specifies (UL-2580, "Batteries for Use in Electric Vehicles") the short-circuit resistance as 20 mΩ at 20°C.

5.3.3.2 RMS: Root mean square.

Intrusion of a foreign object resulting in short circuits is usually simulated in a nail penetration test is carried out by puncturing the cell using a nail or a blunt rod. The specifications vary with the objective of the test and the requirements of the end users. One common approach is to use a nail with a tip radius of 0.9 mm and tip angle of 45° at a constant press speed of 0.1 mm/s until the cell voltage drops by at least 100 mV. However, as described in Section 5.2, the propagation of thermal events varies significantly among different cell designs. More recent test methods have recommended inducing a short circuit of definite resistance at a specific location. Test procedures such as those recommended by the Battery Association of Japan (JIS C8714) for consumer electronic cells involve placing a conductive particle between the layers of a jelly roll obtained after opening a cell at 100% SOC have been found unsuitable for large-format cells. Alternate test procedures involve triggering a short circuit internal to the cell by electrical, mechanical, or thermal means. These tests report higher rates of reproducibility and better control over the type of the short circuit introduced, short-resistance, location, and time of trigger.

5.3.3.3 Crush Testing

The mechanical integrity of the cells and packs used in large-format applications is tested by applying a force equal to 1,000 times the weight of the test article or a maximum of 100 kN to 85% of the initial dimension and the load is held for a duration of 15 minutes, after which the test is continued until the strain reaches 50% of the original dimension. The tests are repeated on different samples held between one flat platen and another with semicircular ridges along each of the three axes of orientation.

5.3.3.4 Hot-Box Test

Whereas some standards such as the SAE J2424 call for an open-ended test to determine the maximum temperature at which the cell remains stable "indefinitely," others such as the one developed by the International Energy Commission (IEC 62660-2) specify a ramp rate of 10°C/min to 130°C followed by storage at that temperature for 30 minutes. An equivalent IEEE 1725 standard prescribes a slower ramp rate of 5°C/min and storage for a longer duration of 60 minutes. The more rigorous test condition at the individual cell level for smaller-format cells in this instance is justified by the likelihood of a battery used in consumer electronics to be more susceptible more often to high temperature exposure compared to well-packaged vehicular batteries, for example, that have a good mechanism to carry heat away from individual cell surfaces. The emphasis instead is on thermal shock testing, where the cells or modules are subjected to thermal cycles between -40°C and +80°C and held at each extreme for a duration of 1 hour. This test captures the heat generation within the module during the high-power cycles and checks for any difficulties within the cells in facilitating quick heat transfer across the jelly roll, which is typically an issue for cells of larger dimensions.

5.3.3.5 Impact Testing

Whereas mechanical impact testing, such as a drop from a prespecified height, is considered a good measure of the individual cell's mechanical durability, these tests provide little information on the safety of large-format cells in a pack. Instead, the focus is on mechanical shock due to vibrations or a vehicle undergoing a crash test. Accordingly, the acceleration at impact is reduced from those prescribed for single-cell testing—these usually range from 25 to 50g for vehicular batteries (SAE J2424, ISO/CD 12405) to 125 to 175g for cell phone batteries under IEEE 1725.

5.3.3.6 Pressure/Humidity Testing

The electrolyte used in Li-ion batteries is very sensitive to moisture. Any failure across the weld on the container or the seal along the pouch leads to exposure of the electrolyte to moisture and triggers chemical reactions that result in swelling and/or pressure accumulation within the cell. Due to the relatively lower maturity in packaging techniques employed to make large-format cells, compared to say, the manufacturing process for the 18650 cells, these issues are more prevalent in larger cells. Immersion testing under saltwater at 25°C for 2 hours is usually recommended for modules and packs, whereas storage at 0.1 atm pressure for 6 hours is commonly used to check for leaks at the cell level. Testing at a higher temperature (55°C) or storage at very low temperatures (–17°C/0°F) are also recommended in order to reduce the testing duration. However, the lack of an inline test with a prespecified metric for leaks or quality of welds during the manufacturing process is an issue currently not addressed by the industry.

5.3.3.7 Overcharge Testing

Whereas the ability of the cell to hold charge in excess of 100% SOC is tested extensively, under various charge rates and durations of exposure to overcharge the different testing bodies have identified limited utility in subjecting large-format cells or modules and packs to overcharge. This is in part due to the dependence on protective circuitry to isolate overcharged cells. After several earlier designs of battery packs have repeatedly shown the need for additional safety measures, protective circuitry for single-cell has been proposed recently for large-format cells in battery packs.

5.3.3.8 Fire Hazard

Li-ion batteries have been widely advertised as hazardous when exposed to fire, owing to the presence of organic solvents and highly reactive chemicals. Flammability characterization of Li-ion batteries vary significantly in reporting the amount of heat generated when a battery is subjected to fire. Testing of individual cell as well as theoretical calculations place the amount of heat generated from a single cell of the order of 150-250 kJ/Ah. The most commonly identified source of fuel in fires related to Li-ion batteries has been the packaging material, which in some instances amounts to as much as 40% of the pack by weight and exceeds the flammability of the batteries themselves by as much as 10 times. With the effect of packaging material that may act as fuel, the energy equivalent of fires reported in battery warehouses and shipment containers has been estimated to be as high as a few megajoules per Ah capacity. The Society of Automotive Engineers (SAE) standards (J2464 and J2929) recommend exposing the battery to an open flame at 890°C for 10 minutes to simulate a fuel fire. UL recommends exposing the battery to a lower temperature (590°C) for longer durations (20 minutes).

Whereas many of these tests provide at best a repeatable qualitative assessment of safety at the individual cell level, further work is necessary to provide more measurable insights, such as the ability to gauge the propensity for propagation of failure from one damaged cell to others in a multicell module or a pack. Many standards such as the SAE J2424, UL 2580, and the FreedomCar EESS Abuse Test Manual (SAND2005-3123) discuss propagation, but there are no well-defined requirements, beyond setting an individual cell within a representative subunit of a pack to get into thermal runaway and monitoring the other parts of the pack for propagation. Differences in trade-offs between performance metrics and safety within a given operating bound for different applications preclude development of quantitative standards for multicell modules. Another key challenge in developing safety standards for large-format cells is the wide range of specifications available in the market for the cells themselves. As the technology matures, a better degree of standardization is expected to contribute to improvements in the test procedures as well as safety of the batteries themselves.

References

[1] Dahn, J., and G. M. Erlich, "Lithium Ion Batteries," in *Linden's Handbook of Batteries*, 4th Edition, T. B. Reddy (ed.), New York: McGraw Hill, 2011, p. 26, Table 26.3.

[2] Arai, H., M. Tsuda, K. Saito, M. Hayashi, and Y. Sakurai, "Thermal Reactions between Delithiated Lithium Nickelate and Electrolyte Solutions," *J. Electrochem. Soc.*, Vol. 149, p. A401, 2002.

[3] Doughty, D. Vehicle Battery Safety Roadmap Guidance, Subcontract Report, NREL/SR-5400-54404, 2012.

[4] Santhanagopalan, S., and Z. Zhang, "Separators for Lithium Ion Batteries," in *Lithium-Ion Batteries: Advanced Materials and Technologies*, Green Chemistry and Chemical Engineering, X. Yuan, H. Liu, and J. Zhang (eds.), Boca Raton, FL: CRC Press, 2011, pp. 197–253.

CHAPTER 6

Applications

The suitability of a battery to any particular purpose can be reduced to one seemingly simple criterion: can it adequately power the application? However, as discussed in the previous chapters, the performance of a Li-ion battery is highly sensitive to temperature. It also degrades through life at a rate dependent on electrical and thermal cycling characteristics. Further, the electrical and thermal cycling to which the device will be subjected over its intended service life may not be defined with certainty. Thus, answering this question can become quite complex. The first step—understanding the application—is of critical importance and is addressed herein. Methods to apply this knowledge to the design of Li-ion battery systems are discussed in Chapter 7.

6.1 Battery Requirements

Understanding an application means understanding the requirements it places on an energy storage system. These demands can vary substantially from one application to the next, but a similar set of data is often required. They typically span electrical, thermal, mechanical, and safety topic areas.

6.1.1 Electrical Requirements

The energy and power demanded of a battery are often defined by one or more electrical duty cycles. When known, it is best to acquire these duty cycles as time histories of either power or current. However, often this level of detail is not known with a good degree of certainty. In place of high-resolution data, the duty cycle can instead be represented by its total duration, average, and RMS power or current, and minimum and maximum power or current coupled with a predetermined duration. Table 6.1 shows these values as computed for an example peak shaving duty cycle shown in Figure 6.1.

A quick look at Figure 6.1 and Table 6.1 reveals that significant information can be lost in such a simple translation. For example, Figure 6.1 reveals that the peak discharge power requirement occurs early in the discharge. Where a conservative designer is given only the information in Table 6.1, he or she may assume the peak power discharge comes at the end of the discharge. Working with such an assumption may result in a larger battery than actually needed. Further, the peak

Table 6.1 Translating Duty Cycle in Figure 6.1 to Average and RMS Power over Duration

Average discharge power	3.9 kW
RMS discharge power	5.1 kW
Peak discharge power	12.4 kW for 1 minute
Total discharge duration	119 minutes
Average charge power	4.2 kW
RMS charge power	5.3 kW
Peak charge power	10.0 kW for 1 minute
Total charge duration	111 minutes

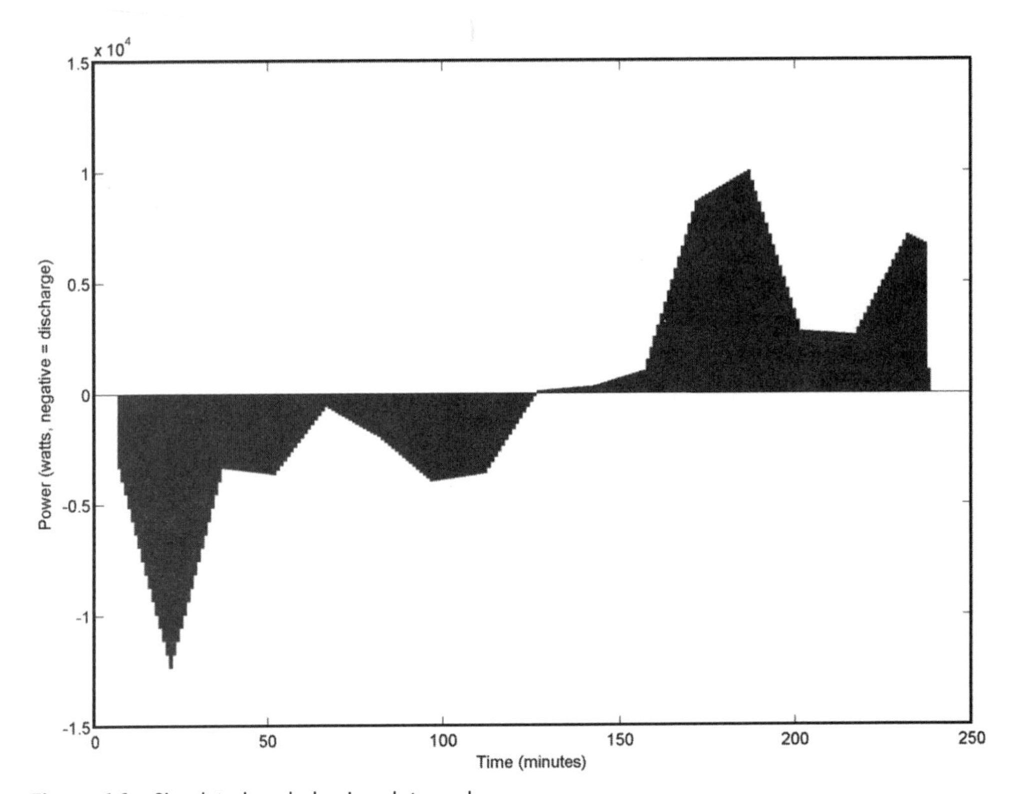

Figure 6.1 Simulated peak shaving duty cycle.

discharge duration is only specified for 1 minute; however, the nature of the profile is such that a battery that meets this peak duration may not be capable of the supplying the actual duty cycle. A more conservative translation may be to specify a peak discharge power requirement of 12.4 kW for 10 minutes.

It is not atypical that the discharge duty cycle be defined succinctly while the charge duty cycle is defined more ambiguously. For example, a discharge profile may be specified per Figure 6.1, and a charge profile may be specified simply by a maxium available power or current and an accompanying duration. This allows the battery system to manage its own charge profile to maximize performance and safety (e.g., implementing a taper charge as the battery reaches its maximum voltage). In other cases, it may be necessary for the battery to absorb all the energy that is provided to it. Care should be taken to clarify when this is the case, as it can have a large impact on the design of the system.

The total number and frequency of occurrence must also be defined. The number and frequency of occurrence affects battery wear in the long term, but also impacts the battery's thermal response in the short term. For example, a given charge-discharge cycle may be easily achievable when performed once per day, but when performed 10 times per day the increased heat generation and reduced cooldown periods may lead to unacceptably high battery temperature.

When multiple duty cycles are specified, the sequence in which they occur is also important. In the short term it can impact thermal response as previously noted (e.g., sequential high-power discharge and charge operations may induce unacceptably high battery temperatures), while in the long term it could affect degradation patterns and the ability to meet performance requirements (e.g., increasing versus decreasing DOD through life).

6.1.2 Thermal Requirements

The thermal environment that the battery must operate in must also be properly defined. Ideally, the correlation of thermal environment to the electrical duty cycle will be specified. Most frequently, this is accomplished by dividing thermal limits into operational and survival limits. For example, the battery may be required to complete its electrical duty cycle in an environment ranging from –10°C to +40°C, while it may be required to survive (but not operate) in a more extreme –30°C to +60°C. Given the reduced power capabilities of batteries at low temperatures, it is also common to couple a less demanding electrical duty cycle to the lowest temperatures of the operational band.

It is important to note that the specification of environmental temperature is different than specifying battery temperature. By specifying environmental temperature we must recognize that electrical operation will generally result in heat generation and increased battery temperatures. At high environmental temperature we must therefore be sure to assess the maximum battery temperature that will be achieved and its negative impacts on degradation and safety. At low environmental temperatures, however, we can leverage the battery's self-heating effect to improve performance.

6.1.3 Mechanical Requirements

While there is, at present, a lack of publicly available data relating mechanical environmental factors, such as vibration and shock, to electrochemical performance and long-term degradation, such exposure is known to be capable of damaging electrical pathways both within and between cells. These factors can also affect balance of systems components such as disconnect relays and charge balancing systems. Further, extreme mechanical abuse such as crush or impact that results in mechanical deformation can create hazardous short circuits within the system. For these reasons it is also necessary to specify mechanical requirements for the system.

Two common mechanical requirements are shock and random vibration. The former corresponds to exposure of the battery to very short duration acceleration events, while the latter concerns longer duration exposure to nonperiodic vibrations. Both shock and random vibration requirements are typically specified by a power spectral density (PSD) plot, direction, and in the case of random vibrations, a duration. PSDs are generally created from the discretization of measured or simulated time histories of acceleration by frequency.

Vibration requirements may also include tolerance to *sine sweeps*, where a sinusoidal vibration pattern is applied that slowly increases or decreases in frequency over time. While such exposure is uncommon in the application itself, testing the battery via a sine sweep is beneficial for identifying the system's natural frequencies. Such testing is also used to develop mathematical models that capture the mechanical failure of a component as a function of the storage and loss moduli.

Impact and intrusion requirements may also be pertinent, particularly for vehicular applications. Specifying such requirements can be challenging due to the unpredictable nature of crush events in the field. Therefore, most specifications are designed around an intended test method. It may be likely that these requirements are coupled with abuse response specifications.

6.1.4 Safety/Abuse Requirements

Requirements for abuse tolerance of a battery vary by application and become most important when human safety is a factor in manned applications. Most frequently, abuse tolerance requirements demand that a battery system be capable of enduring one or more off-nominal conditions while maintaining all temperatures and any gas expulsion below a threshold level and without the presence of fire or explosion. The off-nominal conditions that may be specified include over- and under-temperature, over- and under- voltage, short-circuit exposure, or mechanical deformation. It may also be required that these criteria are met in the presence of one or more systems failures (e.g., the failure of a fuse) to encourage added redundancy.

6.2 Automotive Applications

There are many applications for batteries in the automobile. SLI, start-stop, and HEV applications are discussed briefly herein, although we do not consider them large applications and therefore will not treat them thereafter. PHEV and battery electric vehicle (BEV) applications are discussed here and will be employed for further discussion in this book since we consider these batteries as falling under the category of large-format batteries.

6.2.1 Drive Cycles

Drive cycles—histories of velocity of a given vehicle versus time—are an important input for vehicle simulation and analysis. Real-world drive cycles have been recorded from drivers around the world to characterize the speeds and accelerations that are typically requested of vehicles. Studies of these data have shown that significant variability exists between different drivers. Rather than analyze every real-world drive cycle, which is clearly infeasible, standardized drive cycles are often used instead. As discussed in [1], the relative aggressiveness of standardized drive cycles may depend on the vehicle platform. To this end, the National Renewable

Energy Laboratory (NREL) has developed a drive cycle that produces fuel consumption numbers indicative of median driver aggressiveness for conventional vehicles (CVs), HEVs, PHEVs, and BEVs. A summary of this and other common drive cycles is presented in Table 6.2.

Besides drive cycle specification, the requirements on batteries used in SLI, microhybrid, and power-assist HEV batteries are slightly different. The following sections briefly outline the requirements for these applications before entering a detailed discussion on the PHEV and BEV batteries.

6.2.2 SLI

SLI batteries are commonplace in combustion engine vehicles. They serve primarily for starting the engine; this application requires a high power-to-energy ratio by itself (typically on the order of 6 kW for 6 seconds). This operation is generally expected to be performed on the order of 10 times per day for ~5 years. The ability to perform at low temperatures is critical and typically drives the sizing of the battery. However, this application also demands that the battery buffer the output of the vehicle's onboard electrical generator (alternator) to support operational electrical loads and provide additional energy for key-off auxiliary loads (750W for >10 minutes) and standby loads (15 mA for 30 days). The battery is recharged by the vehicle's alternator when the engine is running. The charging system and vehicle loads are designed to operate at 12V nominally. Lead-acid batteries have dominated this application for decades due to their simplicity and low cost. Typically, these batteries are sized at ~500 Wh (12V and ~40 Ah).

6.2.3 Start-Stop (Micro) Hybrids

Recently, start-stop hybrid (also called microhybrid) vehicles were introduced in the market. These vehicles stop the engine every time the vehicle comes to a stop, to reduce idling periods, thereby reducing fuel consumption and emissions. This

Table 6.2 Common Standardized Drive Cycles

Name	*Abbreviation*	*Also known as*	*Note*
Urban Dynamometer Driving Schedule	UDDS	LA4, FTP-72	Intended to represent urban driving of light-duty vehicles. Average speed of 19.6 mph, 7.45 miles total distance. Note there is also a different UDDS for heavy-duty vehicles.
California Unified Cycle	UC	LA92, UCDS	More aggressive version of UDDS.
Highway Fuel Economy Test	HWFET	HFET	Intended to represent highway driving of light-duty vehicles with an average speed of 48 miles per hour and a total distance of 10.3 miles.
US06	US06	—	High-speed, high-acceleration light-duty vehicle driving. Reaches a peak speed of 80.3 mph.
NREL DRIVE	NREL DRIVE	—	Synthesized from thousands of real-world drive cycles; accurate representation of median driver aggressiveness across multiple light-duty powertrain types.

increases demand on the battery on top of regular SLI applications in two primary ways. First, the number of times the battery is asked to start the engine increases by an order of magnitude. Analysis of real-world drive cycles in [2] indicates that 73 start events per day may be expected for the 95th-percentile driver. Second, the battery must power auxiliary loads when the vehicle is stopped.

While analysis has shown that the total energy requirement imposed by such operations (~56 Wh) is not significantly larger than that of the SLI application, it does impose significantly different cycling requirements on the battery. As can be seen in Figure 6.2, a start-stop hybrid requires a battery to perform numerous but short auxiliary load discharges followed by an engine-start discharge through the battery's operational range. Traditional lead-acid batteries are known to exhibit poor cycle life to such partial-state-of-charge cycles, and are thus ill-suited for this application. Advanced lead-acid batteries and other chemistries—including Li-ion—are good candidates for this application.

6.2.4 Power Assist Hybrids

In power-assist HEVs, a battery performs the functions similar to that in a start-stop hybrid, but also provides power for propulsion via an electric motor/generator attached either directly to the combustion engine or to the driven wheels via a transmission. This enables the combustion engine to be downsized, forcing it to work more frequently at a larger percentage of its maximum power output, and thereby a point of higher thermal efficiency. The battery is charged during deceleration events, capturing kinetic energy that would otherwise be lost as heat in the

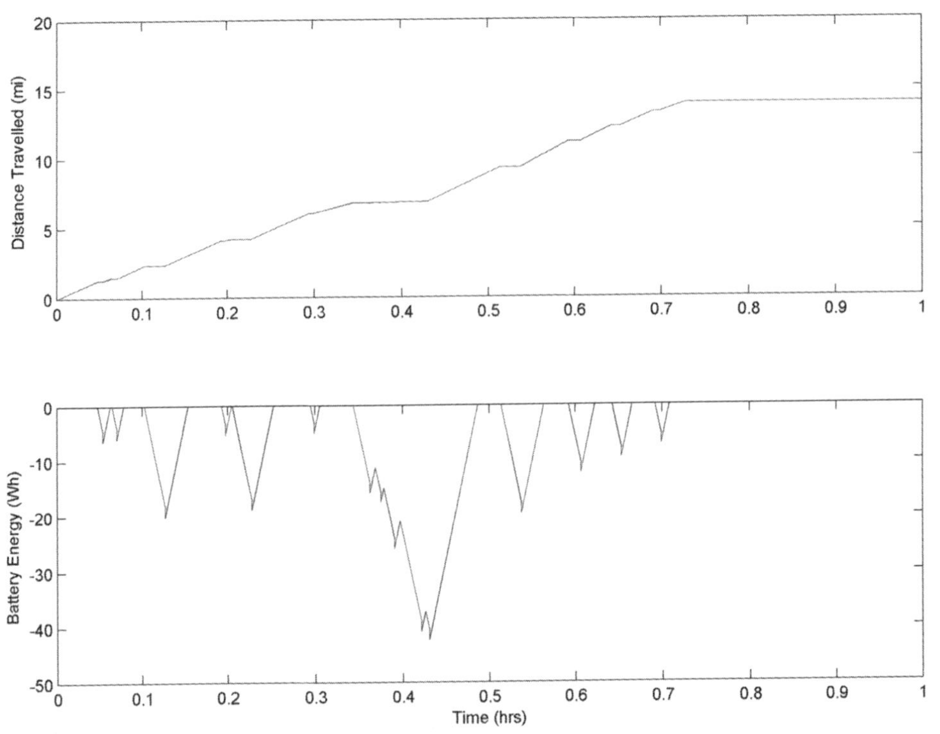

Figure 6.2 Example of a start-stop battery duty cycle and the simulated state-of-charge variation.

vehicle's friction brakes and providing it later for acceleration. Together these effects can greatly decrease per-mile emissions and fuel consumption in comparison to a conventionally powered vehicle.

HEV batteries operate primarily in a charge-sustaining (CS) mode, where the battery SOC will swing up and down briefly during regenerative deceleration and power-assist acceleration events, but is approximately constant when viewed at longer time scales. Note that HEVs do not have the capability of charging their batteries from an external source.

The requirements placed on the battery by such operation modes are much more rigorous than those of start-stop or SLI automotive applications. Drive cycle analysis shows [3] that as little as ~600 Wh of available energy can be sufficient to yield significant fuel economy benefits in light-duty vehicle, but the required power levels can be extremely high in comparison (> 30 kW, translating to effective C-rates near 50). Combined with nearly continuous current flow during vehicle operation, this can result in significant amounts of heat generation. Due to the typical expense and complexity of these high-voltage batteries (~150 to ~400V in light-duty vehicles and up to ~800V in medium- and heavy-duty vehicle), they are often designed to last the life of the vehicle (15 years and more).

As such, high-power, long-life chemistries are required for HEV applications. NiMH has been the most broadly deployed chemistry in HEV applications to date. However, Li-ion batteries are becoming more commonplace in HEVs in recent years, likely due to their higher specific energy, energy density, and cell voltage, as well as their declining price. High-energy capacitor derivatives (ultra-, super-, and asymmetric capacitors) are also being explored for low-energy HEVs due to their high power and long cycle life at larger DODs.

6.2.5 Plug-In Hybrids

A PHEV includes similar drivetrain components as an HEV; however, the size of the electric motor/generator and battery are significantly larger, and a charger is included to enable the battery to be charged from an external source. When the battery SOC is higher than a predetermined value, the vehicle operates in the charge-depleting (CD) mode. Here, higher levels of power and energy are drawn from the battery and SOC continually decreases as the vehicle is driven. If the system can source sufficiently high levels of power (e.g., from regenerative braking when driving downhill), the CD mode can be a purely electric mode of operation. Alternatively, the vehicle may be designed such that the combustion engine turns on to provide high power levels, when requested, in the CD mode. Once the battery is depleted to a predetermined SOC, the vehicle switches to the CS mode and operates like a HEV, with the combustion engine providing the majority of the required driving energy. The response of the battery and vehicle in each mode, as well as the transition between the two, is shown in Figure 6.3. As such, the driving range of a PHEV is not restricted by the energy of the battery.

The powertrain of a PHEV can be configured in one of three different architectures to create either a through-the-road, parallel, or series PHEV (Figure 6.4). In a through-the-road PHEV, the combustion engine powers one pair of drive wheels while the electric drivetrain powers the other pair. A proper control strategy is critical in such a configuration so as to not upset the dynamics of the vehicle. In a parallel

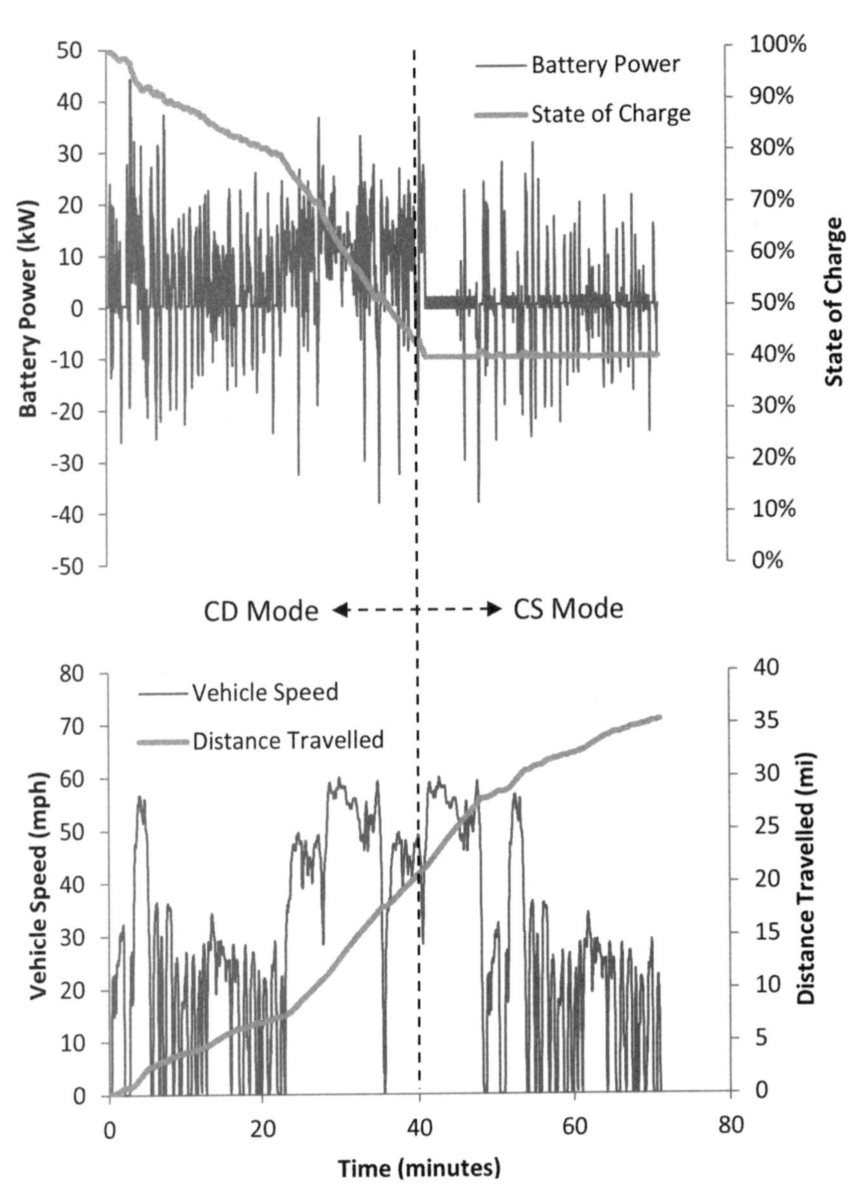

Figure 6.3 Example of a PHEV battery's response in the CD and CS modes.

PHEV, both the combustion engine and electric motor/generator are mechanically connected to the driven wheels. This is typical of how HEVs are constructed, and allows the combustion engine to operate similar to that of a conventional vehicle. In a series PHEV, the combustion engine is connected only to a generator, and the driven wheels are connected only to an electric motor. A series PHEV allows the full decoupling of the speed of the combustion engine from the speed of the vehicle itself, as there is no mechanical connection between the two. This can potentially increase the effective thermal efficiency of the combustion engine. Furthermore, the series PHEV offers the dynamic benefits of a full electric drivetrain even if the combustion engine is providing the electricity. However, the series PHEV suffers from efficiency losses in transmitting the energy from the combustion engine to

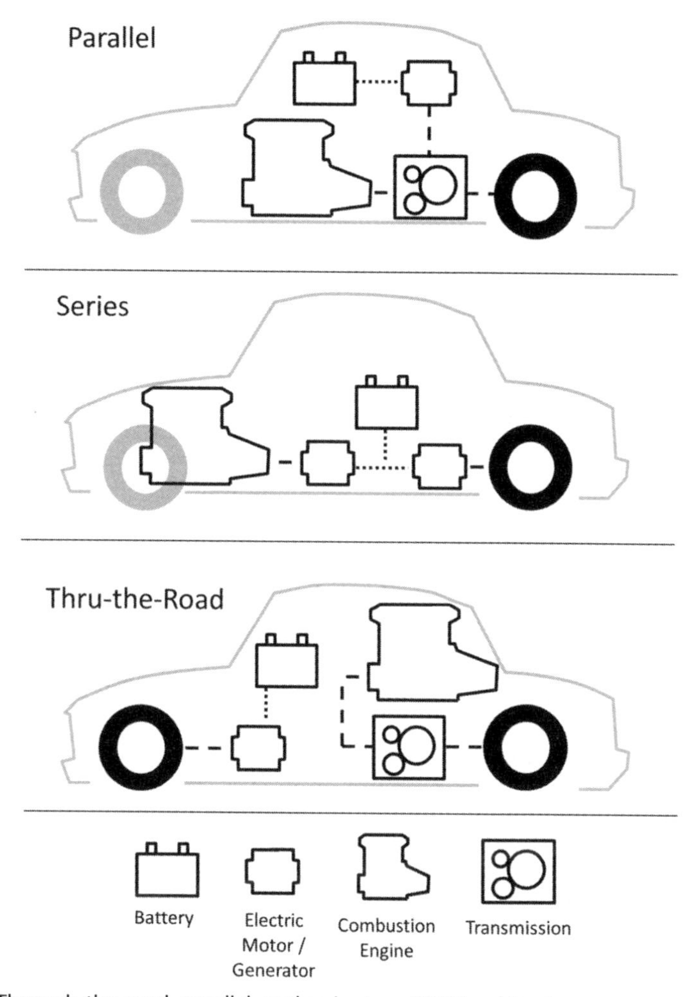

Figure 6.4 Through-the-road, parallel, and series type PHEV architectures.

the wheels, since it must first convert the engine's mechanical energy to electrical energy with the generator and then convert this back to mechanical energy via the electric motor. The series architecture also requires more electrical componentry, and can thus increase the cost and weight of the vehicle.

When a series PHEV is designed such that the battery can provide a large CD range and enough power to meet all of the vehicle's needs without intervention from the combustion engine, it may be referred to as an extended-range electric vehicle (EREV). For example, the 2012 Chevrolet Volt has a battery that is capable of providing a range of 37 miles [4] and supplying enough power to meet all of the vehicle's dynamic requirements without assistance from the combustion engine.

Creating duty cycles for PHEV battery design and testing are challenged by the sensitivity of the cycle to the vehicle platform, system architecture, the control strategy for the battery, and so forth, as with HEVs, but today, designers must also approach the questions of external charge frequency and the distribution of CS and CD mode operations. For example, drivers who charge the battery overnight at home and travel relatively short distances before returning home may find that they can operate their PHEV almost solely in CD mode on one charge a day. This would imply a duty cycle with one charge event per day and little to no CS operation. However, other drivers with long distances between charge events may charge the battery infrequently and operate predominantly in CS mode, or, if they have access to at-work or public charging stations, they may charge the battery several times per day and operate in the CD mode primarily. Which duty cycle is most taxing on the battery will depend on the nature of the battery chemistry, thermal characteristics of the system, and other factors.

As with HEVs, developing a set of duty cycles that cover all worst-case scenarios for a particular PHEV is possible within the course of a vehicle development program, provided access to sufficient vehicular and customer data is available. For the purpose of broader study, the standardized cycles of Table 6.2 may prove useful. However, in either case, a detailed vehicle simulator is necessary to translate the vehicle drive profile into an electrical duty cycle. Freely available software like FASTSim can be readily applied [5] for this task.

The United States Advanced Battery Consortium (USABC) has also developed technology targets for PHEV batteries, as presented in Table 6.4. Note that these values are generalized and that battery requirements will vary by specific vehicle model. Information on the development of these targets can be found in [7]. While it can be seen from these targets that PHEV batteries operate at a lower average

Table 6.3 USABC PHEV Battery Technology Targets

Characteristic at the End of Life of the Battery	*Unit*	*High Power/ Energy Ratio Battery*	*High Energy/ Power Ratio Battery*
Reference equivalent electric range	Miles	10	40
Peak pulse discharge power (10 sec)	kW	45	38
Peak Regen pulse power (10 sec)	kW	30	25
Available energy for CD mode, 10 kW rate	kWh	3.4	11.6
Available energy for CS mode	kWh	0.5	0.3
Minimum round-trip energy efficiency	%	90	90
Cold cranking power at –30°C, 2 sec, 3 pulses	kW	7	7
CD life/discharge throughput	Cycles/MWH	5,000/17	5,000/58
CS HEV cycle life, 50-Wh profile	Cycles	300,000	300,000
Calendar life, 35°C	Year	15	15
Maximum system weight	Kg	60	120
Maximum system volume	Liter	40	80
Maximum operating voltage	Vdc	400	400
Minimum operating voltage	Vdc	>0.55 × Vmax	>0.55 × Vmax
Maximum self-discharge	Wh/day	50	50
System recharge rate at 30°C	kW	1.4 (120V/15A)	1.4 (120V/15A)
Unassisted operating and charging temperature range	°C	–30 to +52	–30 to +52
Survival temperature range	°C	–46 to +66	–46 to +66
Max current (10-sec pulse)	Amps	300	300
Maximum system production price at 100k units/year	$	$1,700	$3,400

From [6].

C-rate than HEV batteries, the PHEV batteries are still often required to operate in the 5 to 10 C-rate range. They may also be required to complete several large DOD cycles approximately on a daily basis, which can make meeting lifetime requirements challenging. The inclusion of both electric and combustion powered drivetrains into one vehicle also makes the volume, mass, and cost requirements more demanding of the battery, necessitating high energy density and specific energy at low cost. This combination of high power, high energy, long life, and low cost can be an extremely difficult combination of metrics to achieve. Thus, PHEV battery requirements may be considered the most challenging of all automotive battery applications.

While lead-acid offers an attractive cost point for this application, its energy density and specific energy are wholly insufficient to meet the battery requirements of PHEVs. NiMH technologies suffer similar challenges, albeit at a higher price point. The Li-ion battery is at present the chemistry of choice for PHEV applications, as it offers the highest energy density and specific energy among today's battery chemistries when designed to best meet the other requirements. However, delivering the required lifetime, cold temperature performance, and cost at the levels requested by the USABC are still a challenge.

6.2.6 BEVs

The drive train of a BEV consists solely of a battery and an electric motor. It operates entirely in the CD mode without assistance from a combustion engine. Thus, while energy is captured from regenerative braking events, the battery SOC is in a state of decline on an average when driving. The energy and power of the battery, as well as the power of the electric motor, is generally required to be much larger than those of PHEVs and HEVs. And unlike PHEVs and HEVs, the driving range of a BEV is determined by the energy of the battery. Once the battery is depleted, it must be recharged from an external source (or replaced with a freshly charged battery, if battery swapping infrastructure is available [8]) before further travel can be completed.

Creating short-term duty cycles for BEVs is somewhat simplified relative to PHEVs and HEVs. As the battery is the only source of propulsion, a vehicle's velocity profile can be translated readily to a battery power request. Interaction from a combustion engine need not be considered. The USABC has provided a test cycle recommended for BEVs, although it is a coarse simplification of the BEV battery's operation.

Alternatively, it is relatively straightforward to create a battery power profile from real-world or standardized drive cycles using vehicle simulation software such as FASTSim, as discussed earlier. However, the sequencing of duty cycles through life must be considered, and is primarily dependent on consumer driving patterns. For example, like the prior PHEV example, some drivers may charge their battery daily and drive relatively short distances. Other drivers may leverage at-work or public charging infrastructure to cycle their battery multiple times per day. For some, this may include the use of fast chargers that can recharge 80% of the battery's capacity in say, 30 minutes, inducing addition wear and thermal stress on the battery.

Further, the climate in which the vehicle is operated is important. Due to the generally larger battery capacity, lower C-rates, and shorter continuous operational periods seen in BEV batteries relative to PHEV and HEV batteries, the environment can become the largest determining factor in determining the average operating temperature of the battery. Auxiliary loads can also impact battery operational requirements, as they must be solely sourced from the battery. As well, cabin heating, ventilation, and air conditioning (HVAC) loads generally dominate here, and are highly dependent on the local climate. It is important to recognize that there is no significant source of waste heat available to heat the cabin in BEVs, as is the case with conventional and hybrid vehicles.

All of these factors can affect the three critical BEV performance metrics: vehicle range, utility factor, and battery wear patterns. With respect to vehicle performance requirements, range often becomes the most important metric. This is due to the needs of consumers based on their travel patterns in addition to their potential range anxiety. Range anxiety effectively adds a safety margin to the vehicle range a consumer demands to reduce the likelihood of becoming stranded when the battery is depleted. Identifying an optimum range for a BEV is challenged in large part due to the lack of available driving pattern data from potential consumers. Such data must provide adequate resolution to accurately calculate the fraction of driving a given consumer will be able to complete with a BEV of a specific range under a specific infrastructure availability scenario (e.g., at-home charging, at-work charging, public fast charging). Most rigorously, year-long trip data including the time, distance, and destination of each trip taken for each specific individual is required.

While such data sources are few and far between, analysis of 3 months of data from each of 398 drivers data collected in the Travel Choices Study [9] can yield some insight into the relation between vehicle range and utility. Figure 6.5 shows how the fraction of these drivers who can complete their original driving range on 70%, 80%, 90%, and 99% of days with a BEV varies with vehicle range. This plot shows that a 100-mile BEV would provide sufficient range on 7 out of every 10 driving days for nearly all of these drivers, and 9 out of every 10 driving days for more than 90% of these drivers. However, it only completely satisfies the driving needs of ~35% of these drivers. This is due to the fact that while these drivers commonly complete fewer than 100 miles per day, they occasionally take much longer trips that greatly exceed 100 miles per day. Accordingly, achieving widespread 100% BEV utility factors will require either extremely long-range vehicles, the ability to extend BEV range conveniently with infrastructure, or significant changes in consumer behavior.

Identifying the optimal BEV range is a question of interest. At the time of writing this book, the majority of the BEVs on the market offer an EPA rated range of ~80 miles. While the consistency across the market may lead one to believe that the automotive manufacturers have decided this range to be optimum, this is not the case. Instead, the manufacturers have designed BEVs to yield a 100 mile or greater range when measured using the UDDS—a requirement of the California Air Resource Board's (CARB) zero emissions vehicle (ZEV) program. Thus, these vehicles have been designed to achieve ZEV credits at the lowest possible vehicle manufacturer's suggested retail price (MSRP).

Increasing BEV range is a major goal of the automotive industry, however. The primary impediments to doing so are battery volume, mass, and cost. The

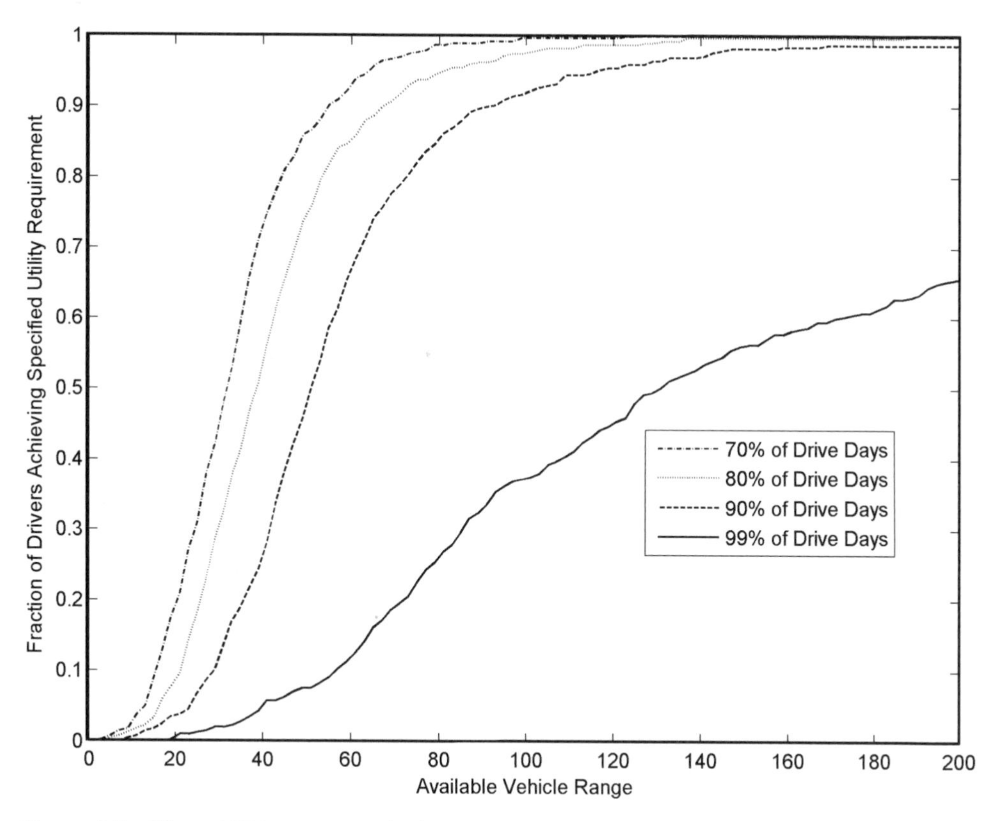

Figure 6.5 Effect of BEV range on vehicle utility for Travel Choices Study drivers.

USABC has recently updated their battery technology requirements that drive future battery development. As presented in Table 6.4, it is clear that the industry has ambitious targets for energy density, specific energy, and battery cost to enable cost-effective, longer-range BEVs. While Li-ion is the best battery technology for meeting these targets at present, batteries will need to undergo significant improvements to reach these goals. Advanced high-energy-density candidates are therefore also under investigation for BEV applications, including solid-state Li-ion, lithium air, and others.

6.3 Grid Applications

In 2011, electricity service in the United States alone was a $371 billion/year industry, delivering approximately 4.1 × 106 GWh of electricity with a peak power of 782 GW [11], and yet there was only approximately ~23 GW of energy storage, of which ~95% was pumped hydro storage [12]. Thus, unlike almost every other industry, the electricity industry operates with practically no warehouse for its product, but rather continuously synchronizes supply and demand.

Increasing the amount of grid-connected energy storage can dramatically change the way the grid operates, improving both reliability and quality of service. There are many different services that energy storage can provide to the electricity grid. The U.S. Department of Energy's (DOE) *Energy Storage Handbook* [13]

Table 6.4 USABC BEV Battery Technology Targets

End-of-Life Characteristics at 30°C	*Units*	*System Level*	*Cell Level*
Peak discharge power density, 30s pulse	W/L	1000	1500
Peak specific discharge power, 30s pulse	W/kg	470	700
Peak specific regen power, 10s pulse	W/kg	200	300
Useable energy density @ C/3 discharge Rate	Wh/L	500	750
Useable specific energy @ C/3 discharge Rate	Wh/kg	235	350
Useable energy @ C/3 discharge rate	kWh	45	N/A
Calendar life	Years	15	15
DST cycle life	Cycles	1000	1000
Selling price @ 100k units/year	%/kWh	125	100
Operating environment	°C	–30 to +52	–30 to +52
Normal recharge time	Hours	<7 hours, J1772	<7 hours, J1772
High rate charge	Minutes	80% ΔSOC in 15 minutes	80% ΔSOC in 15 minutes
Maximum operating voltage	V	420	N/A
Minimum operating voltage	V	220	N/A
Peak current, 30s	A	400	400
Unassisted operating at low temperature	%	>70% useable energy at C/3 discharge rate at –20°C	>70% useable energy at C/3 discharge rate at –20°C
Survival temperature range, 24 hour	°C	–40 to +66	–40 to +66
Maximum self-discharge	%/month	<1	<1

From: [10].

classifies these applications into five different categories as listed in Table 6.5. A detailed description of each specific service can be found in the DOE/ Electric Power Research Institute (EPRI) *Electricity Storage Handbook* [13]. Herein, we list the primary performance metrics for each service—the amount of power required, the target discharge duration for one cycle, and the number of expected cycles per year.

Numerous studies have been performed to quantify the monetary value of these services to the grid [14, 15] in search of avenues to increase the amount of energy storage installed on the grid. It is generally found that a single energy storage system must provide multiple services to justify the cost of the hardware, installation, and maintenance. To aggregate services as such, it is important to consider geographical, regulatory, and ownership requirements of each service, as well as the ability to marry the duty cycles of different applications.

For example, providing distribution infrastructure services requires the energy storage system be physically installed near the distribution assets it serves; regulations may prevent a single energy storage system from collecting revenue for both transmission infrastructure and bulk energy services; customer-owned, behind-the-meter energy storage systems can provide customer energy management service but may not be eligible to participate in ancillary markets.

When designing applications for Li-ion systems, it is additionally important to consider the strengths of competing energy storage technologies. Pumped-hydro storage (PHS) currently makes up the majority of installed energy storage. Its cost is extremely low on a per-kWh basis; however, it is impractical to install small PHS

Table 6.5 Energy Storage Specifications for Electric Grid Applications

Bulk Energy Services	*Power Range*	*Discharge Duration*	*Minimum No. of Cycles/Year*
Electric energy time shift (arbitrage)	1–500 MW	<1 hr	250+
Electric supply capacity	1–500 MW	2–6 hr	5–100
Ancillary Services			
Regulation	10–40 MW	15–60 min	250–10k
Spinning, nonspinning, and supplemental reserves	10–100 MW	15–60 min	20–50
Voltage support	1–10 MVAR	n/a	n/a
Black start	5–50 MW	15–60 min	10–20
Other related uses			
Transmission Infrastructure Services			
Transmission upgrade deferral	10–100 MW	2–8 hr	10–50
Transmission congestion relief	1–100 MW	1–4 hr	50–100
Distribution Infrastructure Services			
Distribution upgrade deferral	0.05 to 10 MW	1–4 hrs	50–100
Voltage support	10 KVAR–1 MVAR	n/a	n/a
Customer Energy Management Services			
Power quality	0.10 to 10 MW	0.17–15 min	10–200
Power reliability			
Retail electric energy time shift	0.001–1 MW	1–6 hr	50–250
Demand charge management	10 kW–1MW	15 min–4 hr	10–20

systems, and it is a geographically limited technology—further expansion of PHS on the grid is constrained by the availability of relevant sites for PHS and resistance due to their environmental impact. Compressed air energy storage (CAES) is similar in that it requires specific geological structures to operate and must be installed as a large system, although related technology that could be installed more flexibly is under development.

Batteries offer the grid a much more flexible means of energy storage that can be installed at a smaller scale with fewer geographical and environmental constraints and yet a limited number of battery installations—primarily lead acid and sodium sulfur—have been completed to date [13, 16]. While the former suffers from sensitivity to high temperature environments and poor cycle life, the latter is largely impervious to high environmental temperatures (sodium sulfur is a high-temperature chemistry that operates above 300°C) and can provide excellent high DOD cycle life [13]. The downside of sodium sulfur batteries is that they are rate-limited: the maximum discharge rate of a sodium sulfur battery is ~C/7. The sodium nickel chloride chemistry, which is very similar, can operate at higher rates, but is still limited to ~C/2.

More recently, a small number of grid-connected Li-ion battery installations have been completed [13]. Relative to alternate energy storage options, Li-ion's primary advantages in the grid energy storage markets are its ability to be installed flexibly in sizes ranging from 1 kW to hundreds of MW, to operate at high C-rates, and to provide a long cycle life. It also offers the benefit of low mass and volume, which, despite what may be viewed as less of a constraint under common perception, can actually be quite important for space-constrained stationary applications. Considering these advantages alongside the technical requirements and economic benefits of the grid energy storage services as shown on Table 6.5, Li-ion batteries can yield several attractive use scenarios for batteries on the grid, as outlined in the next section.

6.3.1 Demand Charge Management and Uninterruptable Power Sources

Demand charges are a feature of many utility rate structures for commercial customers that charge a fee based on peak power rather than total energy. The peak power is often recorded by dividing each month into 15-minute intervals, then selecting the interval yielding the largest 15-minute average power to determine the fee charged. These rate structures often consist of different demand charges for on-peak, midpeak, and off-peak periods that vary by season. Excerpts of an example demand charge rate structure from Southern California Edison (SCE) (serving the Los Angeles, CA area) are shown in Table 6.6. With a behind-the-meter energy storage system, the said system can be employed to discharge the battery during high demand periods, reducing the meter load and thereby the demand charge.

Performed perfectly, this service could be worth more than $232/kW/yr for customers of SCE's TOU-GS-2-B rate schedule. However, it must be recognized that for each additional kilowatt of power added to the system, a greater amount of energy must be added as well, as illustrated in Figure 6.6. Thus, the annualized

Table 6.6 Demand Charges from SCE's TOU-GS-2-B Rate Structure

Charge	*Time*	*Cost*	*Units*
Facilities-related demand charge	All	$13.94	$/kW
Time-related demand charge	Summer on-peak	$16.20	$/kW
	Summer midpeak	$4.95	$/kW

From: [17].

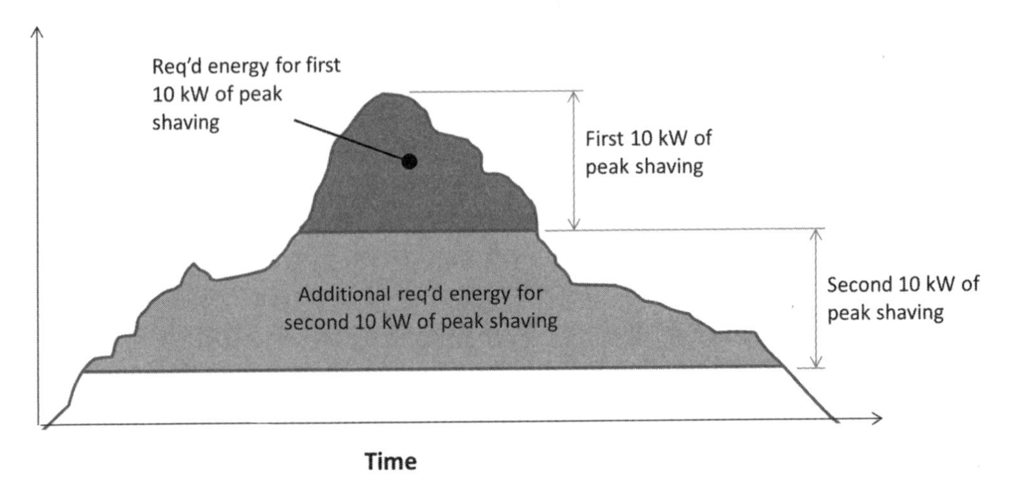

Figure 6.6 Required energy can increase geometrically as the peak reduction power is increased.

return on investment generally decreases as the power of the energy storage system increases. One study of the value of demand charge management (DCM) has found that ES is most cost effective when power is scaled to ~5% to 10% of facility peak power [18].

The reader will note that some energy/time-shifting value is also likely to be accrued, as the battery is likely to discharge during on-peak periods and charge during off-peak periods. However, this value will generally be an order of magnitude smaller than the value of DCM.

Intelligent control and forecasting systems are also necessary to implement DCM services effectively. Battery dispatch for peak reduction must be done in such a way that the battery does not prematurely run out of energy prior to a period of peak demand. Due to the fact that demand charges are calculated on the peak load interval observed over the entire month, only one such mistake during the course of each month is sufficient to eliminate that month's entire financial benefit. Developing an optimal control to command battery dispatch in the presence of an accurate forecast is relatively straightforward. However, developing accurate demand forecasts, or, alternatively, control strategies that require less accurate forecasts, are more challenging.

Implemented aggressively on a customer facility with a diurnal demand cycle, battery power levels will fluctuate considerably and may even oscillate rapidly between charge and discharge in response to facility demand. But battery SOC will generally trend downward until the facility demand is consistently below the peak shaving load target. An example response of a system using a perfect load forecast and an optimal controller targeting a minimum battery SOC of 40% is illustrated in Figure 6.7.

Frequency of such operations will depend on the control strategy, system sizing, forecast horizon, and consistency of the facility load. At minimum, it should be expected that one such discharge will occur per month. This can be the case when facility load is consistently decreasing over the course of the month, or when a highly accurate, long time horizon forecast is available. Alternatively, such cycles may be required on a daily basis. This is usually the case when facility load is consistent from day to day or when facility load increases day to day and only short-term forecasts are available.

If the facility load is not diurnal, and instead dominated by customer-specific load peaks (perhaps due to business-specific processes), duty cycles and cycle frequency will be much different. Indeed, demand pattern variability is high between different customers and the demand charge management value varies accordingly. Thus, battery duty cycles must be considered on a case-by-case basis to support system design.

An uninterruptable power source is any device that provides a reliability service to an individual customer from behind the meter. The value of this service will be specific to the customer's intended use. For example, the telecommunications industry places a high value on keeping its towers operating during power outages, and information technology companies place a high value on keeping their servers operational. Both industries can place a specific dollar value on each minute their services are interrupted. Such industries can use a Li-ion system to totally mitigate the effects of short outages (<1 hour) or as a bridging system to allow

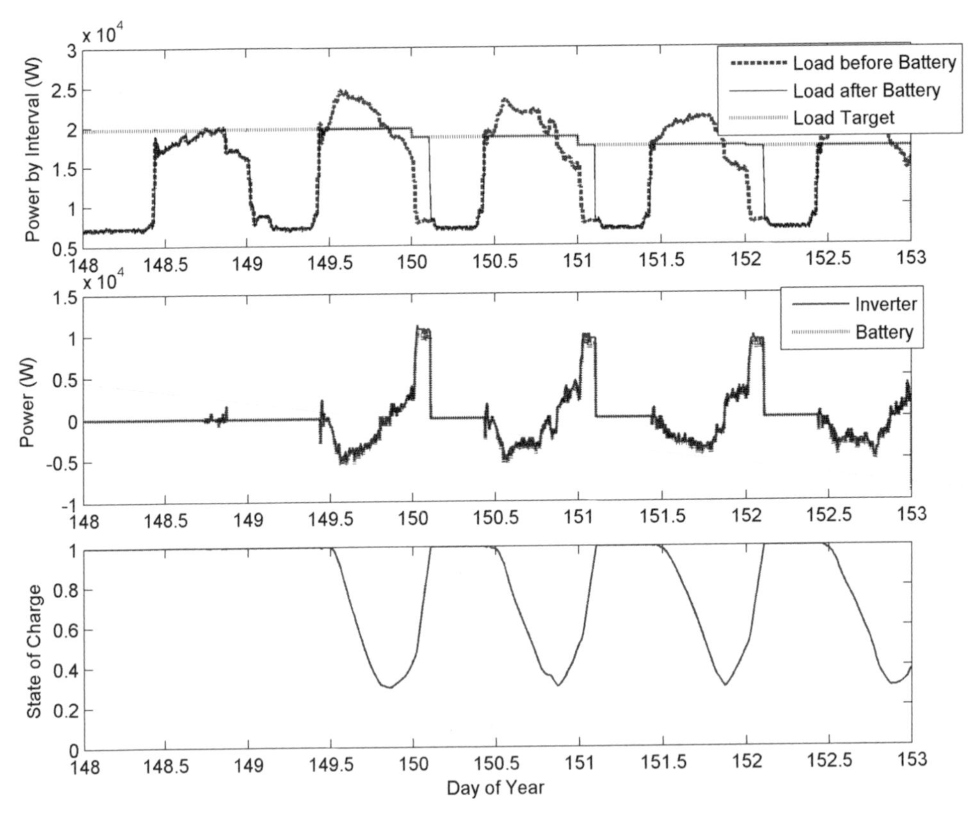

Figure 6.7 Example demand charge management battery response.

seamless transition from grid-supplied power to backup-generator-supplied power for longer outages.

The unanticipated interruption of manufacturing processes, on the other hand, can cause damage to in-process product and equipment. For example, consider the final stages of computer-controlled machining of precision products; loss of power to the machining operation will irreparably damage the nearly finished product and also damage the machining tools. Such industries can thereby place a dollar value on each outage event. Here, Li-ion systems can provide a seamless transition to battery-backup power for a short duration (<1 hour) to allow the safe shutdown of manufacturing equipment. The addition of long duration backup power generation is not needed.

Battery duty cycles will of course be specific to the customer and nature of their requirements. Similarly, cycling frequency will depend on the customer's service territory and quality of service. However, it can generally be assumed that the battery will be required to discharge infrequently at a largely constant power level for a period of 15 to 60 minutes.

Given that both applications are behind-the-meter applications, there is no geographical conflict when pairing the two. Their thermal, environmental, interconnection, and abuse requirements are also similar. Thermal requirements will depend on whether the system is to be installed inside or outside of the customer facility. If the system is installed inside the facility, the designer must identify whether the interior installation site is HVAC-controlled and whether the existing facility

HVAC system will be sufficient to accommodate the needs of and heat generated by the battery. Mechanically, it will be necessary for the system to be tolerant to expected earthquake loads per the California Building Code or International Building Code, if it is to supply UPS service. It is also likely it must comply with the UL's Standard for Uninterruptible Power Supply Equipment UL 1778.

However, pairing the electrical duty cycles such that the demand charge management and uninterruptible power source services do not conflict poses a larger challenge. While it is impossible for the demand for each service to exist simultaneously, the need to provide the services sequentially is problematic. If the battery is fully discharged following an aggressive demand charge management event when a UPS event occurs, UPS service may not be adequately supplied. Conversely, if a single DCM discharge is missed in anticipation of a UPS event, significant DCM value can be lost. Accurate ability to forecast DCM and UPS events is thus extremely important to this pairing.

Understanding the nature of UPS events can help determine whether pairing such applications is viable. For example, outages caused by high service-area-wide demand periods that induce failure of utility equipment, and thereby service interruptions, may be likely to follow DCM discharges. In this case, the pairing of DCM and UPS services may not be practical. On the other hand, if the primary source of outages is due to storms or other exogenous factors, pairing UPS and DCM may be more readily achieved (particularly when the exogenous events are predictable).

Another point that must be considered when pairing these applications is system sizing. As noted above, optimal return on investment for DCM systems generally occur when sized for low power levels relative to the facility's peak demand (<10%). While it is generally unnecessary for a UPS service to back up the entire facility load, it is not guaranteed that the fraction of loads in need of backup will perfectly coincide with the power and energy capabilities of a value-optimized DCM system. Thus, the level of DCM service and number of loads to be backed up must be carefully considered when selecting the power and energy levels of the system.

6.3.2 Area Regulation and Transportable Asset Upgrade Deferral

Area regulation is an ancillary service intended to ensure that the electricity supply precisely matches electricity demand over short time periods. This creates the need to increase (upregulation) or decrease (downregulation) the amount of electrical energy on the grid. It is common for conventional (combustion) generation to provide area regulation alongside a separate energy service by operating at an output less than its maximum capacity for the energy service, thereby leaving headroom for up regulation by increasing its output. For example, a 1-MW gas-powered turbine may elect to operate at 800 kW to provide the energy service, then offer 200 kW for upregulation with its remaining unused capacity. It may also elect to offer 200 kW for downregulation, which would reduce its output to 600 kW to remove energy from the grid. While in theory a conventional generator could offer downregulation to an output of 0 kW, in practice conventional generators have minimum load points below which it becomes impractical to operate. Once these bounds are set, the regulation output of the plant at any given time is set in response to an automatic generation control (AGC) signal provided by the local balancing authority.

Energy storage, on the other hand, can respond to upregulation requests by discharging and downregulation requests by charging. Thus a 1-MW storage system can offer its full 1 MW in either direction (provided the system is not fully charged or discharged at the moment of the request). Compared to conventional generation, it also often has the advantage of fast response. Conventional generation typically has ramp rates on the order of minutes or more; thus, there can be considerable delay between a change in the AGC and a change in output of the generator. Storage typically has response times of seconds or less, and thus can respond to requests by the balancing authority almost instantaneously. This improvement in response time can significantly reduce the total amount of regulation (as measured in megawatts) necessary to balance the grid. In recognition of this fact, recent regulation has implemented a pay-for-performance requirement that will pay fast-responding technologies twice as much as slow-ramping resources for regulation service, making regulation a profitable value proposition for many storage technologies [19].

It is also important, however, to consider the efficiency of the energy storage system. While it will be paid for the energy delivered during upregulation, it must pay for the energy consumed during downregulation. This is above and beyond area regulation payments. Thus, energy storage systems with low round-trip efficiency will be financially penalized.

When providing area regulation services, a battery operates nearly continuously at highly variable power levels, switching often between charge and discharge. Future demands on the battery are largely unpredictable, presenting a challenge to the designer in deciding how the battery will respond to the AGC signal (or similar) in such a manner as to maximize the value of the system. Disparities in the value of upregulation and downregulation value, system efficiency, thermal response, battery degradation, and historical AGC signal data must all be considered. Figure 6.8 shows the impact of one potential strategy for designing an area regulation energy storage control strategy that results in many small SOC cycles and an occasional large SOC cycle.

Transmission and distribution upgrade deferral both pertain to the use of an energy storage system to reduce loads on expensive transmission or distribution equipment that are approaching their load limits, thereby deferring the replacement of said asset(s). For example, consider a transmission line with a100-MW limit. If this limit is exceeded, the transmission line will overload thermally and fail. The transmission planner anticipates that the load on this line is growing at a rate such that it will become overloaded by 1 MW in July next year when demand peaks for the season. The transmission planner has two options: (1) begin and complete a $10M project to upgrade the transmission line before the overage occurs, or (2) find a way to eliminate the 1-MW July overages. Option 2 can be implemented with an energy storage system placed at the consumption end of the transmission line commanded to perform up to 1 MW of peak shaving when transmission line loads exceed 100 MW. The avoided cost of deferring the large-scale transmission upgrade can be determined by the cost that would be passed through to the rate payer for the upgrade. This is calculated as the annual fixed charge rate (0.08 to 0.15, typically) times the total capital investment. In this example, the avoided cost is $1.1M for a 0.11 fixed charge rate. Thus, if an energy system can be procured and operated on-site effectively for the month of July for less than $1.1M, then option 2 is the more cost-effective option.

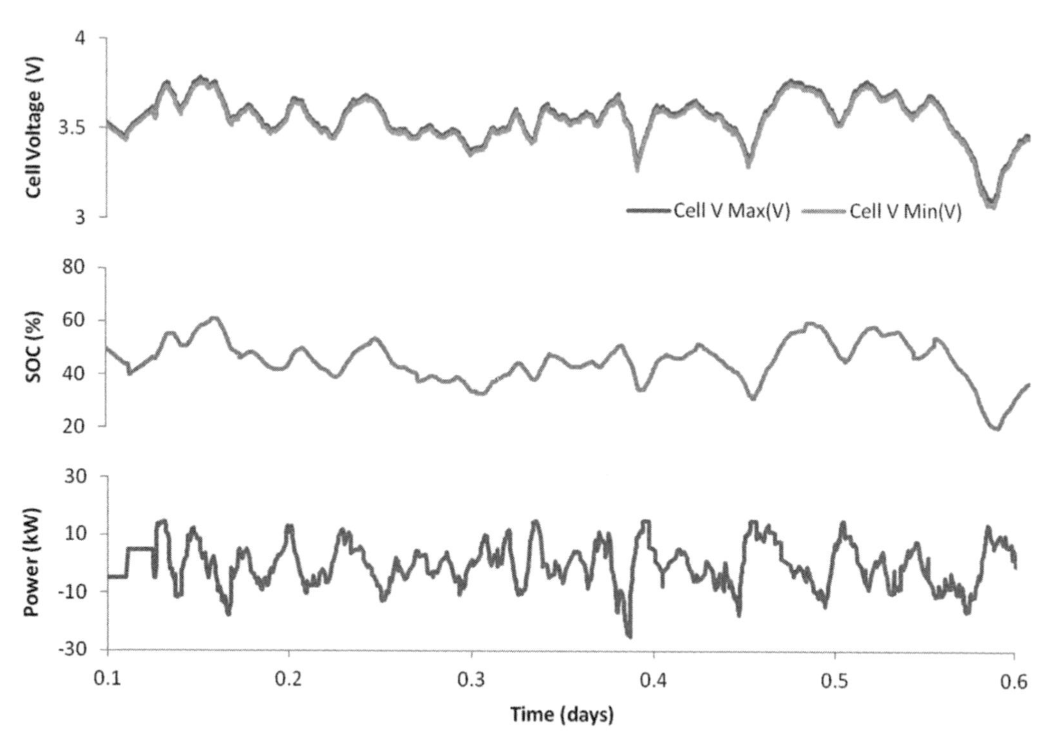

Figure 6.8 Example excerpt of an area regulation duty cycle and response of the energy storage system.

Asset upgrade deferral can be a financially attractive use of energy storage. The need to discharge the system can be quite infrequent, on the order of 10 or fewer instances per year. It can also be relatively easy for some storage technologies to perform, requiring average C-rates of C/2 and lower. However, service value and technical requirements are extremely site-specific. Thus, like the DCM and UPS applications discussed earlier, the system designer must approach each potential installation separately to optimize a system.

Further, care must be taken in attempting multiyear deferrals. In cases where load is expected to grow continually, the peak shaving power and energy requirements will grow as well. Notably, the requirement for energy is likely to increase much faster than the requirement for power (Figure 6.9), which can quickly degrade project economics. Thus, 1- to 2-year deferrals are often most cost effective.

For this reason, it is attractive to employ a transportable storage system for this service. A transportable system would be capable of not only serving different sites in sequential years, but could potentially serve multiple sites in the same year. While identification of sites where this is possible is challenging, it could potentially double the profitability of the system. For such transportable systems, energy density and specific energy become more important performance metrics, as does mechanical robustness to survive the transport environment. These requirements give Li-ion a strong advantage over competing technologies like flow and lead acid batteries (energy density and specific energy), as well as sodium sulfur batteries (mechanical robustness, C-rate).

Combining area regulation and asset upgrade deferral is somewhat straightforward. Geographic considerations must be given to the asset upgrade deferral

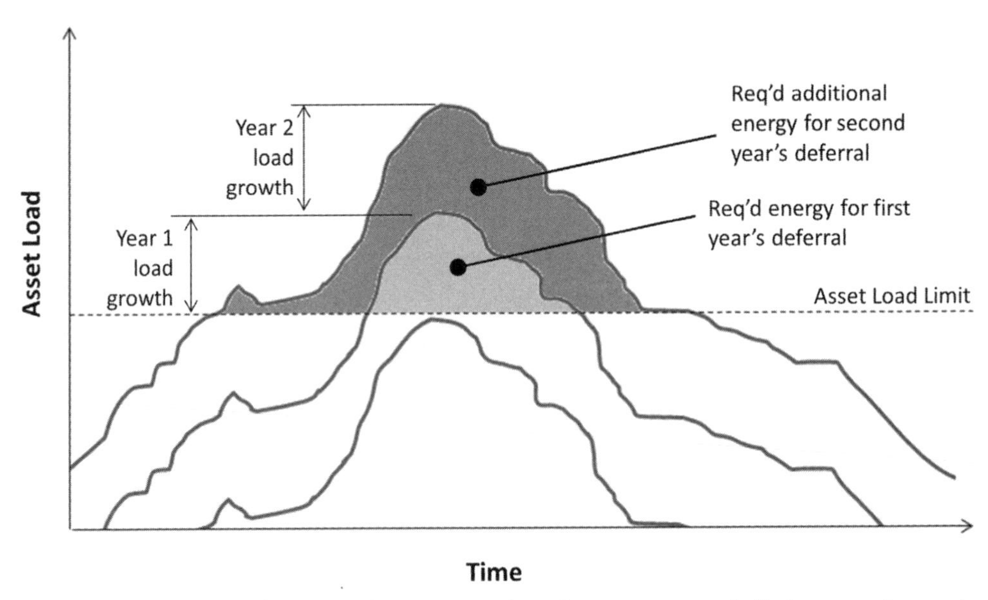

Figure 6.9 The required amount of energy stored can increase geometrically from year to year in multiyear asset deferral scenarios if asset load growth is constant or accelerating.

service with the restriction that only site is acceptable for also providing regulation services. The system will likely be a stand-alone unit sized >100 kW packaged in a shipping container to facilitate transportation that is installed and operated outdoors. Thermal requirements will be driven by the operating environment and the area regulation service. The continuous operation, and the fact that the inverter will often be installed in the same container, will likely require an active cooling system. Mechanical requirements will be designed around the shipping environment, number of sites per year to be served, and total service life of the unit.

Logistically, a third party could own and operate the energy storage system, selling ancillary services to the market and asset deferral services to the transmission operator. Assuming that the device is to be installed and operated from a single location for 10 to 12 months at a time, it could provide regulation services at all times that the asset to be serviced is below its contracted load limit. When a near term forecast (1 to 2 days) predicts that the load limit will be exceeded, the energy storage would begin to curtail its regulation services to prepare (charge) in advance of the overload condition. When the overload condition begins, the battery would enter peak shaving mode, discharging the battery to keep the asset below its overload limit. Once the threat has passed, the battery would return to regulation service.

Where the frequency and duration of asset upgrade deferral events are small, the system has the opportunity to collect considerable revenue from regulation in addition to the asset upgrade deferral benefit. However, it must be recognized that the asset upgrade deferral service must have extremely high reliability, lest the asset be allowed to fail and significant additional cost be incurred. Thus, highly accurate forecasts must be available and considerable margin for error (via oversizing system energy and power) are recommended.

6.3.3 Community Energy Storage

In community energy storage (CES), it is envisioned that a large number of battery systems sized from 10 to 100 kWh and 10 to 100 kW are installed on distribution feeders throughout a utility's service area. The edge of the grid location, ability to aggregate many units to provide large capacity services, and ownership by the utility enable CES to provide many different services, including transmission and distribution upgrade deferral, regulation, reserves, voltage support, electric supply capacity, and more. Thus there are many opportunities to optimize the value of the system, but doing so is not straightforward. Many of the same challenges associated with combining different applications as noted previously must be addressed. Duty cycles and system requirements will vary based on the selected pairing of applications. Several developmental CES deployments have been initiated to investigate these matters.

In addition to the ability to serve so many applications, one potential advantage of CES is the ability to leverage technical developments and cost reductions achieved by the automotive industry. The system size and technical requirements for CES batteries are often similar to those of BEV batteries. Thus, as the BEV market grows, so will interest in and deployment of CES.

One challenge with CES, however, is the need for extremely low-maintenance systems, due to the need to minimize cost and the large number of units that may be deployed in a geographically dispersed fashion. This concern often arises in discussions of CES lifetime and thermal management. Long lifetimes are desired to minimize hardware replacement costs—while active cooling systems could help extend life, they also come with their own maintenance requirements. Early studies of CES thermal management show that installations that behave like a greenhouse can lead to high battery temperatures that are detrimental to battery wear. However, they also suggest [20] that simple passive thermal management techniques such as shading the units from solar irradiation or vaulting the batteries below grade with a strong thermal connection to ground temperatures can be adequate when paired with intelligent restraints on electrical operation (Figure 6.10).

While early CES installations have anticipated the benefits of the vaulted configurations, some actually behave more like a greenhouse, amplifying the effects of

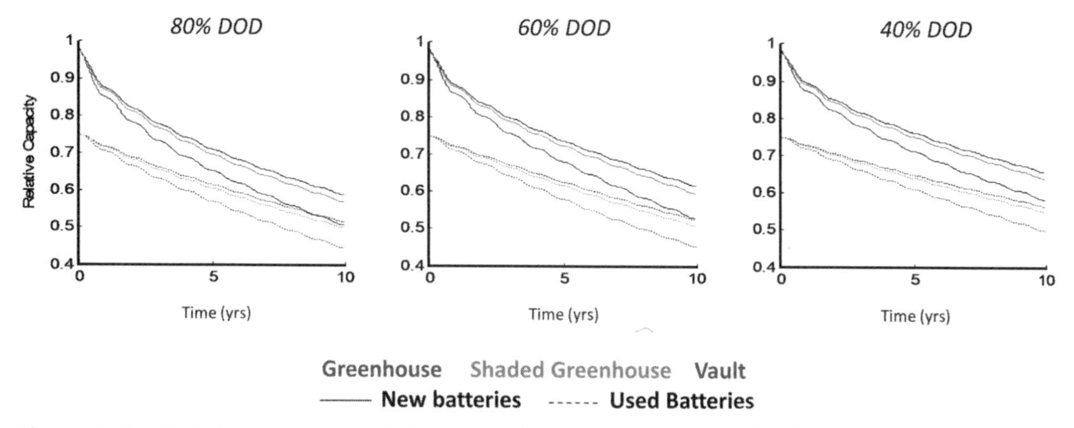

Figure 6.10 Variation of simulated battery wear for three different passive CES thermal configurations (greenhouse, shaded greenhouse, and ideal vault) in Phoenix, AZ.

solar radiation and insulating the battery from the soil. Further, vaulted installations must carefully manage water penetration, are less flexible to relocating the unit, and are of high cost.

6.3.4 Other Grid-Connected Applications

The use of energy storage on the grid is still evolving, as is the grid itself, with rapid growth in distributed, variable-generation, and smart-grid technologies. Certainly the technical requirements of grid-connected energy storage and the ways to identify the best-suited systems will also continue to evolve. For example, increased penetration of wind energy may incite the need for regulation-like services at wind farms. Similarly, increased popularity of behind-the-meter solar power could incite utilities to require behind-the-meter storage to smooth power and voltage fluctuations. It will therefore be important for the battery system designer to stay abreast of changes in this exciting new market as the penetration of energy storage grows.

References

[1] Neubauer, J., and E. Wood, "Accounting for the Variation of Driver Aggression in the Simulation of Conventional and Advanced Vehicles," presented at the SAE 2013 World Congress & Exhibition, April 16-18 2013, Detroit, MI; SAE Technical Paper 2013-01-1453; NREL Report No. CP-5400-58609; and CP-5400-57503.

[2] Tartaria, H, O. Gross, C. Bae, B. Cunningham, J. Barnes, J. Deppe, and J. Neubauer, "USABC Development of 12 Volt Battery for Start-Stop Application," EVS27, Barcelona, Spain, 2013.

[3] Gonder, J., A. Pesaran, D. Howell, and H. Tataria, "Lower-Energy Requirements for Power-Assist HEV Energy Storage Systems—Analysis and Rationale," presented at the 27th International Battery Seminar and Exhibit, Fort Lauderdale, FL, March 18, 2010.

[4] http://www.fueleconomy.gov/feg/Find.do?action=sbs&id=32655, accessed 11/19/2013.

[5] http://www.nrel.gov/vehiclesandfuels/vsa/fastsim.html, accessed 11/19/2013.

[6] USABC Requirements of End of Life Energy Storage Systems for PHEVs, http://www.uscar.org/commands/files_download.php?files_id=156 , accessed 11/19/2013.

[7] Pesaran, A., T. Markel, H. S. Tataria, and D. Howell, "Battery Requirements for Plug-In Hybrid Electric Vehicles: Analysis and Rationale," presented at the 23rd International Electric Vehicles Symposium and Exposition (EVS 23), Sustainability: The Future of Transportation, December 2–5, 2007, Anaheim, CA.

[8] Neubauer, J., and A. Pesaran, "A Techno-Economic Analysis of BEV Service Providers Offering Battery Swapping Services," NREL Report No. PR-5400-58343, presented at SAE 2013 World Congress, April 17, 2013, Detroit, MI.

[9] Traffic Choices Study—Summary Report, Puget Sound Regional Council, April 2008, http://psrc.org/assets/37/summaryreport.pdf.

[10] Neubauer, J., A. Pesaran, C. Bae, R. Elder, and B. Cunningham, "Updating United States Advanced Battery Consortium and Department of Energy Battery Technology Targets for Battery Electric Vehicles," *J. Power Sources,* Vol. 276, pp. 614–621, 2014.

[11] http://www.eia.gov/electricity/annual/, accessed 11/9/2013.

[12] Energy Storage Activities in the United States Electricity Grid, Electricity Advisory Committee, May 2011, http://www.doe.gov/sites/prod/files/oeprod/DocumentsandMedia/FINAL_DOE_Report-Storage_Activities_5-1-11.pdf.

[13] Akhil, A., et al, "DOE/EPRI 2013 Electricity Storage Handbook in Collaboration with NRECA," Sandia report SAND2013-5131, July 2013.

[14] *Electricity Energy Storage Technology Options: A White Paper Primer on Applications, Costs and Benefits*, 1020676, Electric Power Research Institute (EPRI), Palo Alto, CA, December 2010.

[15] Eyer, J., and G. Corey, Energy Storage for the Electricity Grid: Benefits and Market Potential Assessment Guide, SAND2010-0815, Sandia National Laboratories, February 2010.

[16] Doughty, D., et al, "Batteries for Large-Scale Stationary Electrical Energy Storage," *Electrochem. Soc. Interface,* Fall 2010, pp. 49–53.

[17] www.sce.com, accessed 11/9/2013.

[18] Neubauer, J., and M. Simpson, "Optimal Sizing of Energy Storage and Photovoltaic Power Systems for Demand Charge Mitigation," Electrical Energy Storage Application & Technologies, San Diego, CA, October 2013.

[19] http://www.ferc.gov/whats-new/comm-meet/2011/102011/E-28.pdf, accessed 11/9/2013.

[20] Neubauer, J., et al, "Analyzing the Effects of Climate and Thermal Configuration on Community Energy Storage Systems," Electrical Energy Storage Application & Technologies, San Diego, CA, October 2013.

CHAPTER 7

System Design

With the possible exception of flow batteries, the building block of electrochemical energy storage systems is the single cell. Energy storage systems may range from just a single cell to a few cells, such as in consumer electronic devices, to truly large-scale systems, such as in grid and automotive applications that combine tens to thousands of electrochemical cells into a single system. Design of an energy storage system builds upon topics discussed in earlier chapters, including:

- Determination of application requirements and possible duty cycles (Chapter 6);
- Selection of appropriate electrochemical technologies (Chapter 1);
- Selection and/or design of electrochemical cell (Chapter 2);
- Thermal, life, and safety characterization (Chapters 3, 4, 5).

In this chapter, we assume that a suitable electrochemical cell has been selected from available technologies. Ideally, factors that were considered during cell selection included matching the power-to-energy ratio of the cell with the application, anticipated lifetime, cost and business relationship with the supplier, product uniformity, safety and compatibility of cell form factor, and packaging with the system.

Scaling up from the cell to the system, design aspects discussed in this systems chapter include:

- Sizing beginning of life (BOL) excess energy and power in order to meet energy and power requirements at end of life (EOL);
- Establishing the number of cells and electrical topology;
- Establishing the thermal management system topology, design, and control set points;
- Mechanical packaging of cells and modules within the pack;
- Design of electrical management systems, including cell balancing, relays, contactors, switches, and fuses;
- Battery management system (BMS) controller tuning:
    - Estimation of battery state of charge and state of power.
    - Allowable power limits. For Li-ion, it is particularly important to establish acceptable charge rate at cold temperatures to avoid Li plating,

as well as derating the power limits at high temperatures to avoid overheating.
- Cell balancing strategy.

- Supervisory controller tuning (e.g., establishing how much of the battery's BOL energy and power to use and still achieve acceptable lifetime, including under reasonable worst-case assumptions such as a hot environment with regular fast charging);
- Validation of the technology versus the original business requirements;
- Warranty and life-cycle management (anticipated years of life, infant mortality of cell technology, probability of premature failure, maintenance of inventory for replacing failed units), decommissioning (reuse, recycling, disposal).

Similar to cell design, systems design involves evaluating trade-offs in cost, performance, lifetime, and safety at each step of the process. Topics discussed in this chapter include electrical systems design, thermal and mechanical design, electronics and electrical control systems, the design process, and design standards. The chapter concludes with two design case studies: one for an automotive application and one for a stationary, grid-tied application. The automotive case study develops a semiempirical life prediction model and uses the model to calculate trade-offs in life with performance and cost. The grid case study investigates the application of energy storage for so-called demand charge reduction for a commercial electricity customer with a large amount of on-site photovoltaic (PV) power generation. The customer's goal is to reduce their monthly utility bills as much as possible.

7.1 Electrical Design

Analysis of the energy storage application, discussed in Chapter 6, provides anticipated power versus time duty cycles that the energy storage system must meet. For energy intense applications, the power level will be low but the time duration of discharge and charge will be long. For power-intense applications, the power level will be high, with the power levels and discharge/charge pulse time dictating energy requirements.

7.1.1 Power/Energy Ratio

As a rule of thumb, for 10 years of life, the system is sized with on the order of 20% to 30% excess energy and 30% to 70% excess power relative to requirements derived from duty cycles. This provides an allowance for degradation throughout lifetime. Stationary storage applications have some flexibility to periodically add capacity throughout life to maintain performance as degradation occurs. Mobile applications do not. Excess energy and power also allow some margin for difficult operating conditions such as at cold temperatures.

These power and energy requirements—including excess—establish the power-to-energy (P/E) ratio of both the application and cell. Regardless of how many cells are combined in series and how many cells are combined in parallel, the P/E ratio of both cell and pack stay the same. For a system designer, the search for cells with the desired P/E ratio begins. Other electrical attributes to document during the cell market survey include capacity (Ah), nominal, minimum and maximum cell voltage, expectations for life and thermal behavior/requirements.

7.1.2 Series/Parallel Topology

The pack nominal, minimum, and maximum voltages must be matched to the system(s) to which it is connected. AC transformers cannot directly be used to scale the DC power supplied by an electrochemical energy storage system. DC-DC converters can scale the pack's voltage to a different system requirement; however, this is often more expensive than simply finding a different electrochemical cell that better meets requirements. The desired nominal voltage of the battery pack, V_{pack}, together with the nominal cell voltage V_{cell}, establishes the number of cells that must be combined in series, n_s, according to the following relationship:

$$V_{pack,nom} = V_{cell,nom} \times n_s \tag{7.1}$$

Equation (7.1), written for nominal cell/pack voltages, can be similarly written for minimum and maximum cell/pack voltages. With the number of series cells now known, we can find the number of cells required in parallel, n_p, to meet the total energy requirement, $E_{pack,total}$ (Wh)

$$E_{pack,total} = V_{cell,nom} \times Q_{cell} \times n_s \times n_p \tag{7.2}$$

where Q_{cell} (Ah) is the cell's capacity.

Regarding series/parallel topologies, series/parallel combinations fall into two main categories at the extreme. Generally, parallel cells are connected together as supercells, with supercells then connected in series. This is the so-called P before S topology as shown in Figure 7.1(a). The opposite, S before P topology is shown in Figure 7.1(b).

The *benefit of P before S* is that a parallel group of cells provides some capability for stronger cells to help weaker cells. A parallel group of cells can potentially partially operate with a single cell failed in open circuit but not with a shorted cell. A further benefit of the P before S topology is that balancing circuitry connections must only be made once per each parallel group rather than to every individual cell for S before P configurations.

The *benefit of S before P* configurations arises when contactors or active switches are placed at the ends of the series strings, as in Figure 7.1(c). In this switched S before P topology, any one series string can be electrically removed from the system in case of a failure in that string.

The choice of S before P or vice versa is essentially a trade-off in complexity of electrical wiring to each cell (for cell voltage measurement and balancing) versus anticipated cell failure rates and need for fault tolerance (determined from system requirements and anticipated rates of cell premature failure). Beyond several hundred cells, extreme large-scale systems usually have some combination of nested parallel/series/parallel connections (e.g., Figure 7.1(d)) with cost, complexity, fault tolerance, and safety dictating the configuration.

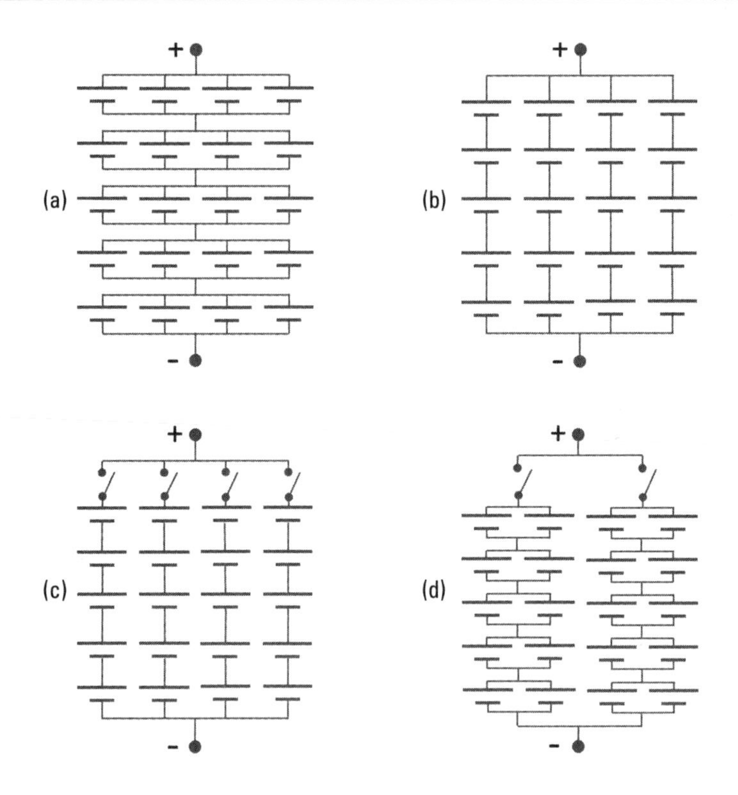

Topology	Cell groups to balance	Useable energy with failed cell
(a) 4P-5S	5	~0
(b) 5S-4P	20	~0
(c) 5S-4P/switched	20	75%
(d) 2P-5S-2P/switched	10	50%

Figure 7.1 Example of series/parallel topologies, each with identical energy and power capabilities.

The four topologies in Figure 7.1 have trade-offs in the number of connections required for cell voltage monitoring/balancing versus capability to operate with partial functionality in case of a cell failure.

Regarding safety, a limiting factor for the maximum number of cells that can be combined in a parallel group is the fault tolerance to an internal short in a single cell. In a parallel connection, the healthy cells will discharge their energy through the internally shorted cell, exacerbating consequences of the failure. So long as one ensures that a single cell can absorb all energy from its neighbors in case of internal failure, it is appropriate to make a hard connection between parallel cells in a P before S topology. If there is concern about this failure mode, the parallel group of cells should be designed with fuses, thin fused busbar connections such as in Figure 7.2(a), or separated into S before P subgroups.

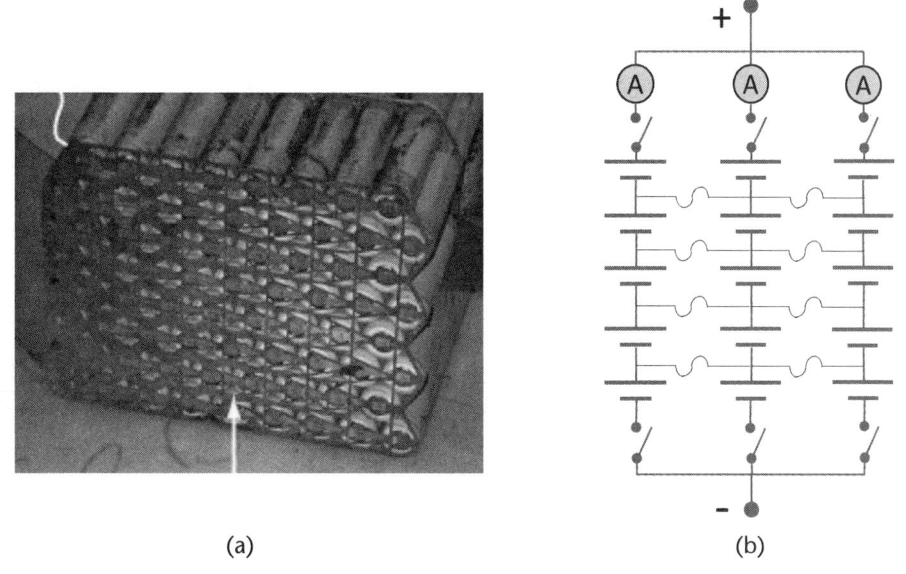

Figure 7.2 Fused busbar arrangements for parallel groups of cells: (a) passive design (source: [1]), and (b) fail-safe design with passive and active features. (Source: [2].)

The fail-safe topology shown in Figure 7.2(b) is a design both to detect cell internal shorts and mitigate their consequences on the system [2]. This design uses fused connections between parallel cells. In contrast to series connections in a pack, parallel connections for balancing need not use highly conductive leads. The design also employs active elements, placing current sensors and contactors at the top of series strings. Current imbalance between series strings indicates an internal cell failure. Contactors can then be opened to isolate the failed string. Further details are given in [2].

7.1.3 Balance of Plant

A schematic of a Chevrolet Volt automotive battery pack is shown in Figure 7.3. This PHEV uses 288 cells, 15-Ah capacity, and a 3.75V nominal in a 3P-96S configuration. This T-shaped pack combines several modules of different sizes to fit within available volume inside the vehicle's crash protected zone. Discussion of balance of plant electrical systems that are common with other large battery packs is given below.

Module sizing. Modules are sized to have voltage <50V to minimize risk of electrical shock to humans during manufacturing, maintenance, refurbishment and disposal.

Manual service disconnect (MSD). This manual switch is located roughly at the middle of the pack. The intent of the MSD is to provide a simple, error-proof method to de-energize the pack's main terminals before performing maintenance on the pack. With the MSD disconnected, the pack cannot be externally shorted across its main terminals, nor can the electrical connections arc even if contactors are closed.

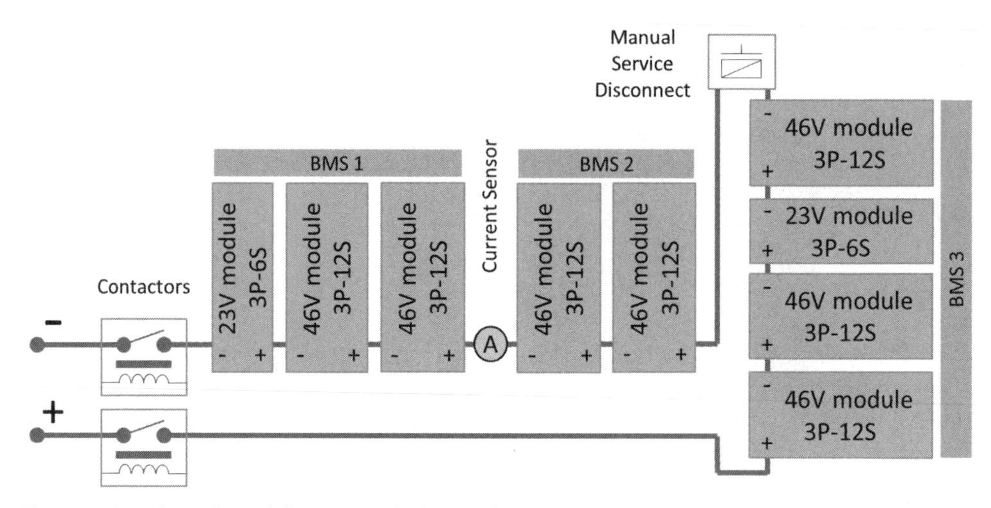

Figure 7.3 Chevrolet Volt battery pack electrical configuration.

Main pack fuse. The main fuse is often colocated with the MSD. The fuse may have slow-blow design characteristics to tolerate brief unintended pulses but to protect in case of a continuous high current draw in case of short or system malfunction.

Contactors. Two contactors (relays) are located near the pack terminals. Both the positive and negative leads are independently switched to provide fault tolerance. The contactors are actively switched on/off by the BMS, but can also be switched off if the BMS signal is wired in series with a passive safety interlock loop. This low-voltage loop ensures that operational/safety critical items (MSD, high-voltage electrical leads to/from pack, etc.) are correctly engaged. Another feature of high-voltage contactors is precharge resistors that match voltages of both sides of the switch before closing. This prevents large inrush currents and possible arcing across the switch to avoid damage to the contactors, fuses, batteries, and application. Only DC contactors may be used for battery packs as AC contactors rely on the changing current direction to break the switch open.

Ground fault detection. For fault tolerance, neither the positive nor negative terminal of the battery pack is connected to ground. Connection of either terminal with ground is an abnormal condition that exposes the system to risk of shorting should the other side also connect with ground. A ground fault detection system (not pictured in Figure 7.3) alternately connects a large resistor (possibly the precharge resistor) from the positive terminal to ground, looks for leakage current across the resistor, opens that connection, connects the negative terminal to ground through a large resistor, and looks for leakage current across that resistor. If either leakage current exceeds a determined threshold, a ground fault is declared by the BMS. A ground fault indicates a breakdown in electrical isolation somewhere in the system, either in wiring or the cell walls and holders (as some cells have containers that are electrically connected to one of the two cell terminals).

Thermal system. Discussed in Section 7.2.

Battery management system. Discussed in Section 7.4.

7.2 Thermal Design

The thermal management system must maintain the battery within the cell's temperature operating limits for performance, life, and safety. See Chapter 3 for discussion of thermal management techniques and methods to measure and model heat generation rates. Chapter 5 provides important thermal considerations with respect to safety. When designing the thermal system, its effectiveness must be traded off with the extra cost, volume, and mass it adds to the energy storage system. It is rare that a thermal management system is sized to handle worst-case heat generation rates on a continuous basis.

Figure 7.4 shows cell-level RMS and average current statistics—important for determining heat generation within cells—for a PHEV40 vehicle. The PHEV40 simulations are for a 17.6-kWh, 292V nominal pack with 270 20-Ah cells in a 3P-90S configuration. The figure aggregates simulation results for several hundred possible drive cycles in a cumulative distribution of the RMS and average currents. Given the sharp rise in currents in the right side of the plot, it is reasonable to design the thermal management system to handle the continuous heating of perhaps an 80th percentile drive cycle with 60A or 3C RMS current. For the other 20% of drive cycles that exceed the thermal management system's continuous cooling capability, the system can rely on thermal mass to dampen cell heating during the transient, sometimes short-duration drive cycles. If the cell temperature rises too far, the BMS can derate the power limits communicated to the application supervisory controller so it can limit charge/discharge rates with the impact of reducing heat generation within the cells.

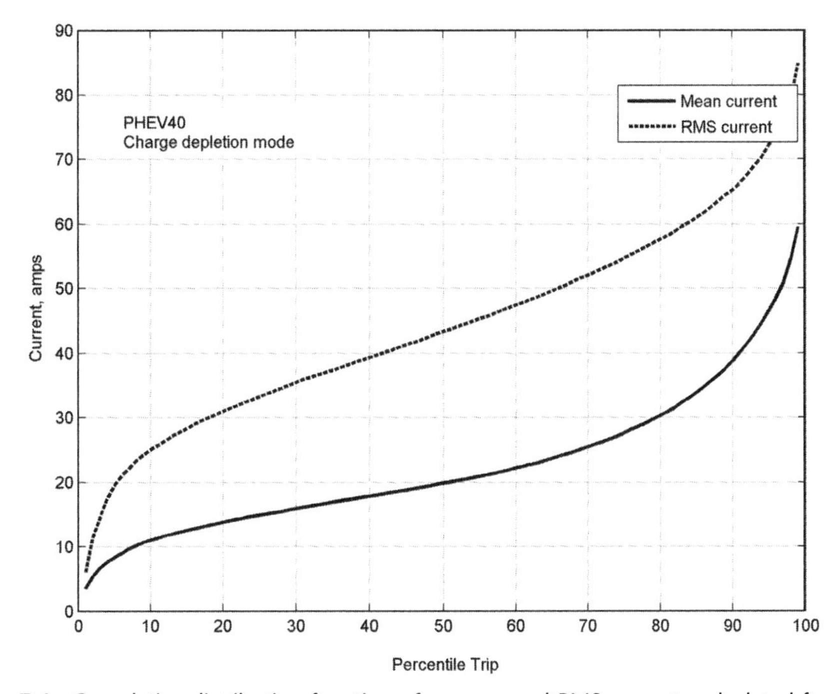

Figure 7.4 Cumulative distribution function of average and RMS currents calculated for several hundred PHEV40 drive cycles. (Source: Neubauer et al. [3].)

When selecting the cooling/heating working fluid, air cooling is preferable to liquid cooling if its performance is acceptable. Air cooling is less expensive and simpler. Liquid cooling is more complex, more prone to leaks, yet cools cells at a higher rate and more uniformly due to large heat capacity of liquid relative to air. Another strategy to reduce temperature rise is to choose a cell with lower resistance or higher P/E ratio. This reduces the heat generation rate for a given duty cycle. Including a vapor-compression system to chill the air or liquid provides further benefits to lifetime, especially in hot ambient temperatures. An example of trade-offs with lifetime for four candidate thermal strategies is shown in Figure 7.5. The longer battery life possible with a chilled liquid cooling system can either be used to extend the life of a given system with a fixed size or to reduce the system size required for a fixed lifetime requirement, say 10 years. For the PHEV, a chilled liquid cooling system that costs less than $500 can pay for itself by reducing the size of the battery, assuming $300/kWh Li-ion battery cost and Phoenix, AZ ambient conditions. Or if the battery size is held fixed, the battery with the chilled liquid cooling can operate for 2 to 5 additional years compared an air cooled system that uses ambient air [5].

As an example of battery thermal management, Figure 7.6 shows a schematic of the Chevrolet Volt system. A thin plate with microchannels for liquid cooling is placed in between every other cell of the 288-cell battery pack. The liquid supplied to the cells is controlled in three modes: (1) heating mode, via an electrically powered resistive heater in-line with the liquid coolant for warm-up during winter, (2)

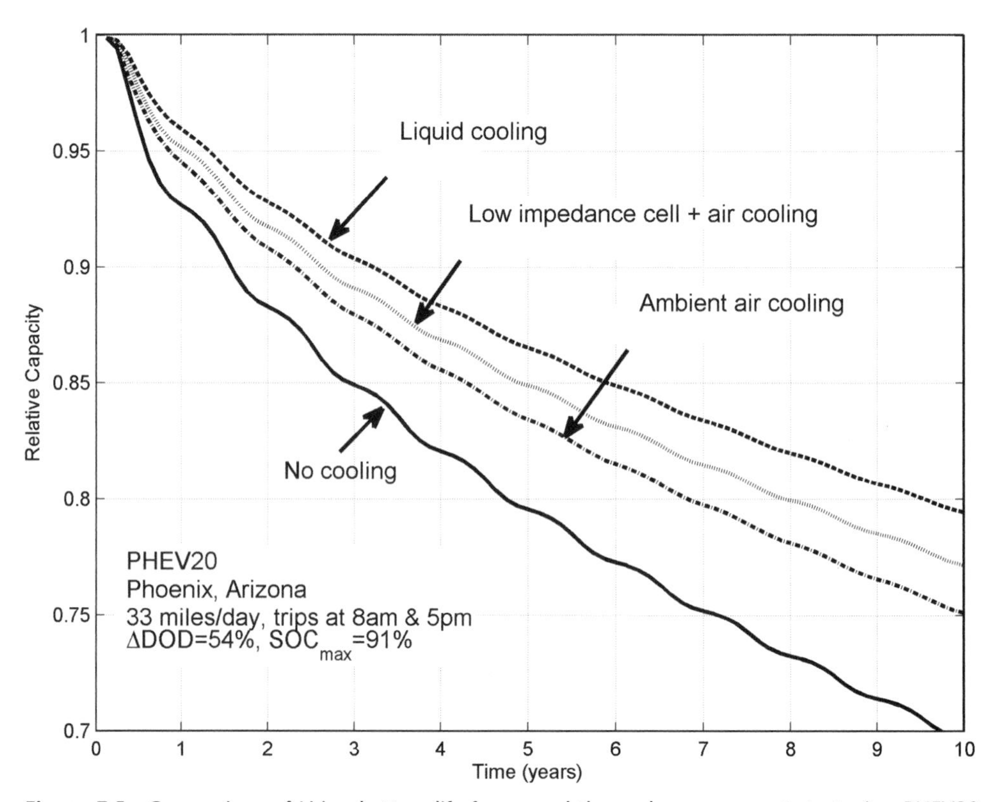

Figure 7.5 Comparison of Li-ion battery life for several thermal management strategies. PHEV20 vehicle with graphite/NCA Li-ion battery. (Source: Smith et al. [4].)

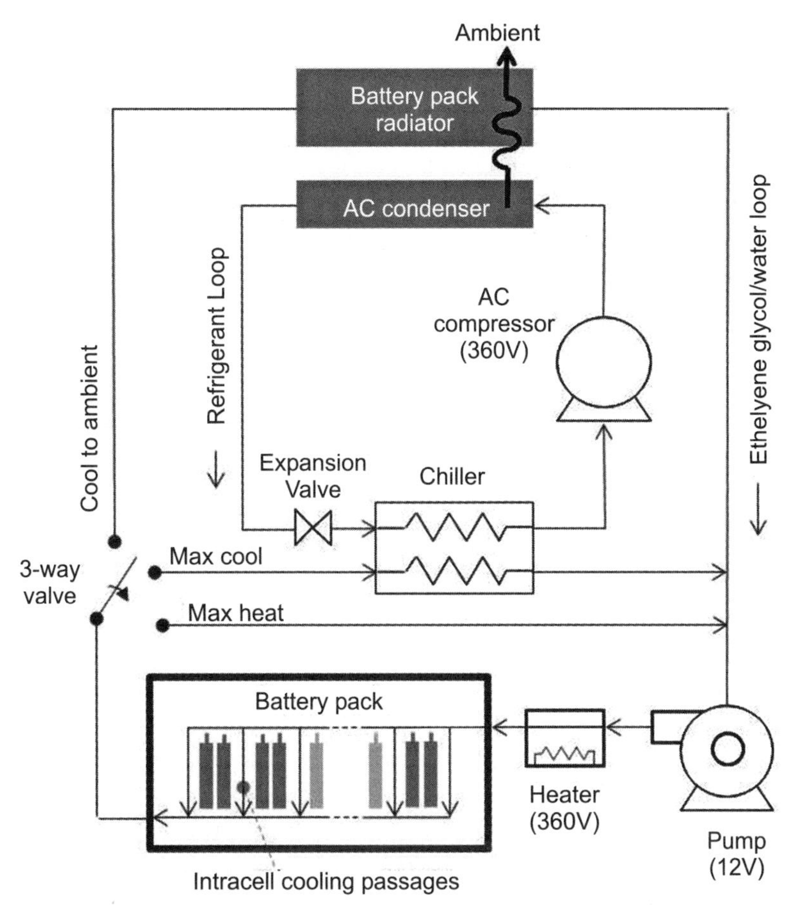

Figure 7.6 Chevrolet Volt battery thermal system.

an ambient cooling mode where the battery liquid coolant is circulated through a radiator at the front of the vehicle when the battery is hot and the outside temperatures are sufficiently cooler than the battery, and (3) a chilled cooling mode where the liquid is cooled in a chiller. The chiller is part of a HVAC vapor-compression system that also cools the passenger cabin through a separate evaporator.

7.3 Mechanical Design

Mechanical enclosures for cells and packs are designed to meet requirements for safety, durability, cell lifetime, and thermal management. As a brief overview of such requirements, cell holders and module structures must:

- Maintain compression on cells with prismatic form factor. For prismatic cells—both hard can or soft pouch packaging—a small amount of external pressure, 10 to 20 kPa, suppresses electrode creep and delamination and is generally beneficial for life. Cylindrical cells maintain pressure on the jelly roll without the need for external compression.

- Allow flow paths for thermal management in designs employing intracell cooling, or conduction paths for systems employing fin or cold-plate cooling.
- Be mechanically rigid, holding cells in place for the lifetime of anticipated worst case vibration loads.

Pack enclosures must:

- Be designed with outer casing capable of withstanding expected mechanical abuse (e.g., a person standing on the pack or a person or object falling onto the pack).
- Be tolerant to dropping the pack from some height (e.g., a forklift) or, in the case of vehicles, be able to withstand crash events. The mechanical requirements of automotive packs varies greatly depending on whether the pack is inside the protected zone (e.g., passenger vehicles) or not (e.g., delivery trucks).
- Be able to route cell vent gasses to a safe exit and contain possible cell failures.

7.4 Electronics and Controls

In some industries, the BMS is referred to as a battery electronic control module (BECM). At the highest level, the BMS receives commands from the application supervisor to power on/off and close/open contactors and reports back the battery's operational state. The various additional functionalities of a BMS are discussed next.

7.4.1 Roles of Battery Management

Figure 7.7 summarizes roles of a BMS. Since the battery is a passive device that responds to whatever load the application attempts to source/sink, the application supervisor is responsible for active control of the load within operating limits reported by the BMS. Estimation algorithms are needed to determine battery energy and power limits using measured current, voltage, and temperature signals.

In addition to estimating and reporting operating limits, the BMS controls the battery thermal management and cell balancing. The BMS must also detect faults and log abnormal events including cell/pack over/under voltage/current/temperature, ground fault/loss of electrical isolation, gas detection, sensor failures, communication watchdog timer, and checksum errors.

7.4.2 BMS Hardware

Numerous topologies are possible for BMS and balancing systems. For detailed discussion, we refer the reader to [6]. Technical considerations for BMS hardware include input/output accuracy, voltage reference stability, operating temperature range, communications speed, tolerance to electro-magnetic interference, memory,

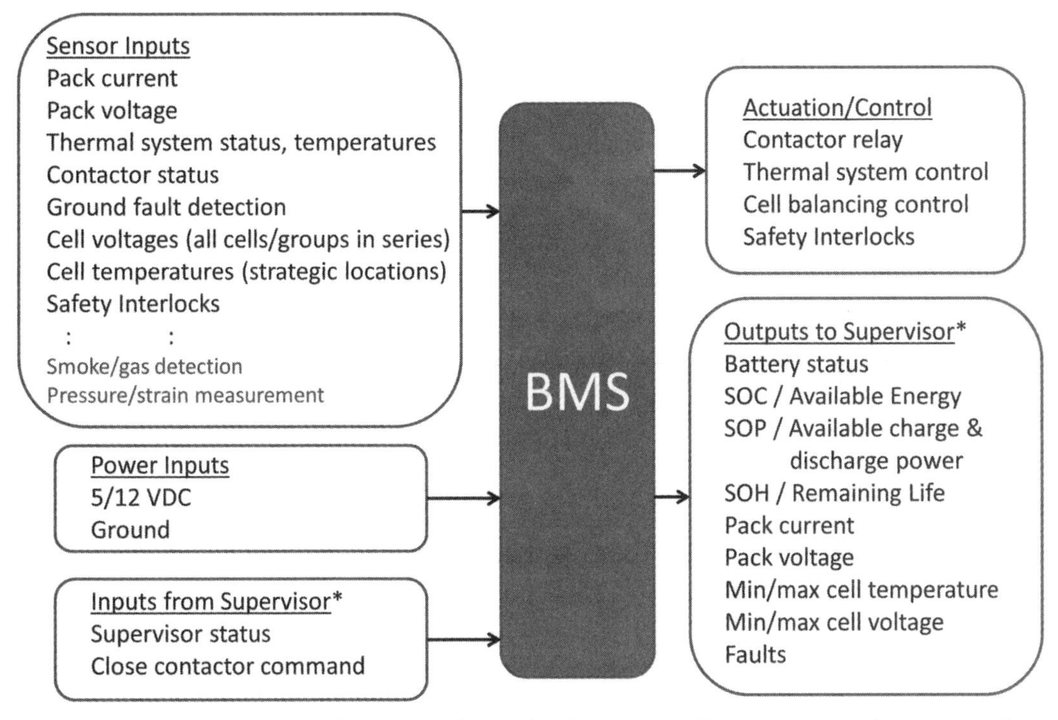

Figure 7.7 BMS inputs and outputs.

and computation speed. For Li-ion systems, the BMS works hand-in-hand with the cell balancing system. If balancing is co-located within the BMS, it must be able to dissipate heat generated by balancing resistors and circuitry.

The BMS may be arranged in master/slave configurations to minimize number of and length of wires travelling between the BMS and cells. At one extreme is the master-only BMS configuration, using a fully centralized BMS that has sensor measurement, cell balancing, and computation all taking place in one box for the entire pack. At the other extreme is a slave-only BMS configuration, using distributed cell-level processors, each accomplishing cell sensing, computation and balancing for its individual cell or group of parallel connected cells. Perhaps a true slave-only configuration is not strictly possible, as there must be some means for coordinating cell balancing as well as carrying out pack-level objectives such as closing contactors, detecting faults and communicating with application supervisory controllers.

The particular master/slave hardware architecture is chosen to minimize cost, volume and mass, facilitate maintenance or module replacement, and to maximize performance and lifetime of the pack. Amongst available hardware options, architectures are chosen to avoid long distances of wires and number of wiring connections. As a rule of thumb from the automotive industry, each single wire connection in an automobile costs $1. Additionally, each send/receive node for CAN communications costs around $0.50.

7.4.3 Cell Balancing

NiMH, NiCd, and Pb acid chemistries all have internal overcharge protection mechanisms in the form of a redox shuttle. The redox shuttle is some molecule that can be reversibly oxidized at the positive electrode, diffuse through the electrolyte, and be reduced at the negative electrode, all at a potential slightly higher than the typical end-of-charge voltage. Chemistries with redox shuttles may be overcharged at a gentle rate with the result that a series string of cells with mismatched capacity or SOC all eventually reach 100% SOC. The unused charge generates heat inside the cell.

Even for Li-ion, which presently has no proven redox shuttle, extremely well matched cells may require no balancing and still achieve long life. This topology is sometimes used for satellite batteries where the risk of balancing system electronics failure is a concern and cells can be well matched prior to building the battery system. Cells must be precisely matched for their capacities and self-discharge rates. The additional cell prescreening and aging tests are expensive, as is the increased scrap rate for cells that fail qualification. At the opposite end of the spectrum, low-cost replaceable consumer electronics devices that only need a few years of life may not require cell balancing.

But in general, most Li-ion systems require balancing; passive balancing is the most common choice. If a cell reaches too high a SOC (or voltage, though SOC is preferred) relative to its neighbors, resistors in parallel with each cell are switched on to dissipate some energy. Passive balancing systems generally only balance during charging because balancing during discharge is a waste of energy. Passive balancing is cheaper than active balancing; however, it has the disadvantage that it leaves unused energy stranded in healthy cells. With cells perfectly matched at full charge, during discharge weak cells having low capacity and/or high resistance will be the first to reach their minimum permissible voltage, causing end of discharge for the entire string of cells. Stronger cells that have not yet reached their minimum voltage will have stranded energy left in them. This unused energy obviously has larger consequences for energy—rather than power—applications. In contrast, active balancing systems have some provision to shuttle energy from strong to weak cells during discharge and charge. For a system with mismatched cell capacities, all the energy of the system can be used minus some losses due to inefficiencies.

Li-ion cells can be well matched in the factory, perhaps with just 1% to 2% capacity mismatch when a pack is constructed. Even with this tight level of matching, cell imbalance will grow throughout life. Figure 7.8 shows an extreme case of cell imbalance growth throughout life. In this simulation, cells are matched within ±1.2% imbalance at BOL. Daily cycling for this high annual mileage case causes cells in the center of the large EV battery pack to heat up more significantly than cells close to the pack exterior. After 10 years, the strongest cell holds some 14% more capacity that the weakest cells. Temperature gradients across the pack are responsible for roughly half of the imbalance growth by year 10. Variation of aging processes within individual cells is responsible for the other half. At year 10, the system shown in Figure 7.8 will have 76% of BOL capacity if it uses active balancing or 65% of BOL capacity if it uses passive balancing, as it will be limited by the weakest cell. The simulated level of imbalance varies depending on use scenario. For example, packs smaller than this BEV75 or with stronger thermal management

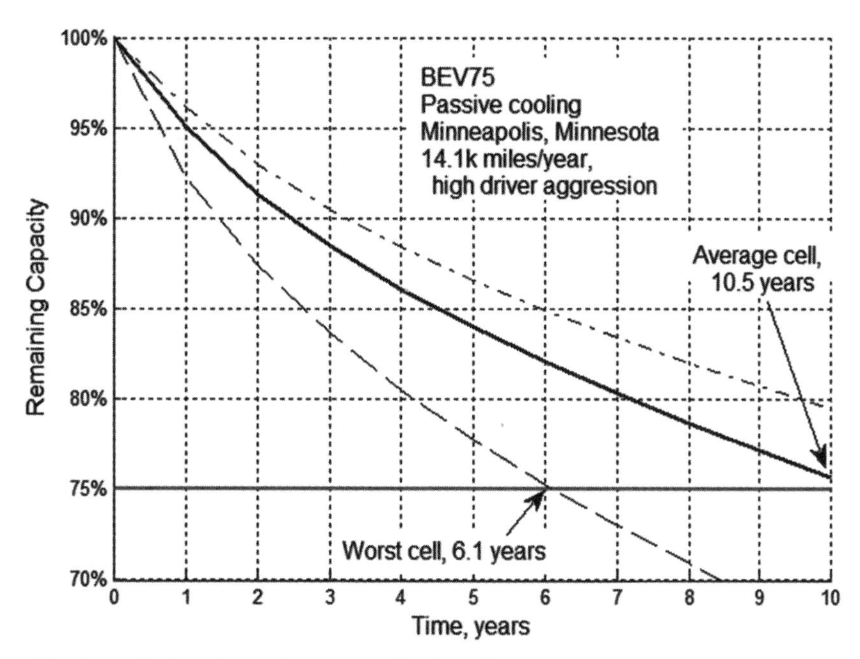

Figure 7.8 Pack lifetime dependence on cell aging dispersion and balancing system. With active balancing, average cell aging dictates pack life. With passive balancing, worst-case cell aging dictates pack life. (Source: [5].)

may have less temperature imbalance leading to less capacity imbalance over the lifetime.

Whether this stranded energy at year 10 is an issue depends on the application. In years 1 to 3, the stranded energy is much less. Active balancing therefore has questionable value for short lifetime consumer devices, but may be worthwhile for long term capital investments that need to last 5 to 20 years. Also, the magnitude of imbalance growth shown in Figure 7.8 strongly depends on the cell quality, BOL matching and thermal gradients across the pack. Hotter cells generally age the fastest, though it is also possible for coldest cells to age the fastest if the pack encounters frequent low temperature operation, especially charging. For more information on cell imbalance growth, see [7].

7.4.4 State Estimation Algorithms

Estimation algorithms are needed to deduce state of charge (SOC), state of power (SOP), and state of health (SOH) from current, voltage, and temperature measurements [31]. The algorithms may be rule-based or model-based. For example, a rule-based algorithm might estimate initial SOC using a lookup table of SOC = $f(V_{oc})$ when the battery is at rest and terminal voltage equals open-circuit voltage, V = V_{oc}. During charge and discharge however, the rule $V = V_{oc}$ does not hold, and the algorithm might switch to a coulomb counting calculation:

$$SOC(t) = \int_0^t \frac{-I(t)}{Q} dt + SOC_0 \tag{7.3}$$

where Q is the capacity of the battery and SOC_0 is the initial SOC estimate, and $I(t) < 0$ indicates battery discharge. The issue with coulomb counting, however, is that the SOC estimate will eventually diverge from the proper SOC due to inevitable current sensor measurement error. To correct for current sensor error, it is better to use current and voltage measurements simultaneously to constantly adjust the SOC estimate. This is the concept behind a state estimator or observer, one class of which is the Kalman filter. In general, model-based algorithms are preferred to rule-based, as they are formulated to provide a smoothly changing, best-possible estimate within anticipated sensor error bounds and are more accurate across a wide operating range.

To estimate SOC, SOP, and SOH in a model-based fashion, multiple algorithms are needed. A set of algorithms is suggested in Figure 7.9.

1. A *reference model* is needed that predicts battery current/voltage dynamics and relates them to the internal states of the system. All other algorithms in Figure 7.9 make use of the battery reference model. The more accurate the model, the more accurate the state estimates.
2. The *state estimator* or observer uses the reference model in a predictor-corrector fashion to converge on estimates of the model states, one of which is SOC. (Recursive regression algorithms are an alternative for calculation of model states.)
3. For *limits calculation,* the *reference governor* inverts the reference model to find allowable current or power levels that do not violate some battery constraint, usually minimum and maximum voltage limits. These current/power limits are reported to the supervisory controller for controlling the load. (It is also possible to use model-predictive control algorithms to calculate limits.)
4. Health changes in the battery may be deduced through slow adaptive tuning of parameters, namely resistances and capacity in the reference model. Not discussed here, *online parameter identification* algorithms include recursive regression [8] and/or augmentation of state estimators to jointly estimate model states and parameters [9]. Parameter estimation should be carried out several orders of magnitude slower (~months) than state estimation (~seconds) to avoid instability between the two estimates.

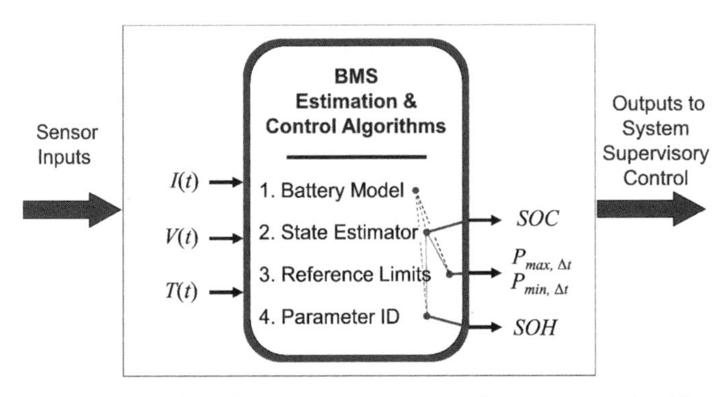

Figure 7.9 Models and algorithms for SOx estimation, x = charge, power, health.

The above algorithms take both linear and nonlinear forms. This section illustrates basic concepts using linear models and estimators. The linear algorithms can then be easily extended to more accurate nonlinear ones.

7.4.5 Battery Reference Model

Battery reference models typically take the form of equivalent circuit models that mimic battery current/voltage dynamics, as shown in Figure 7.10. The drawback of the circuit model is that SOC is the only internal electrochemical state of the model. The resistive/capacitive states of the model provide no physical insight into battery internal processes, but are useful in predicting proximity of battery terminal voltage to voltage limits. Research is underway to develop electrochemical models fast and accurate enough for online application [10–13]. For most state estimation algorithms, the reference model should be in state variable form. A continuous time linear state variable model takes the form

$$\dot{\mathbf{x}}(t) = \mathbf{A}\mathbf{x}(t) + \mathbf{B}\,\mathbf{u}(t) \tag{7.4}$$

$$\mathbf{y}(t) = \mathbf{C}\,\mathbf{x}(t) + \mathbf{D}\,\mathbf{u}(t) + \mathbf{y}_0 \tag{7.5}$$

Equation (7.4) is called the state equation and (7.5) is called the output equation. In these equations, **x** is a vector of model states, **y** is a vector of the model outputs, and **u** is a vector of model inputs.

The model must be strictly proper in order to ensure stability of the algorithms. This requires that current be the model input, $\mathbf{u}(t) = I(t)$, and voltage be the model output, $\mathrm{y}(t) = V(t)$. The opposite model would constitute an improper system in control theory. Figure 7.10 defines the model matrices $\mathbf{A},\mathbf{B},\mathbf{C},\mathbf{D},\mathbf{y}_0$ for a typical equivalent circuit model. Though not necessarily required, if possible it is convenient to arrange the A matrix such that it has diagonal structure. In this way, the model's eigenvalues appear directly on the diagonal, $[0, \lambda_1, \lambda_2]$ in Figure 7.10. This facilitates stability analysis and simplifies integration of the ordinary differential equations with time and/or conversion from continuous to discrete time. A

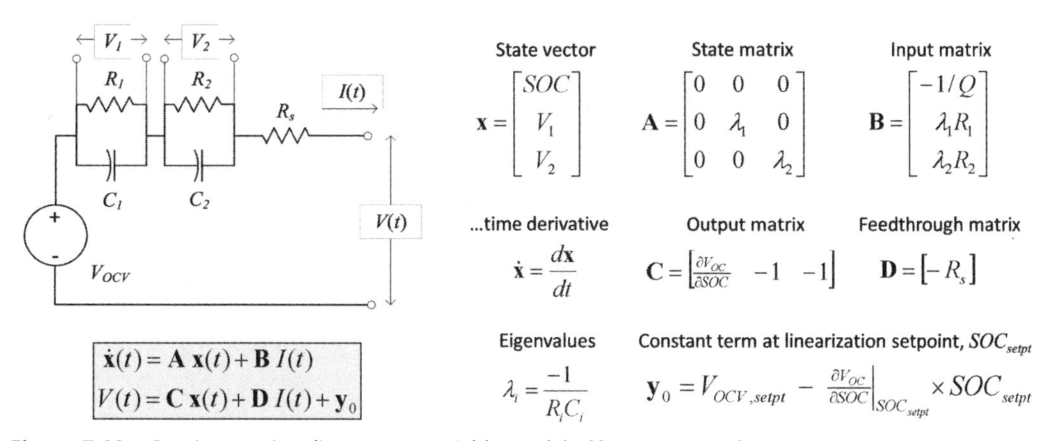

Figure 7.10 Continuous time linear state variable model of battery equivalent circuit.

necessary requirement for the model to have observable states is that the A matrix can only contain one eigenvalue equal to zero. This free integrator term is associated with SOC. The coulomb counting equation (7.3) is a free integrator with an eigenvalue of zero. This sometimes requires clever manipulations so that all other states of the model have negative real eigenvalues such that their states converge to zero if the battery is left at rest. With battery dynamics controlled by diffusion and reaction processes, the eigenvalues of a battery model have no imaginary (oscillatory) part. They are purely real and have quantities less than or equal to zero.

7.4.6 State Estimator

Shown in Figure 7.11, a state estimator operates using principles of feedback control. At each instant in time, a small correction is made to the model state equation. This correction is proportional to the error between the model predicted voltage and the actual measured voltage. The hat symbol (e.g., $\hat{\mathbf{x}}$) indicates an estimated quantity. To analyze convergence criteria, we substitute the measured voltage minus the model voltage times the gain into the state equation

$$\dot{\hat{\mathbf{x}}} = \mathbf{A}\hat{\mathbf{x}} + \mathbf{B}\mathbf{u} + \mathbf{L}(\mathbf{y} - \mathbf{C}\hat{\mathbf{x}} + \mathbf{D}\mathbf{u} + \mathbf{y}_0) \tag{7.6}$$

Collecting the terms for $\hat{\mathbf{x}}$, we find the eigenvalues of the system with the estimator in the loop are the eigenvalues of the matrix $\mathbf{A} - \mathbf{LC}$. For the estimator to be stable and converge, these eigenvalues must all be less than zero. With the A matrix of the battery model formulated in diagonal form and the C matrix a row vector, it is straightforward to find a diagonal L matrix that satisfies this condition. If feedback control is to be applied based on the state estimates, it is good practice for control stability to separate the time scales of the estimation and control functions by an order of magnitude. It is general practice in controls to make estimator eigenvalues faster than controller eigenvalues; however, given the weak observability of battery internal states, it may be advisable to do the opposite: slowly adjust state

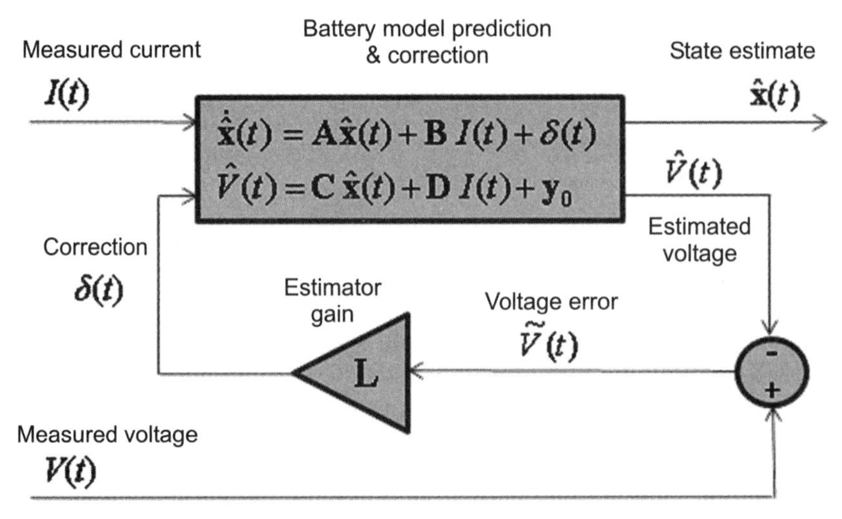

Figure 7.11 Continuous time linear state estimator.

estimates and add appropriate safety margins to the fast control limits to account for estimator error.

The requirement for *observability* of the system is that the observability matrix,

$$\mathbf{O} = \begin{bmatrix} \mathbf{C} \\ \mathbf{CA} \\ \mathbf{CA}^2 \\ \vdots \\ \mathbf{CA}^{n-1} \end{bmatrix} \tag{7.7}$$

have full rank. This condition is what imposes the requirement on the battery reference model that it is **A** matrix only have one eigenvalue equal to zero, as is the case in Figure 7.10.

On a digital controller, the state estimator is implemented in *discrete time* rather than continuous time. The estimator operates based on sensor measurements and calculations made at the present time step, k, and updates the states that are carried forward for operations at the next time step, $k + 1$. First, the output equation is evaluated, then the state equation is updated

$$\hat{\mathbf{y}}(k) = \mathbf{C}^d\ \hat{\mathbf{x}}(k) + \mathbf{D}^d\ \mathbf{u}(k) + \mathbf{y}_0 \tag{7.8}$$

$$\dot{\hat{\mathbf{x}}}(k+1) = \mathbf{A}^d\,\hat{\mathbf{x}}(k) + \mathbf{B}^d\ \mathbf{u}(t) + \mathbf{L}^d(\mathbf{y}(k) - \hat{\mathbf{y}}(k)) \tag{7.9}$$

The discrete time model uses different matrices for its state and output equations than the continuous model. The matrices can be transformed from continuous to discrete time by zero-order hold, Tustin, or any number of other digital control methods. In the case where the **A** matrix of the continuous time model is in diagonal form, an exact solution to the continuous time ordinary differential equation may be used and the continuous to discrete transformation for time step or sampling time, T_s, is simply given by

$$\mathbf{A}^d = \exp(\mathbf{A}\,T_s) \tag{7.10}$$

$$\begin{aligned} \mathbf{B}^d &= \mathbf{A}^{-1}\left(\exp(\mathbf{A}T_s) - \mathbf{I}\right)\mathbf{B} \\ &= diag\left(T_s \quad \tfrac{1}{\lambda_1}(e^{\lambda_1 T_s} - 1) \quad \tfrac{1}{\lambda_{n-1}}(e^{\lambda_{n-1} T_s} - 1)\right) \end{aligned} \tag{7.11}$$

$$\mathbf{C}^d = \mathbf{C} \tag{7.12}$$

$$\mathbf{D}^d = \mathbf{D} \tag{7.13}$$

The condition for convergence of the discrete time state estimator is that the eigenvalues of the matrix $\mathbf{A}^d - \mathbf{L}^d\,\mathbf{C}^d$ be contained inside the unit circle on a plot of real versus imaginary parts, or in other words, the absolute value of the complex eigenvalues must be less than 1.

In Figure 7.10, the battery model is presented as a simple linear model. In reality, the battery is a *nonlinear system*. It is quite common in control theory to use linear control techniques even for nonlinear systems, although stability is not always guaranteed and must be thoroughly tested. To capture nonlinearities of the system, the $\mathbf{A},\mathbf{B},\mathbf{C},\mathbf{D},\mathbf{y}_0$ matrices of the battery model can be scheduled as functions of known nonlinearities of the system, generally temperature and SOC. In this form, the model is called a linear parameter varying (LPV) system. There are numerous references to LPV systems in the control literature [32]. For discrete time implementation, the continuous time LPV matrices may be transformed to discrete time at each time step using (7.10) to (7.13).

Control and estimation theory provides many extensions to the simple linear state estimation algorithms introduced here. These are summarized in Table 7.1. Of these, the unscented Kalman filter is perhaps the most popular for battery estimation [9, 14]. It handles nonlinearities well, has fast computation time, and generally has smaller steady-state error compared to the extended Kalman filter.

7.4.7 Current/Power Limits Calculation

One of the functions of the BMS is to report the charge and discharge power available to/from the battery. Several versions of available power may be needed to satisfy the supervisory controller depending on the time scale of the control. Available power may be specified as (1) an instantaneous limit, (2) a pulse limit available for Δt seconds into the future, and/or (3) a continuous limit. Regarding (3), if energy is the constraining factor, the continuous limit may be calculated based on the energy available divided by the desired charge or discharge time window. If high temperature is the constraining factor, the continuous power may be constrained by the

Table 7.1 Overview of State Estimation Algorithms

Algorithm	*Linear or Nonlinear System*	*Assumed Sensor and Model Process Noise*	*Comments*
State estimator	Linear	N/A	(7.8) and (7.9)
Kalman filter	Linear	Gaussian	Method for tuning gain matrix L based on assumed levels of sensor noise and model error.
Extended Kalman filter	Nonlinear	Gaussian	Extends the linear Kalman filter to nonlinear systems using a linearization of the model at the present state estimate to provide an update to the gain matrix L at each time step.
Unscented Kalman filter (also called the sigma point Kalman filter)	Nonlinear	Hybrid	Rather than relying on a linearization at the present state estimate, samples the model at several possible values of the present state, providing a more reliable estimate in case the present state estimate is too far off in an area of strong nonlinearity.
Particle filter	Nonlinear	Non-Gaussian	Samples the model at numerous possible values of the state to determine the most probable value. High computational cost, but achievable with a simple reference model.

ability of the thermal system to accommodate the cells' heat generation rate as a function of power.

Prediction of instantaneous and pulse power limits requires a model to estimate how much power is available into the future without hitting a cell internal or external limit (e.g., maximum/minimum voltage limits). In the case of pulse power, this is for some predetermined time window extending into the future. In vehicles, for example, power limits may be for just 2-second duration to accommodate pulses of charge current received during regenerative braking or for 10-second duration to accommodate discharge power needed for acceleration of the vehicle.

We introduce a simple reference governor method to estimate limits using the state variable model of the battery and system states estimated by an onboard algorithm, each discussed previously. The reference governor inverts the model at the present states to find the limiting current (model input) that hits the constraint limit Δt seconds into the future. This algorithm relies on (approximate) linear behavior of the battery over the short time period. Since current, rather than power, is the model input in Figure 7.10, we calculate limits in terms of current. Power limits may be approximated by multiplying charge/discharge current limits by the maximum/minimum cell terminal voltage. If these assumptions of linearity are too restrictive, model predictive controls (MPC) provide a means to calculate the limits.

The reference governor seeks to enforce a general constraint in the model output, $y_{min} < y(t) < y_{max}$ and calculate for the supervisory controller a limiting input, in this case $\mathbf{u}(t) = I(t)$, such that at some Δt seconds into the future $y(\Delta t) = y_{lim}$. The constraint is usually minimuim and maximum cell terminal voltage. Inverting the model, we find the limiting current

$$I_{\mathrm{min/max},\Delta t} = [\mathbf{C}\mathbf{A}^{*}\mathbf{B} + \mathbf{D}]^{-1}\left\{(y_{\mathrm{lim}} - y_0) - \mathbf{C}e^{\mathbf{A}(\Delta t)}\hat{\mathbf{x}}\right\} \tag{7.14}$$

where

$$\mathbf{A}^{*} = diag\left(\Delta t \quad \tfrac{1}{\lambda_1}(e^{\lambda_1(\Delta t)} - 1) \quad \tfrac{1}{\lambda_{n-1}}(e^{\lambda_{n-1}(\Delta t)} - 1)\right) \tag{7.15}$$

Equations (7.14) and (7.15) assume the **A** matrix has diagonal form.

As a final consideration, either the BMS and/or the supervisory controller may need to derate these power limits to achieve acceptable lifetime. Often the battery cooling system is the limiting factor, not electrochemical limits, that one unable to continuously cool the cells to an acceptable temperature necessary for long life.

7.5 Design Process

The design process of complex systems must be carefully managed to coordinate the efforts of multiple teams and fields of expertise. The "V" design cycle [15], shown in Figure 7.12, provides a logical order for phases of the design process. Design phases proceed from left to right. Most complex items such as systems appear at the top. Less complex items, such as components, appear at the bottom. The process starts from the top left side of the "V." The first phase is business analysis and requirements definition for the system, followed by the trickle-down of requirements and design of subsystems and components. Once individual components (sensors,

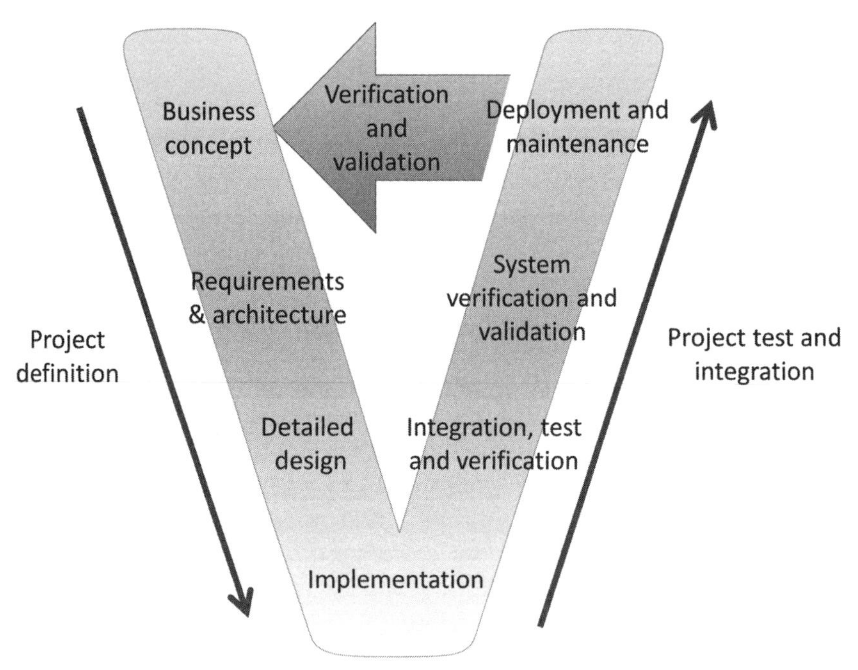

Figure 7.12 "V" design cycle. (Adapted from [15].)

electrical/thermal controls, packaging, etc.) are developed and ready, on the right side of the "V" they are integrated and tested as subsystems and finally complete systems.

Verification testing is performed to ensure each component and system meets its stated requirements. Validation testing is performed to assess the overall system's effectiveness in meeting its original goals. Battery lifetime testing is a particularly time consuming endeavor, requiring anywhere from 6 to 24 months of testing before a new product can be fielded. For emerging technologies, it may be desirable to quickly field prototype units in limited quantities as part of a demonstration phase of the business plan. The performance and reliability of the demonstration units may be tracked for a year or more before making final decisions to deploy at scale. Given the low cost of data telemetry and storage, it is advisable to track energy storage system performance down to the individual cell level so that infant mortality rates or problems with individual cells may be assessed.

7.6 Design Standards

Mature industries using energy storage have established design standards. These include consumer electronics, aerospace, and automotive industries. These design standards provide an evaluation method to assess the safety level of possible failures to hardware, software, or system. These standards generally define the technical state of the art. Failure of the system to meet the technical state of the art exposes the manufacturer to liability in the case of an accident.

An example standard in the automotive realm is ISO 26262, "Electrical System Functional Safety Standard for Road Vehicles" [16]. The standard addresses possible hazards caused by the malfunctioning of electronic and electrical systems. Recognized throughout the automotive industry, ISO26262 defines the so-called "technical state of the art" for electrical systems such as throttle-by-wire and battery pack systems. The basic tenant of ISO 26262 hinges on assigning an automotive safety integrity level (ASIL) for a given hazardous event. The ASIL can have a value of A, B, C, or D, with A as the least severe and D as the most severe in terms of hazard to the user and any other exposed individual.

The ASIL is a product of values assigned from within three categories: Severity, Exposure and Controllability:

$$\text{ASIL} = \text{Severity} \times \text{Exposure} \times \text{Controllability}$$

Possible values for each of the three categories are shown in Table 7.2.

The most severe hazard level, ASIL D, results from combination of worst classifications of each category:

$$\text{ASIL D} = (\text{S3}) \times (\text{E4}) \times (\text{C3})$$

If any one category is reduced, then the ASIL drops one level. For example, if (C3) is reduced to (C2), the hazard becomes ASIL C. At ASIL A there is no safety issue and only standard quality management processes are applied.

ISO26262 considers the entire product life cycle from development, to operation, to decommissioning EOL. Note that ISO 26262 does not apply to electrical shock, flammability, shorting, or thermal abuse events. Relevant standards are discussed in Chapter 5. Multiple safety standards covering abuse, have adopted EUCAR hazard levels, summarized in [17].

7.7 Case Study 1: Automotive Battery Design

In this case study, we highlight some design trade-offs encountered by the automotive system designer whose goal is to meet battery performance and lifetime requirements at minimum cost. In order to explore these trade-offs, we first develop a battery life predictive model and regress its parameters to cell-level experimental aging datasets. This life model captures the dominant aging characteristics of

Table 7.2 Classifications for ISO 26262 Hazard Safety Levels

Severity	*Exposure*	*Controllability*
S0 No injuries	E0 Incredibly unlikely	C0 Controllable in general
S1 Light to moderate injuries	E1 Very low probability (injury could happen only in rare operating conditions)	C1 Simply controllable
S2 Severe to life-threatening (survival probable) injuries	E2 Low probability	C2 Normally controllable (most drivers could act to prevent injury)
S3 Life-threatening (survival uncertain) to fatal injuries	E3 Medium probability	C3 Difficult to control or uncontrollable
	E4 High probability (injury could happen under most operating conditions)	

Source: [16].

Li-ion batteries. Temperature is the most important aging stress factor and, given its importance, this case study provides simple models to predict battery temperature for different ambient environments and thermal management strategies. Last, the lifetime of a PHEV40 battery is explored for several temperatures, excess energy sizes, and charge control strategies.

7.7.1 Life Predictive Model

Using methods introduced in Chapter 4, a semiempirical lifetime model is fit to aging data from the literature for the graphite/NCA Li-ion chemistry. For the NCA chemistry, main aging features observed in the data are

1. SEI growth proportional to square root of time, with rate accelerated by high temperature, *T*, and open-circuit voltage, V_{oc};
2. Loss of active sites with cycling, accelerated with high *DOD*;
3. Electrolyte oxidation at the positive electrode, accelerated by high *T* and V_{oc}.

As an aside, we note that similar aging behavior can also be observed for the graphite/NMC chemistry. For the graphite/FeP chemistry, electrolyte oxidation is negligible due to the lower operating voltage of the FeP positive electrode.

The life model assumes cell resistance growth due to calendar- and cycling-driven mechanisms to be predominantly additive,

$$R = a_0 + a_1 t_{life}{}^{1/2} + a_2 N. \tag{7.16}$$

Cell capacity is assumed to be controlled by either loss of cycleable *Li* or loss of active sites, respectively.

$$Q_{Li} = b_0 - b_1 t_{life}{}^{1/2} \tag{7.17}$$

$$Q_{sites} = c_0 - c_2 N \tag{7.18}$$

where the capacity measured at the cell terminals is the lessor of Li capacity, Q_{Li}, or active site capacity, Q_{sites},

$$Q = \min(Q_{Li}, Q_{sites}) \tag{7.19}$$

In (7.16) and (7.18), the variable N denotes number of charge/discharge cycles. This definition is clear for simple constant current cycling; however, the model requires minor modifications to accommodate complex charge/discharge cycling profiles such as experienced in electric-drive vehicles. From the fatigue literature, the rainflow algorithm is used to break up a complex depth of discharge time history, $DOD(t_{cyc})$ into an array of individual macro- and microcycles, ΔDOD_i. The algorithm also tracks whether each cycle was a full discharge and charge cycle or a single-ended charge or discharge, storing this information in an array N_i with each element having a value of either 1.0 for a full cycle or 0.5 for a half cycle. Miner's rule is used to combine the degradation effects of the various magnitude cycles. With these assumptions, the variable, N, in (7.16) and (7.18) is replaced with

$$N = \frac{\sum_i N_i}{\Delta t_{cyc}} t_{life} \tag{7.20}$$

and the life model is rewritten as

$$R = a_0 + a_1 t_{life}^{1/2} + a_2 t_{life} \tag{7.21}$$

$$Q_{Li} = b_0 - b_1 t_{life}^{1/2} \tag{7.22}$$

$$Q_{sites} = c_0 - c_2 t_{life} \tag{7.23}$$

$$Q = \min(Q_{Li}, Q_{sites}) \tag{7.24}$$

Two different time scales are implied by (7.20), where

- t_{cyc} represents the short-time cycling history encountered by the cell at one state of life;
- t_{life} captures the long-time changes in battery resistance/capacity health with life.

In a modeling environment, t_{cyc} is the time step of performance models (e.g., simply listed as "t" in Figure 7.10, usually with units of seconds, and representing a unit period of cycling or storage at one state of life of the battery). In contrast, t_{life} is time step of the life model, with units of days, months, or years, representing changes in states of life.

Degradation rate constants $k_i = \{a_1, a_2, b_1, c_2\}$ depend on the aging condition. They can either be taken directly from data for one aging condition, interpolated between data sets for several aging conditions or, following our previous work [4], mapped as rate laws that are functions of known stressors to the battery. Neglecting high C-rate and low temperature operation, we find degradation rates for the NCA chemistry are well described with functional dependence on T, V_{oc}, and DOD as follows:

$$a_1 = \frac{a_{1,ref}}{\Delta t_{cyc}} \int_{t_{cyc}} \begin{matrix} \exp\left[-\frac{E_{a,a_1}}{R_{ug}}\left(\frac{1}{T(t)} - \frac{1}{T_{ref}}\right)\right] \times \exp\left[\frac{\alpha_{a_1} F}{R_{ug}}\left(\frac{V_{oc}(t)}{T(t)} - \frac{V_{ref}}{T_{ref}}\right)\right] \\ \times \left[\left(1 + \frac{\max(\Delta DOD_i)}{\Delta DOD_{ref}}\right)^{\beta_{a_1}}\right] \end{matrix} dt \tag{7.25}$$

$$a_2 = \frac{a_{2,ref}}{\Delta t_{cyc}} \int_{t_{cyc}} \exp\left[-\frac{E_{a,a_2}}{R_{ug}}\left(\frac{1}{T(t)} - \frac{1}{T_{ref}}\right)\right] \times \exp\left[\frac{\alpha_{a_2} F}{R_{ug}}\left(\frac{V_{oc}(t)}{T(t)} - \frac{V_{ref}}{T_{ref}}\right)\right] dt \times \sum_i \left[N_i \left(\frac{\Delta DOD_i}{\Delta DOD_{ref}}\right)^{\beta_{a_2}}\right] \tag{7.26}$$

$$b_1 = \frac{b_{1,ref}}{\Delta t_{cyc}} \int_{t_{cyc}} \exp\left[-\frac{E_{a,b_1}}{R_{ug}}\left(\frac{1}{T(t)} - \frac{1}{T_{ref}}\right)\right] \times \exp\left[\frac{\alpha_{b_1} F}{R_{ug}}\left(\frac{V_{oc}(t)}{T(t)} - \frac{V_{ref}}{T_{ref}}\right)\right] \times \left[\left(1 + \frac{\max(\Delta DOD_i)}{\Delta DOD_{ref}}\right)^{\beta_{b_1}}\right] dt \tag{7.27}$$

$$c_2 = \frac{c_{2,ref}}{\Delta t_{cyc}} \int_{t_{cyc}} \exp\left[-\frac{E_{a,c_2}}{R_{ug}}\left(\frac{1}{T(t)} - \frac{1}{T_{ref}}\right)\right] \times \exp\left[\frac{\alpha_{c_2} F}{R_{ug}}\left(\frac{V_{oc}(t)}{T(t)} - \frac{V_{ref}}{T_{ref}}\right)\right] dt \times \sum_i N_i \left[\left(\frac{\Delta DOD_i}{\Delta DOD_{ref}}\right)^{\beta_{c_2}}\right] \tag{7.28}$$

In (7.25) through (7.28), E_a, α, β, and $k_{i,ref}$ are fitting parameters, R_{ug} is the universal gas constant; F is the Faraday constant, and $T_{ref} = 298.15\text{K}$, $V_{ref} = 3.6\text{V}$, and $\Delta DOD_{ref} = 1$ are arbitrary constants included for convenience of comparing $k_{i,ref}$ to standard aging conditions. Degradation rates are calculated for a representative duty cycle period, Δt_{cyc}. Given the prominence of calendar aging of Li-ion chemistries, it is important that the representative duty cycle include not only periods of cycling, but also rest periods such as overnight while an EV is parked. Degradation rates should be calculated for a repeatable unit of aging time. A good practice is to use one full day or one full week of data to represent Δt_{cyc}.

7.7.2 Fitting Life Parameters to Cell Aging Data

Equations (7.21) to (7.28) constitute the life model. Modifications may be required to match aging behavior of other cells and/or chemistries. This model has 19 fitting parameters:

$$p = \{a_0, b_0, c_0, a_{1,ref}, a_{2,ref}, b_{1,ref}, c_{2,ref}, E_{a,a_1}, E_{a,a_2}, E_{a,b_1}, E_{a,c_2}, \alpha_{a,a_1}, \alpha_{a,a_2}, \alpha_{a,b_1}, \alpha_{a,c_2}, \beta_{a,a_1}, \beta_{a,a_2}, \beta_{a,b_1}, \beta_{a,c_2}\}$$

Here, we give brief examples of the model fitting process. The NCA aging data used here is not all for the same cell design; hence the aging model results presented later are meant to illustrate design trade-offs and not the behavior of a specific cell or vehicle battery design.

Figure 7.13(a) shows examples of equations, $R = a_0 + a_1 t_{life}^{1/2}$, separately fit to individual storage aging tests at 20°C, 40°C, 60°C, and 50%,100% SOC [18]. We refer to these individual fits as local models, as they each represent just one aging

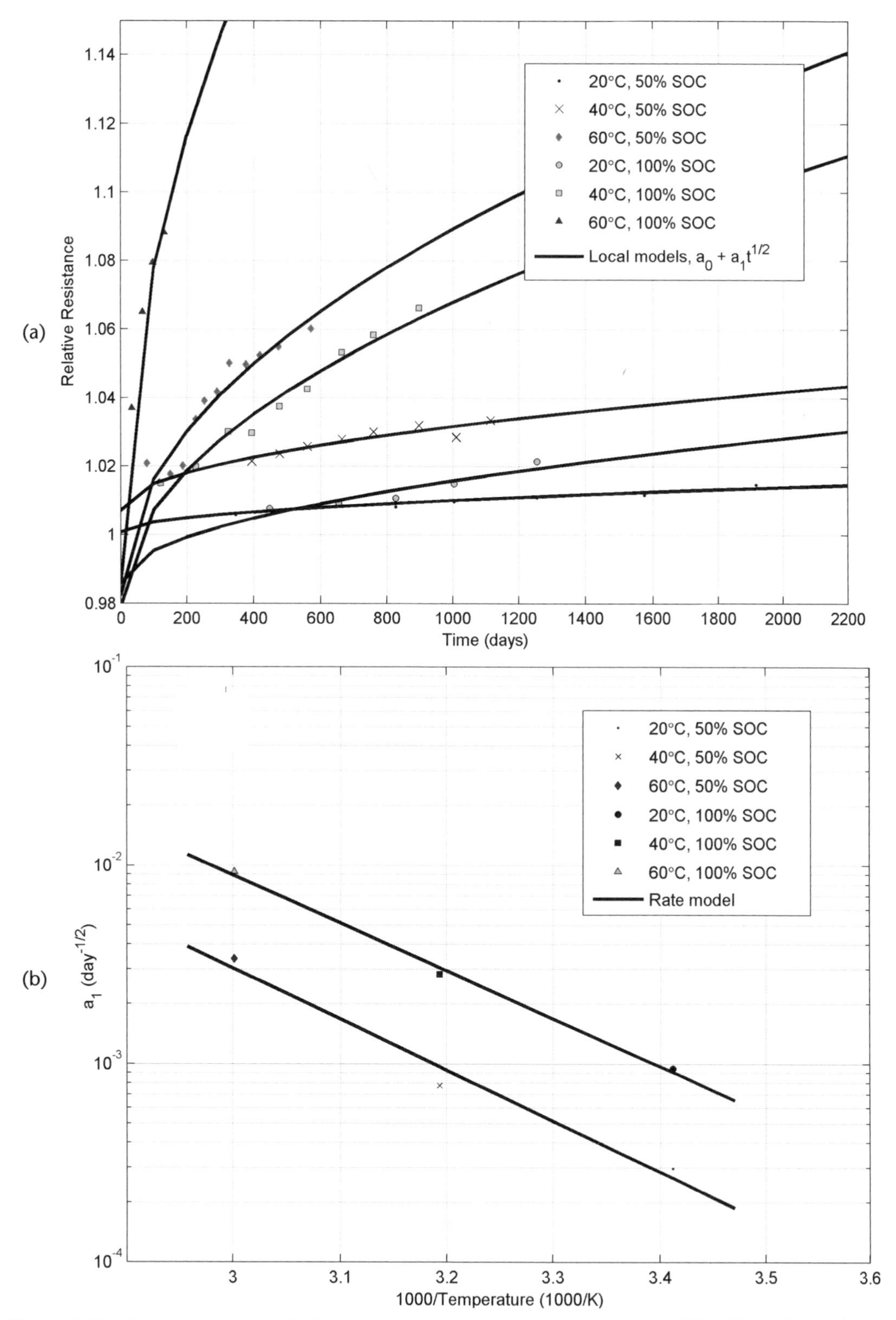

Figure 7.13 Resistance growth during storage at multiple temperatures and S_{OC}s [18]. (a) Local models with separate a_0 and a_1 coefficients fit to each data set. (b) Rate model for a_1 dependence on T and V_{oc}. (Source: NREL/Smith [11].)

condition. The coefficient a_0 takes on slightly different values due to manufacturing variations across the six cells, with apparent ± 2% resistance differences at BOL. In contrast, the degradation rate a_1 varies by more than an order of magnitude depending upon the temperature and SOC. Values of a_1 for the six local models are plotted in Figure 7.13(b) versus the inverse of temperature. The storage terms of the rate law, (7.25), with $\Delta DOD_i = 0$, can be fitted to match these six aging conditions, with best fits provided by $E_{a,a1}$ =70,700 J mol^{-1} K^{-1} and α_{a1}=0.062. Not shown, the coupling of (7.25) with the cycling ΔDOD term is weak, suggesting $\beta_{a_1} = 1$ is reasonable for (7.25). Instead, cycling dependent resistance growth is captured using the a_2 term of the resistance growth model equation (7.21).

Figure 7.14(a) shows values of a_2 resulting from local model fits with equations $R = 1 + a_1(T,V_{oc})\, t_{life}^{1/2} + a_2 t^{life}$, for aging conditions at various DODs, $EOCV$s, and cycles per day [19]. In fitting the rate law equation (7.26), we estimate the time history of open-circuit voltage, $V_{oc}(t)$, based on $SOC(t)$, as it varies greatly across these cycling conditions. In particular, the cells experiencing 1 cycle per day dwell at high SOCs (equivalently, high values of V_{oc}) for long periods of time during their daily cycling. In contrast, the four cycles per day test cases do not dwell for long at high SOCs. The fitted rate-law for a_2 is shown as the solid line in Figure 7.14(b), with values of E_{a,a_2} = 35,000 J mol^{-1} K^{-1}, $\alpha_{a_2} = 0.056$, and $\beta_{a_2} = 2.17$ in (7.26). Figure 7.14(b) compares the final global model for resistance growth, using (7.21), (7.25), and (7.26), to all aging conditions.

Models for capacity fade may be regressed in a similar manner to that described for resistance growth. Figure 7.15 shows the results of a global model for capacity fade. The 1 cycle/day cases are controlled by calendar fade/Li loss equation (7.22), while most of the 4 cycle/day cases are controlled by cycling fade/site loss equation (7.23). Fitted values of rate law coefficients to reproduce this data set [19, 22] are E_{ab_1} =35,000 J mol^{-1} K^{-1}, $\alpha_{b_1} = 0.051$, $\beta_{b_1} = 0.99$ and E_{a,c_2} = 35,000 J mol^{-1} K^{-1}, α_{c_2} = 0.023, and $\beta_{c_2} = 2.61$.

Lastly, reference degradation rates are matched to NCA Li-ion batteries designed for automotive application and tested under automotive cycling conditions [11]. This includes HEV-specific cycling presented by Belt [23] and PHEV-specific cycling by Gaillac [24]. Reference values representing those data sets well are a_0 = 1.0, $b_0 = 1.04$, $c_0 = 1.0$, $a_{1,ref} = 1.123 \times 10^{-3}$ day$^{-1/2}$, $a_{2,ref} = 1.967 \times 10^{-4}$ day^{-1}, $b_{1,ref}$ = 1.794×10^{-4} day$^{-1/2}$, and $c_{1,ref} = 1.074 \times 10^{-4}$ day^{-1}.

7.7.3 Prediction of Battery Temperature in Vehicle

Ambient conditions of a geographic region will strongly influence the average lifetime temperature of any battery that lives outside, such as a vehicle battery. Average lifetime temperature is perhaps the most important factor in determining battery life. The thermal network model shown in Figure 7.16 captures the impact of ambient temperature and solar radiation on the vehicle battery. Table 7.3 provides typical parameters for the model, representing a Toyota Prius passenger cabin and battery located in the car's hatchback trunk. Solar radiation on a vehicle parked in full sun can increase the lifetime average battery temperature by 1°C to 3°C relative to a vehicle experiencing no solar radiation. This number varies with the solar intensity of the geographic location and battery thermal mass. Compared to

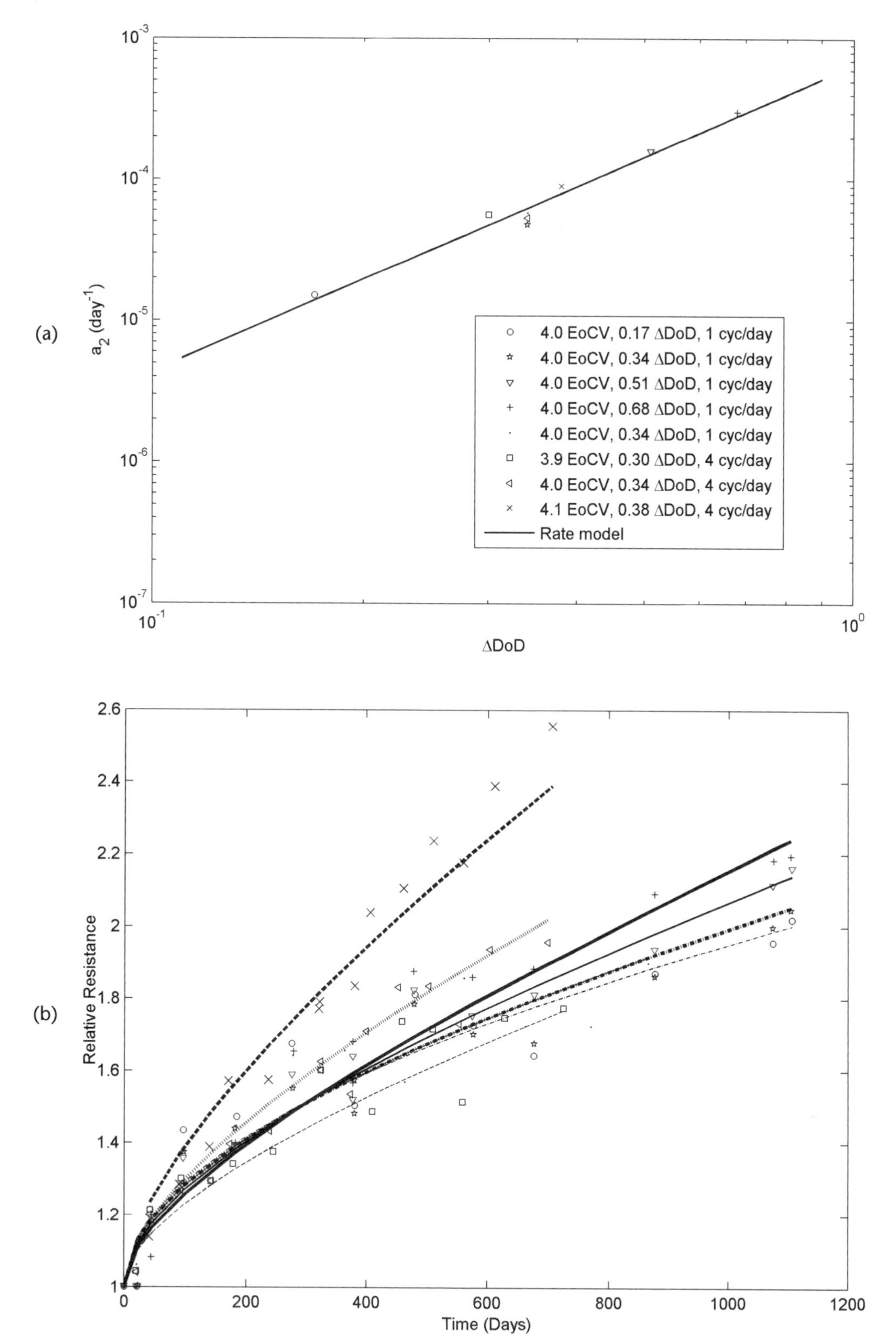

Figure 7.14 Resistance growth at 20°C under multiple cycling conditions [19]. (a) Rate model for a_2 dependence on V_{oc} and ΔDOD. (b) Global model for resistance growth compared to entire data set. (Source: NREL/Smith [11].)

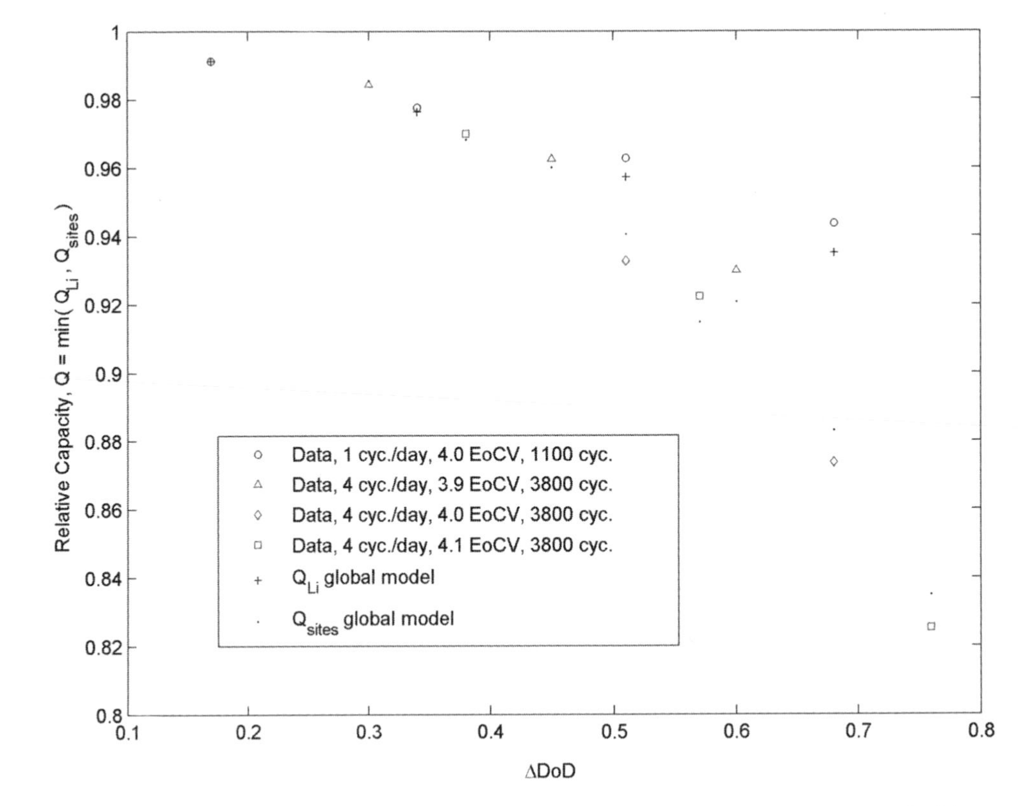

Figure 7.15 Global model for capacity fade versus data from [19]. (Source: NREL/Smith [11].)

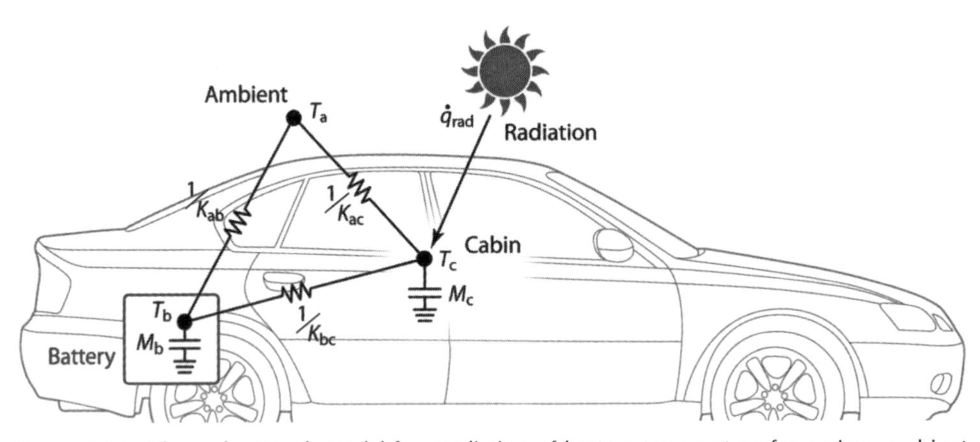

Figure 7.16 Thermal network model for prediction of battery temperature for a given ambient environment. (Source: NREL/Smith; artist: Dean Armstrong [20].)

large BEV batteries, small HEV batteries will experience larger temperature swings during the daytime due to their smaller thermal mass.

By simulating the vehicle thermal network model using ambient temperature and solar radiation data for different geographic locations, one can predict battery temperature and lifetime due to passive heat transfer. Figure 7.17 shows capacity fade for PHEV batteries in two climates. For the hot, Phoenix, AZ environment, solar radiation can reduce battery lifetime by 20% if the car is continuously parked

Table 7.3 Thermal Network Model Parameters for Toyota Prius

	HEV	*PHEV10*	*PHEV40*	*BEV75*
	*NiMH**	*Li-ion*†	*Li-ion*†	*Li-ion*†
M_b (J/K)	35,600	42,970	146,590	182,000
K_{ab} (W/K)	0.6498	0.4641	1.049	4.343
K_{ac} (W/K)	22.6	22.6	22.6	22.6
K_{bc} (W/K)	0.4663	0.3331	0.7527	3.468
M_c (J/K)	101,800	101,800	101,800	101,800
εA_c (m^2)	0.77	0.77	0.77	0.77

From: [20, 21].

*Nickel metal hydride battery. Parameters fit to test data from a 2005 Toyota Prius HEV.

†Parameters adjusted to account for larger thermal mass and surface area of PHEV and BEV battery packs.

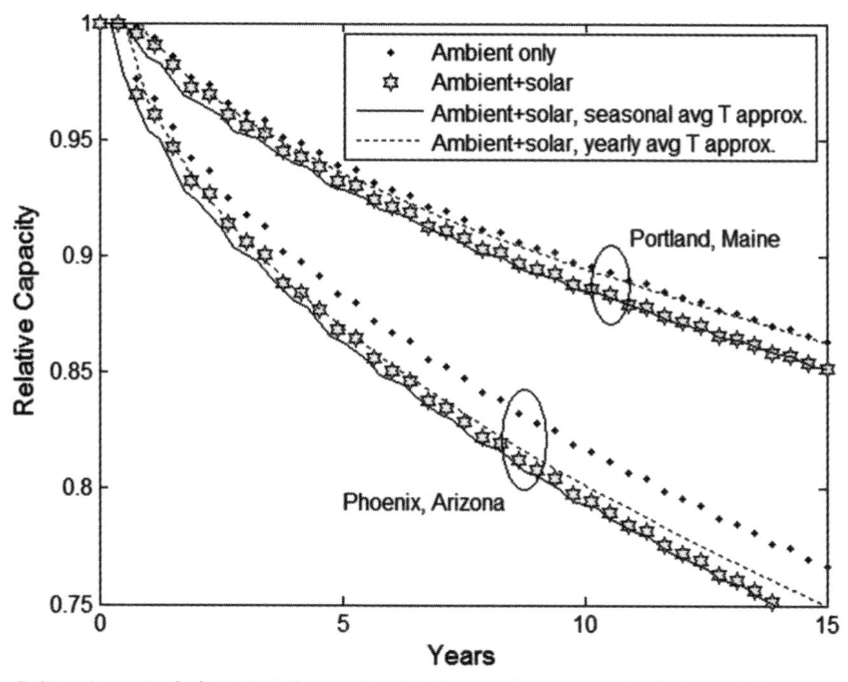

Figure 7.17 Capacity fade in U.S. hot and cold climates for a parked vehicle with battery at 90% SOC. (Source: NREL/Smith [20].)

in the sun. The figure also shows simplifying assumptions for approximating temperature fluctuations of a given ambient environment. A good approximation of the passive thermal environment is to only consider four ambient temperatures throughout the year, each representing the average temperature of winter, spring, summer, or fall. In the automotive industry, it is common to conduct accelerated four-season aging tests in this manner, where the test article is subjected to cycling at four levels of ambient temperature.

Adding to the passive thermal environment model, the impact of different active thermal management strategies was previously shown in Figure 7.5. These simulations augment the passive model by adding the effects of battery heat generation due to cycling and cooling due to active thermal management. This is accomplished by introducing a term for heat generation minus cooling to the battery temperature node, T_b, in Figure 7.16. The results given in Figure 7.5 are for Phoenix, AZ climate with the PHEV40 driven 12,000 miles/year. In this hot environment, a thermal management strategy that uses a refrigeration system to condition chilled liquid to cool the battery can extend battery life by 40% compared to using outside ambient air to cool the battery. Compared to oversizing the battery to achieve the same lifetime, this chilled-liquid thermal management can cost up to $500 more than the oversized air-cooled battery and still pay for itself, assuming $300/kWh battery cost.

7.7.4 Control Trade-Offs Versus Lifetime

In addition to determining thermal system strategy and requirements, additional questions for the systems designer include how much excess energy the battery must be sized with and how the battery should be best controlled within whatever degrees of freedom are acceptable to the end user. Typical expectations for automotive battery lifetime are on the order of 10 years and 100k to 150k miles with 20% to 30% capacity loss.

Here, we present a simple example for a PHEV with 40 miles of all-electric range. If the vehicle consumes 285 Wh/mile on average, then 11.4 kWh of useable energy are required to propel the PHEV for 40 miles. Assumptions follow:

- This simplified life analysis assumes all driving occurs in charge depletion mode, the worst case.
- The typical U.S. driver travels approximately 12,500 miles per year. With this annual mileage, the battery would undergo, on average, 12.5k miles/year divided by 365 days per year, divided by 40 miles/day, or 0.856 cycles per day.
- Time spent at high SOC shortens battery life. For the SOC time profile, we assume the vehicle starts its day with the battery at full charge, $SOC = SOC_{max}$. One 20-mile trip is taken at 8 a.m., depleting the battery to $SOC = SOC_{max} - \Delta DOD/2$. A second 20-mile trip is taken at 5 p.m., depleting the battery to $SOC = SOC_{max} - \Delta DOD$. For the nominal case, the battery is recharged between 10 p.m. and midnight to $SOC = SOC_{max}$, where it remains until the next morning's drive.
- We set a minimum allowable SOC limit of 10% due to the inability of the battery to provide sufficient power required for acceleration below this level. EOL is reached either when remaining total capacity drops below 70% or the battery can no longer provide 70% of the PHEV40's nameplate 11.4-kWh useable energy without dropping below 10% SOC.

Figure 7.18 shows capacity degradation versus time for the PHEV40 operated to several different ΔDODs, under the assumptions of 30°C average lifetime temperature, 95% SOC_{max}, and 12.5k miles/year. At 60% ΔDOD or 67% excess energy, the battery lasts 10 years. At 50% ΔDOD or 100% excess energy, the battery lasts 10.7 years, but costs 20% more due to the additional excess energy required by this more conservative design.

These results are highly specific to the use pattern. For NCA and other Li-ion chemistries, battery life is dependent on:

- Temperature;
- Maximum *SOC* (or equivalently *EOCV*);
- Time spent at combined high *SOC* and high temperature;
- *DOD* and number of cycles.

Figure 7.19 presents a sensitivity analysis of each of these factors on lifetime. The nominal case is 30°C, 95% SOC_{max}, 60% ΔDOD and charging time delay of 4 hours. Perturbing the nominal 30°C temperature by +5°C and –5°C changes the nominal 10-year life by –2.2 years and +3.2 years, respectively. For the moderate to warm temperature cases investigated here for the PHEV40, life is controlled by calendar limitations—(7.22) rather than cycling limitations—(7.23). The calendar life model equation (7.22) has weak sensitivity to *DOD*, but strong sensitivity to all other factors included in Figure 7.19. Given that the cost, mass, and volume of the battery are all inversely proportional to the usable *DOD*, it makes sense to maximize usable *DOD* as much as possible. For the PHEV40, this implies a design should use as much of the battery as possible at *BOL*, and compensate for

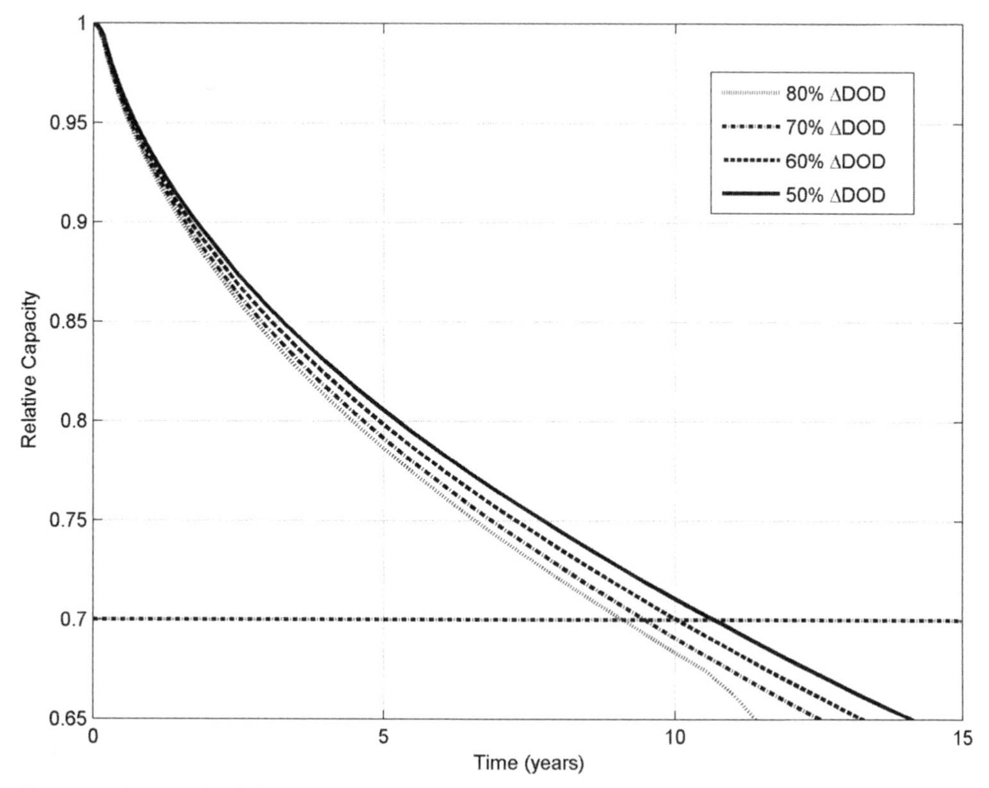

Figure 7.18 Depth of discharge impact on PHEV40 battery life assuming 30°C, 95% SOC_{max}, and 12.5k miles/yea-r.

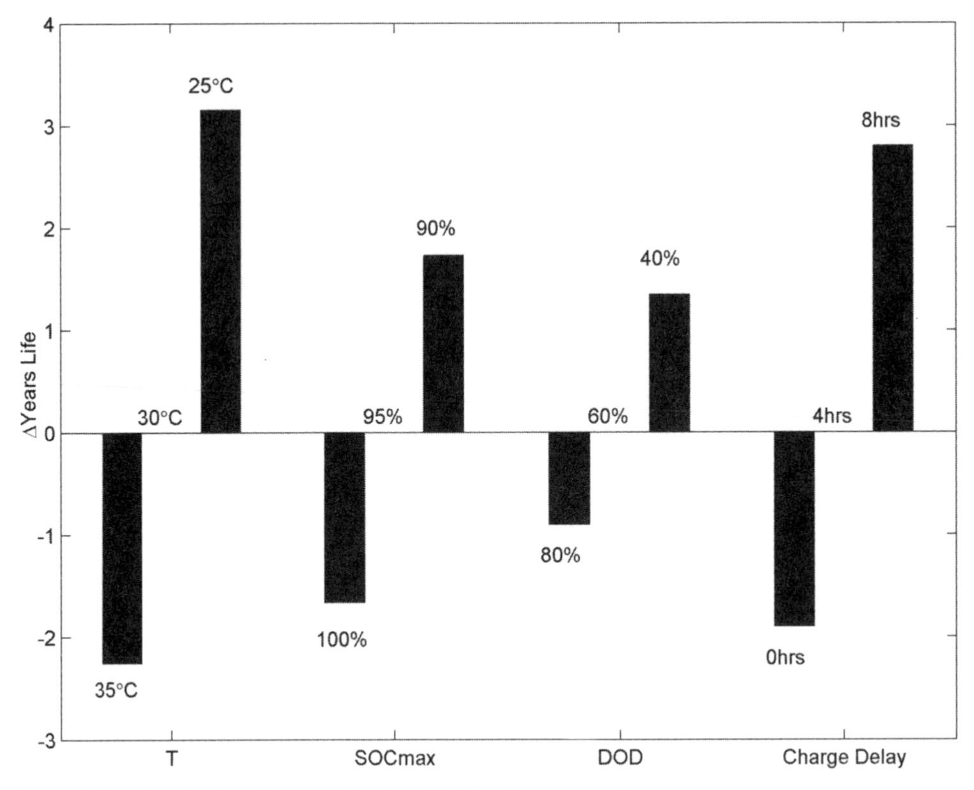

Figure 7.19 Sensitivity of PHEV40 battery life to aging assumptions.

this degradation by employing a combination of effective thermal management, reduction of maximum SOC, and delayed charging, all of which show substantial lifetime benefits in figure 7.19. In particular, delayed charging (or limiting the charge rate) costs almost nothing to implement and is very effective in extending calendar life. Limiting maximum SOC also costs nothing to implement, however it limits the usable energy of the battery. An effective life control strategy will increase SOC_{max} during winter months to compensate for reduced range due to sluggish cold temperature performance. The strategy might also gradually increase SOC_{max} throughout lifetime to regain a portion of the lost usable capacity.

As a final note, the PHEV40 design explored here generally experiences no more than one deep discharge and charge cycle per day. This is because the average U.S. driver travels only 33 miles in a single day. Calendar life controls the overall system life. For HEV and PHEV10 designs, many more cycles per day may be encountered and usable DOD may need to be restricted compared to this example to ensure that cycle life does not limit the overall system life.

7.8 Case Study 2: Behind-the-Meter Peak-Shaving of a Large Utility Customer

In this case study we investigate the application of energy storage for demand charge reduction (i.e., peak shaving) for a large electricity customer with a large amount of on-site PV power generation. The end user's goal is to reduce their monthly utility bills as much as possible in the presence of a budget constraint. Therefore, our objective is first to identify a set of system properties that best meets the customer's goal, then to define a technical specification that will drive the detailed design and validation of the cell, modules, and overall system.

7.8.1 End User Needs and Constraints

The customer in this case is NREL, a research laboratory that studies and develops efficient, renewable energy technologies including wind and solar power as well as energy storage and a myriad of other topics. More than 2,000 staff work at its 327-acre Golden, CO campus [25]. The large campus and nature of its work offer many viable locations for installing an energy storage system; thus, physical size and interconnection requirements are not a primary concern. As a stationary system, mechanical robustness to shock and vibration are also of little concern. Safety will need to be addressed, as the system may be installed in close proximity to working employees. Thermal must also be addressed, as the system may be installed outdoors and Golden is subject to high levels of solar irradiance, which can result in high battery temperatures affecting safety and degradation. However, of primary concern are the system electrical requirements, which must be derived from a techno-economic analysis to meet the end user's primary objective: cost savings.

7.8.2 End User Load Profile and Rate Structure

True to its name, NREL "walks the talk" with multiple LEED-rated buildings and 4,400 kW of onsite PV generation [26]. The shape of the campus load profile is thusly dominated by the PV power profile. Figures 7.20 and 7.21 show the campus demand for 30 and 7 days, respectively. Here it can be seen that peaks in the demand history are most often due to the intermittency of PV power. As these peaks are relatively short in duration, the load profile appears a good fit for the use of storage as a demand charge mitigation tool—large reductions in peak power can be had with a relatively small amount of energy storage.

We will assume the utility rate structure described in Table 7.4 applies, which offers opportunity for substantial savings from demand charge reduction (as discussed in Chapter 6). Note that this rate structure includes no time-of-day dependence for demand charges or electricity costs. Rather, demand charges vary only by season, and electricity costs are constant year round at $0.0473/kWh. As such there is no added value in cycling the battery for energy shifting once monthly peak loads have been minimized, simplifying control. Further note that in comparison to more aggressive demand charge structures (see, for example, those of San Diego Gas & Electric, or of Southern California Edison as referenced in Chapter 6), these demand charges offer much less opportunity for savings. Thus, we can expect similar combinations of storage and facilities to offer better economics in select markets.

7.8.2.1 Preliminary Techno-Economic Sizing Analysis

With the load profile and utility rate structure in hand, we can proceed with a coarse sizing analysis to estimate the battery power and energy level that is most

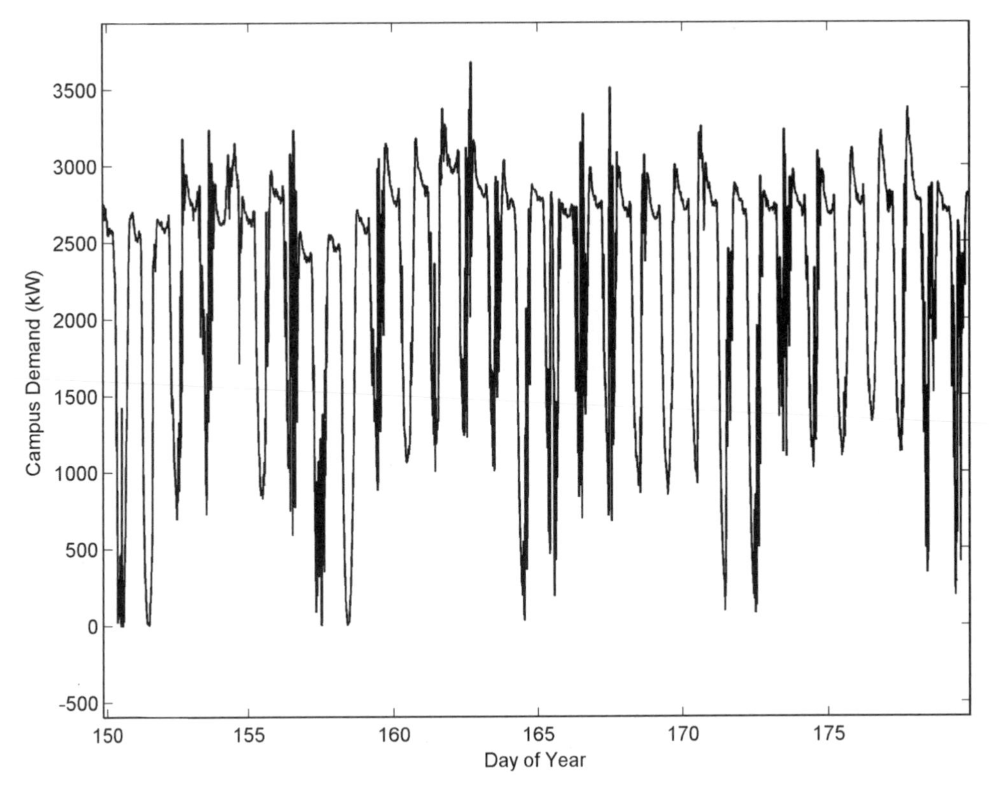

Figure 7.20 Thirty days of 15-minute load data for the NREL campus.

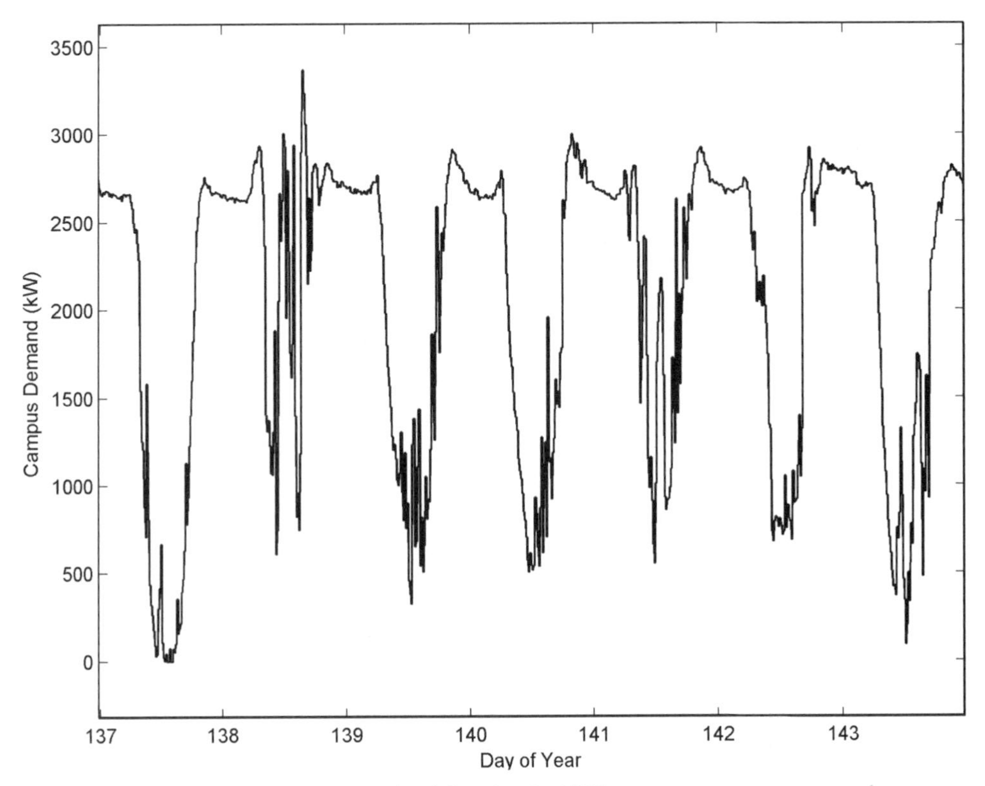

Figure 7.21 Seven days of 15-minute load data for the NREL campus.

Table 7.4 Demand Charges from XCEL's SG Rate Structure

Charge	*Time*	*Cost*	*Units*
Facilities-related demand charge	Summer (June–September)	15.80	$/kW
	Winter (October–May)	12.84	$/kW

From: [27].

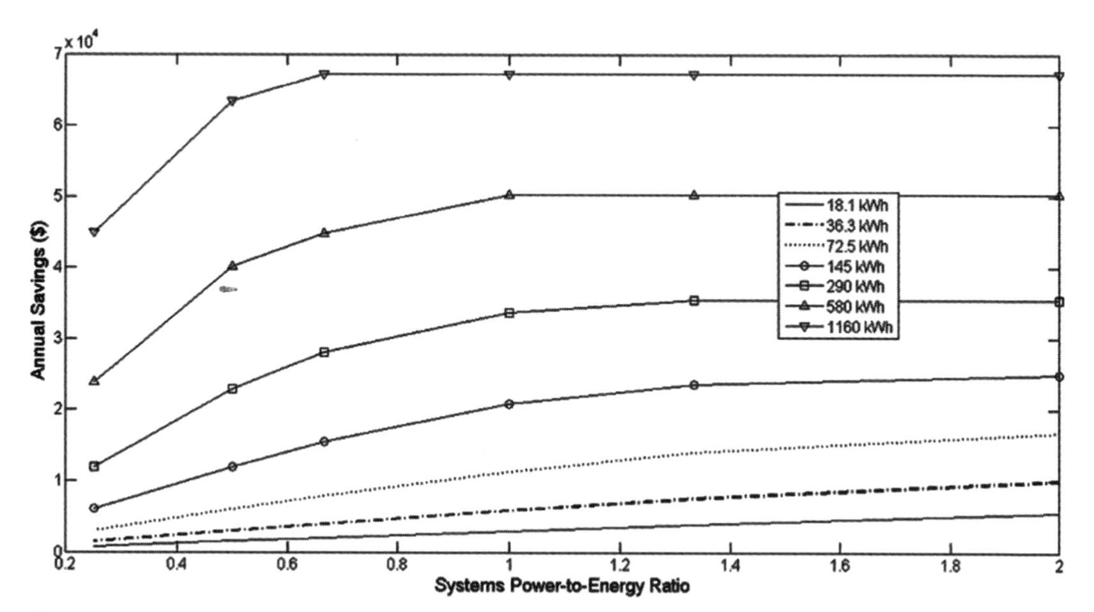

Figure 7.22 Annual savings as a function of system energy and power-to-energy ratio.

economically advantageous for this customer. For this we apply NREL's Battery Lifetime Analysis and Simulation Tool for Behind-the-Meter Lite (BLAST-BTM Lite). This tool is designed specifically for such tasks: it applies a simplified model of energy storage (kWh accounting) in combination with an optimal peak shaving controller to a user-supplied load profile and utility rate structure, then computes the electrical response and cost savings of a specified energy storage unit.

What BLAST-BTM Lite lacks in fidelity, it gains in computational efficiency, making it perfect for a preliminary investigation of a broad range of storage options as is needed here. With this tool we can quickly simulate 42 different energy storage units spanning seven different energy levels and six different power-to-energy ratios. Note that in such full-factorial studies, it is often best to employ a power-to-energy ratio instead of specifying both battery power and energy levels directly. The latter approach may result in unrealistic systems when a broad scope of parameters is sought (e.g., systems requiring unrealistically high C-rates).

Annual savings for the simulated systems are shown in Figure 7.22. Clearly, higher energy and higher power systems lead to greater annual savings. However, increasing energy and power both require increased capital investment in the battery and inverter, respectively. Assuming that total system cost scales approximately with the energy and power levels of these two components at a rate of $500/kWh and $500/kW (additive), we can estimate a simple payback period (estimated total system cost divided by the calculated annual savings) for each system as shown in Figure 7.23.

While this approach includes many approximations, and certainly a more rigorous consideration of savings and costs will be required, it provides a telling looking into the interplay of system power and energy for sizing such a system. Clearly, the best return on investment (minimum payback period) occurs when small amounts of energy are installed and the power-to-energy ratio is high. However, this also results in relatively small annual savings, which may not justify the project given the reality of certain flat or fixed costs that do not scale strongly with energy or power (e.g., system design and analysis or permitting). On the other hand, as energy levels are increased, return on investment falls even after optimizing the power-to-energy ratio (Figure 7.24).

A better approach where a budget constraint is in play may be to view annual savings against total system cost. Applied with a budget of $110,000, we find that a 72.5-kWh system with a 2:1 power to energy ratio appears near optimal for maximizing annual revenues at $16,000/yr and an annual return on investment of ~14%.

7.8.2.2 Electrical Analysis

Before inspecting the electrical response of this battery in any meaningful way, it is important to make our simulation more realistic. In particular, we seek to improve the nature of our electrical model, thermal model, and control model as discussed below. To implement these changes, we switch our analysis tool from BLAST-BTM Lite to the full BLAST-BTM software.

Electrical model. We upgrade the kWh accounting model to a zero-order equivalent circuit model. At the moment, higher-order dynamics are still not required, but we do wish to include the relation of *OCV* to *SOC* and add a resistive element that is sensitive to both *SOC* and temperature. In this manner we can apply minimum cell voltage limits to observe the impact of thermal response on available energy and calculate the amount of heat generated by operating the battery.

Thermal model. We upgrade from complete omission of temperature effects to a lumped capacitance thermal model of the system consisting of battery, inverter, container, soil, and ambient nodes parameterized representative of a standard 20' containerized storage unit. Heat dissipated by both the battery and inverter is assumed to be dumped to the container (which is thermally connected to ambient). Combined with the effects of solar irradiation, the container temperature can thereby greatly exceed ambient temperature—an important effect to account for. We specify a max battery temperature constraint of 50°C so that the battery will not respond to charge or discharge commands when it exceeds this level. Typical meteorological year data for Golden, CO is employed for ambient temperature and solar irradiance histories [28].

Control model. In the preceding analysis we have assumed the ability to perfectly forecast future load to the end of each month. It is unlikely this can be achieved in practice, and reducing the accuracy and/or time horizon of this forecast will significantly affect cycling behavior and resultant savings. To represent more realistic forecasts, we will assume a 48-hour forecast is computed once per day, and that the forecast for each 15-minute interval load has a random error in the

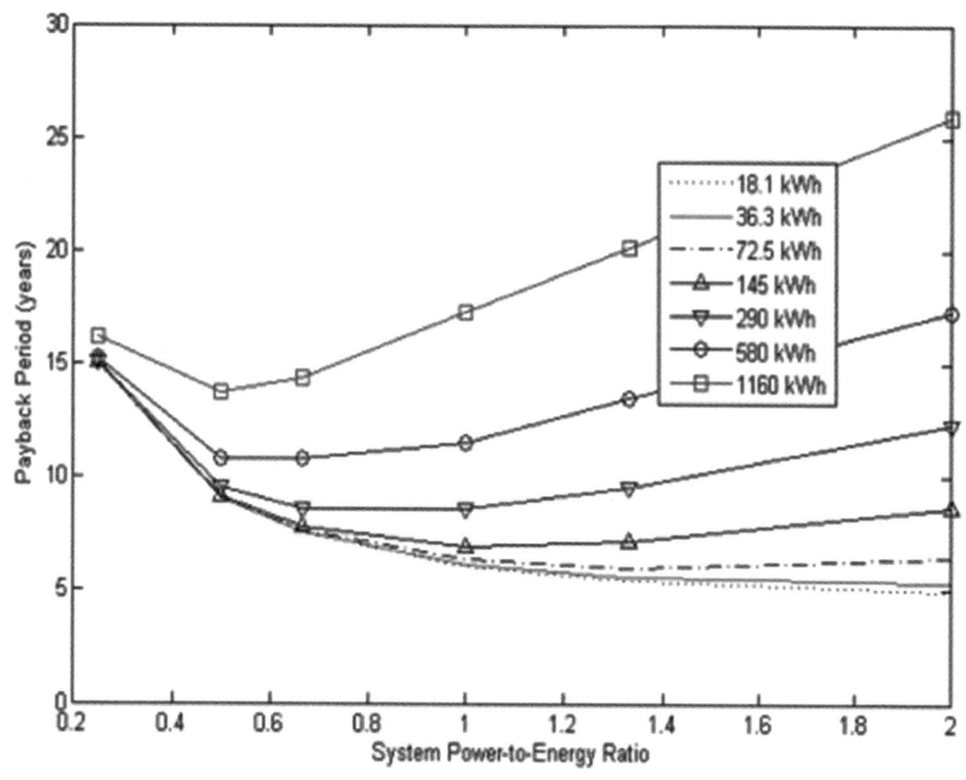

Figure 7.23 Simple payback period as a function of system energy and power-to-energy ratio.

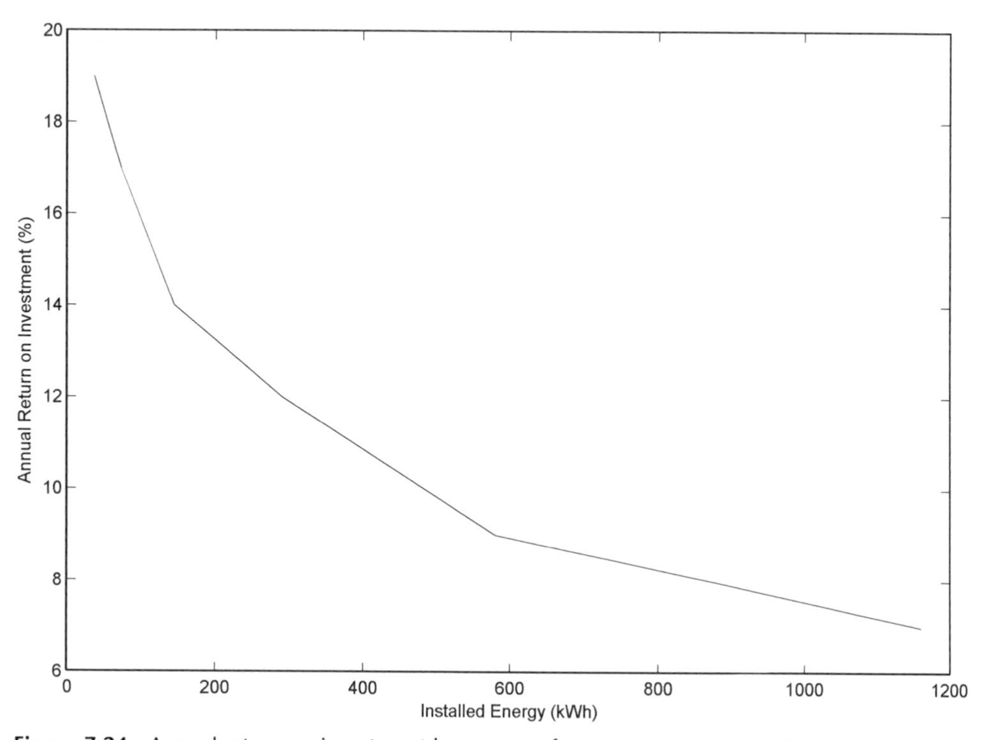

Figure 7.24 Annual return on investment by energy, after power-to-energy ratio optimization.

range of +/–65 kW (+/–2.5% of the average annual load). Our previous simulations also assumed that the anticipated battery response perfectly matched the actual response by utilizing identical models for control and simulation. Herein we will keep the simple kWh accounting method for computing peak shaving load targets, a lower fidelity model than the equivalent circuit model employed for simulation. In this manner, we begin to address the difference in anticipated and actual battery response.

Auxilliary loads. An auxiliary load of 300W is assumed herein to represent control and management functions. We also include the ability to account for heating and cooling loads when present.

As our forecast error element is random in nature, we must run the simulation multiple times for the same system. Resultant value is found to fluctuate between $12,500 and $14,500 per year, averaging $13,600 per year. This represents a decrease of ~15% of the value predicted by our more simplistic simulations. Further inspection reveals that almost all of this deduction is due to the presence of forecast error.

As shown in Figure 7.25, the electrical duty cycle is affected as well. Under our perfect forecast assumption, the battery duty cycle is dominated by a small number of consistently deep discharges to our target *EOD SOC* (20%) over the course of the year. However, in our more realistic simulation, cycling frequency increases as does the variation in depth of discharge. Sometimes the algorithm will overpredict loads, sets an overly conservative load target, and thus result in a small peak load reduction and smaller battery *DOD* than targeted. At other times it underpredicts loads, sets an overly aggressive load target, then discharges the battery past the target *SOC* (defined as the minimum *SOC* that we would prefer that the battery reaches on any given day) in an attempt to achieve that load target. On occasion, this results in the battery discharging to 0% *SOC* and the metered load exceeding the load target. Figure 7.26 shows an example of this behavior, which can result in increased cost to the user if it sets the max load for the month.

7.8.2.3 Thermal Analysis

Over the course of a year, battery temperature is predicted to range from –18°C to +47°C, averaging to 11°C. Ambient, container, and battery temperatures are shown in Figures 7.27 and 7.28 for the maximum and minimum temperature days, respectively. Several important factors are evident from these figures. First, we see that the container temperature can rise drastically above ambient due to its exposure to solar irradiance. Second, battery temperature is only minimally affected by the container temperature fluctuation as the battery has significantly more thermal mass than the rest of the container, and the container has a strong thermal connection to ambient. Additionally, we see that both the battery and container temperature are strongly affected by heat generation of the battery and inverter when the system is charging or discharging (which drives the maximum battery temperature in Figure 7.27).

This predicted maximum temperature may be close to or above the operational range of some Li-ion cells, leaving little to no margin for cases where atypical hot spells occur where the campus loads induce more aggressive battery cycling or where heat generation is increased in later years from resistance growth of the

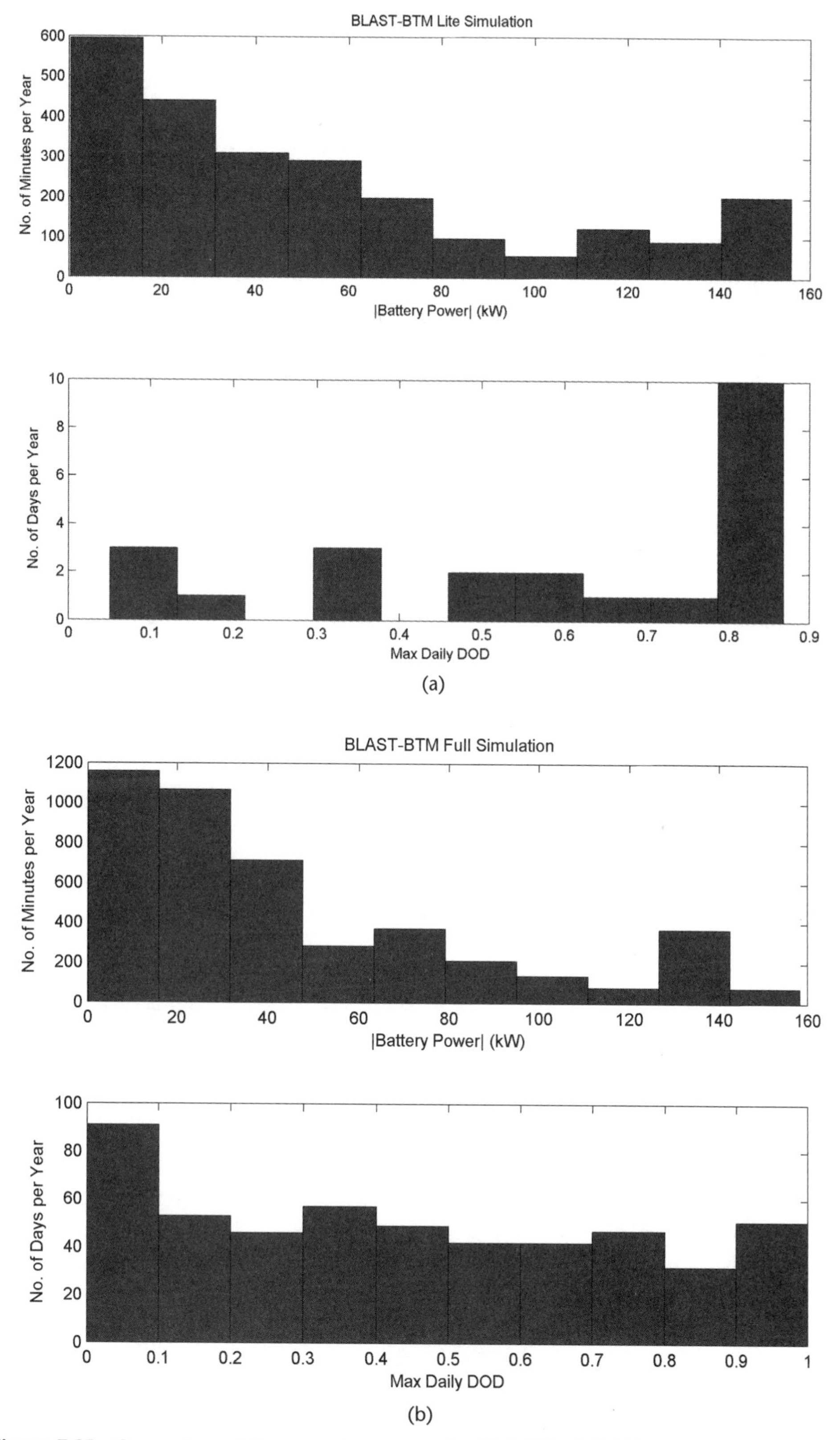

Figure 7.25 Comparison of the annual response of a 72.5 kWh, 145 kW energy storage system minimizing demand charge on the NREL campus in two different simulations.

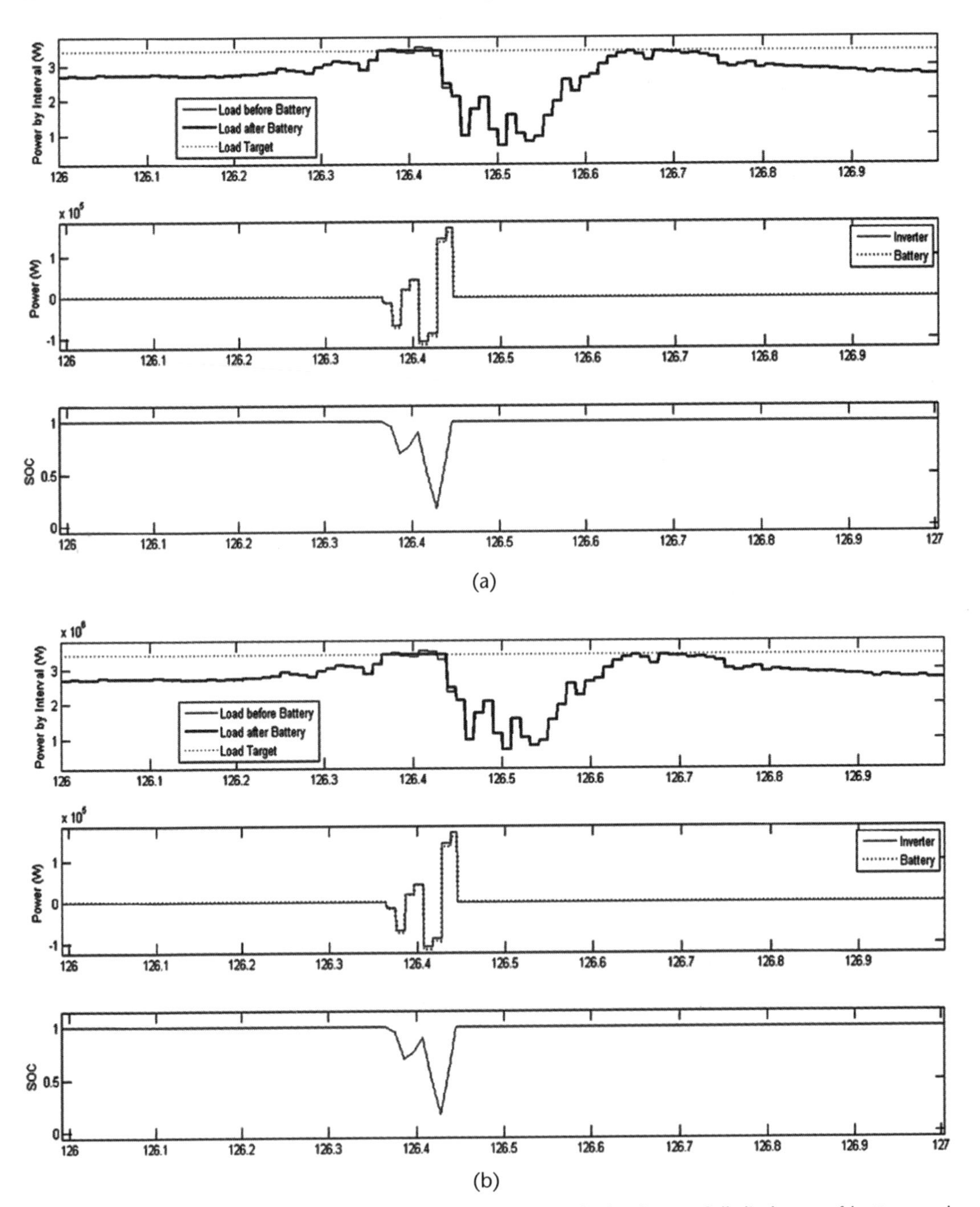

Figure 7.26 Example of algorithm underpredicting facility loads, leading to full discharge of battery and increase in load target.

battery. Similarly, while we may have seen limited impact of minimal temperatures on performance herein, these same factors could amplify the impact of temperature under conditions we have not yet assessed. For these reasons, evaluation of various thermal management strategies should be considered.

Having noted that container temperatures are strongly correlated with solar irradiance, we first investigate locating the system in a shaded area. We simulate this

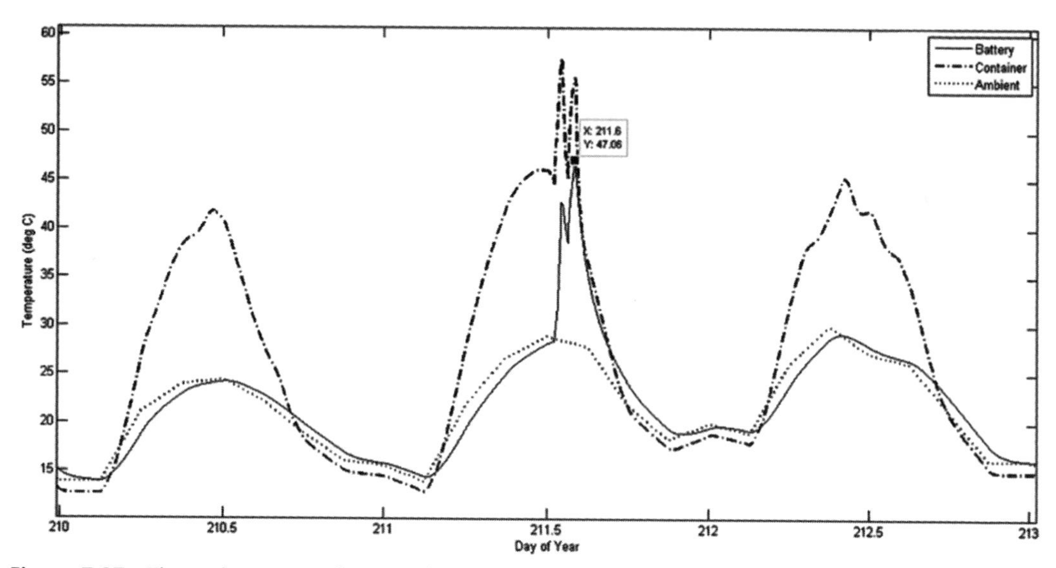

Figure 7.27 Thermal response of system: hot, no battery thermal management.

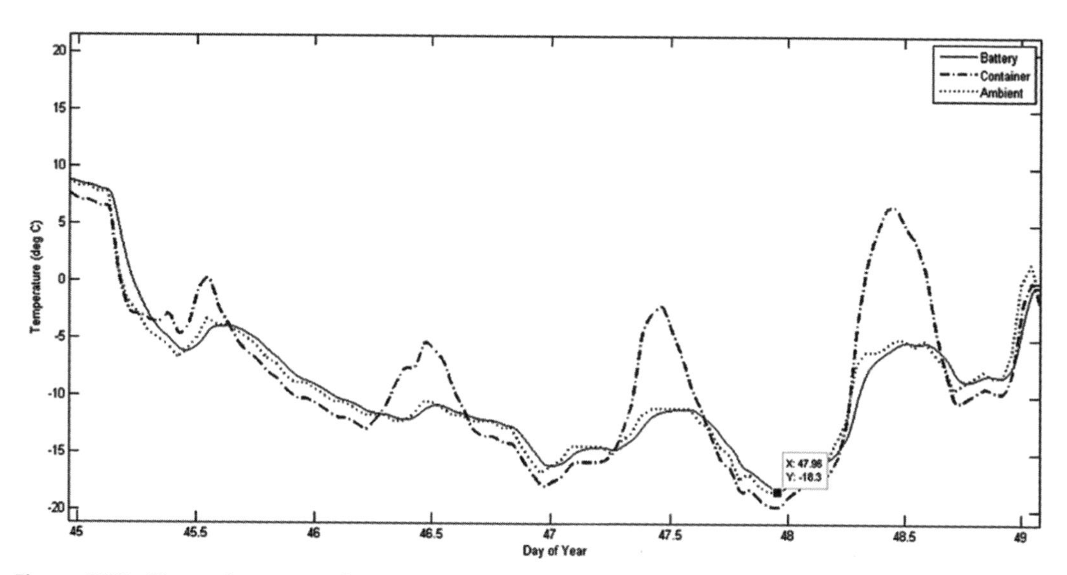

Figure 7.28 Thermal response of system: cold, no battery thermal management.

by reducing the solar irradiance by 50% (allowing for diffuse solar radiation). Such an approach will incur little to no extra cost where shaded sites are available, such as a parking structure or under existing solar panels. We find, however, that while this approach has a strong effect on container temperatures, it has little impact on maximum or average battery temperature. As discussed previously, heat generation by the battery and inverter is the primary driver of maximum temperature.

Similarly, we can explore the addition of a cooling system to the container such as a conventional air conditioning unit. This approach may suffer similar issues as the shading approach—without a stronger thermal connection between the container and battery (that may require a large increase in volume), cooling the container may not have a large effect on battery temperatures. To investigate, we simulate a container air conditioning system with the ability to 10 kW of heat from the container and reject it to the environment with a coefficient of performance of 2.5. Set to activate when container temperatures exceed 30°C, we find that average and maximum battery temperatures fall by approximately 2°C and 4°C, respectively. Ultimately, battery temperature continues to closely track ambient temperature when in standby, and heat generation continues to drive peak temperatures when operational.

Alternatively, we can investigate use of a dedicated battery cooling system rather than a container cooling system. Such a system may employ liquid cooling paths direct to the cells and a refrigerant-based system for removing heat from said liquid, as is employed on advanced vehicles like the Chevrolet Volt [29, 30]. For our system we will assume only half of the total heat rejection capability as with the container air conditioning unit, but otherwise operationally similar—5 kW of heat rejection at a coefficient of performance of 2.5, active at and above 30°C. Removing this heat directly from the battery results in a much larger 10°C reduction in maximum battery temperature to less than 37°C. An example high-temperature thermal response of this system is shown in Figure 7.29. Note that even when the system is actively charging and discharging—as evidenced by the spike in container temperature late in day 184 driven by inverter heat dissipation—the cooling system is capable of constraining the battery temperature to less than 33°C.

In addition to superior control of battery temperature, the direct battery cooling system consumes half of the energy of the container cooling system. However,

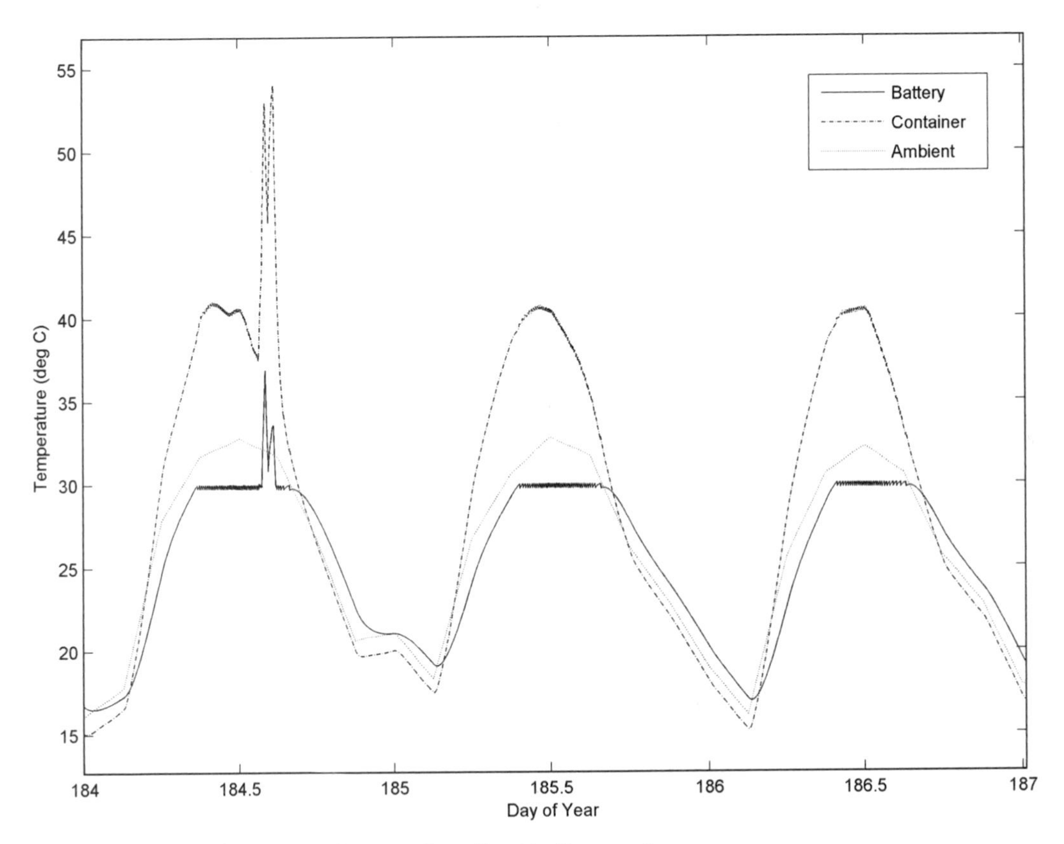

Figure 7.29 Thermal response of system: hot, direct battery cooling.

other components installed in the container may require temperature control as well (inverter, computerized controls, etc.), thus container temperatures cannot be wholly ignored. For this purpose, however, simply shading the container may be sufficient. Accordingly, we will elect to install the unit in a shaded location with a direct battery cooling system.

7.8.2.4 Degradation Analysis

Now we proceed to analyzing the long-term performance of the system. We will now extend the duration of our simulations to 10 years and activate the life model [4] included in BLAST-BTM. This model will calculate the change in battery resistance and capacity at the end of each day, considering the current, voltage, *SOC*, and temperature history that has transpired. As the battery ages, this will result in less energy available for reducing peak loads, and thus utility bill savings may decline. Resistance increases will also amplify the heat generated by the system and possibly raise maximum temperatures. There are three key factors to adjust that can affect these behaviors (see Table 7.5).

7.8.3 Baseline

We begin with a baseline scenario where we employ a 100% maximum *SOC*, 20% target *SOC*, and cooling system active at and above 30°C. As seen in Figure 7.30, 10 years of operation of such a system is calculated to result in ~17% resistance growth and ~22% capacity loss. Note that we also see the degradation rates are sensitive to ambient temperature, as they are higher in the summer than the winter.

Annual savings (Figure 7.32) fluctuate but show no strong trend year to year. Recall that we expect savings to decline. The lack of a clear downward trend implies that the effect of our applied random forecast error overwhelms the economic effects of aging in this scenario.

Temperature data in Figure 7.31 reveals that average temperatures are constant year to year. This is due to the fact that average temperatures are a strong function of ambient conditions rather than operational ones. Maximum temperatures, on the other hand, are strongly coupled to how the battery is operated, coinciding with the occurrence of high-rate, large *DOD* discharges late in the afternoon of a hot summer day. While we do see some trend of increasing temperature year after

Table 7.5 Key Factors to Adjust Degradation

Control Parameters	
Maximum *SOC*	Reducing the maximum *SOC* can reduce battery degradation, but also reduces the amount of energy available for peak shaving.
Target *SOC*	Raising the target *SOC* will reduce the annual battery throughput, maximum *DOD*, and the amount of energy available for peak shaving, and increase the margin for poor load forecasting.
Thermal Parameters	
Cooling thermostat	Lowering the temperature at which the cooling system is active will decrease battery temperature, and thereby battery degradation, but will increase auxiliary loads.

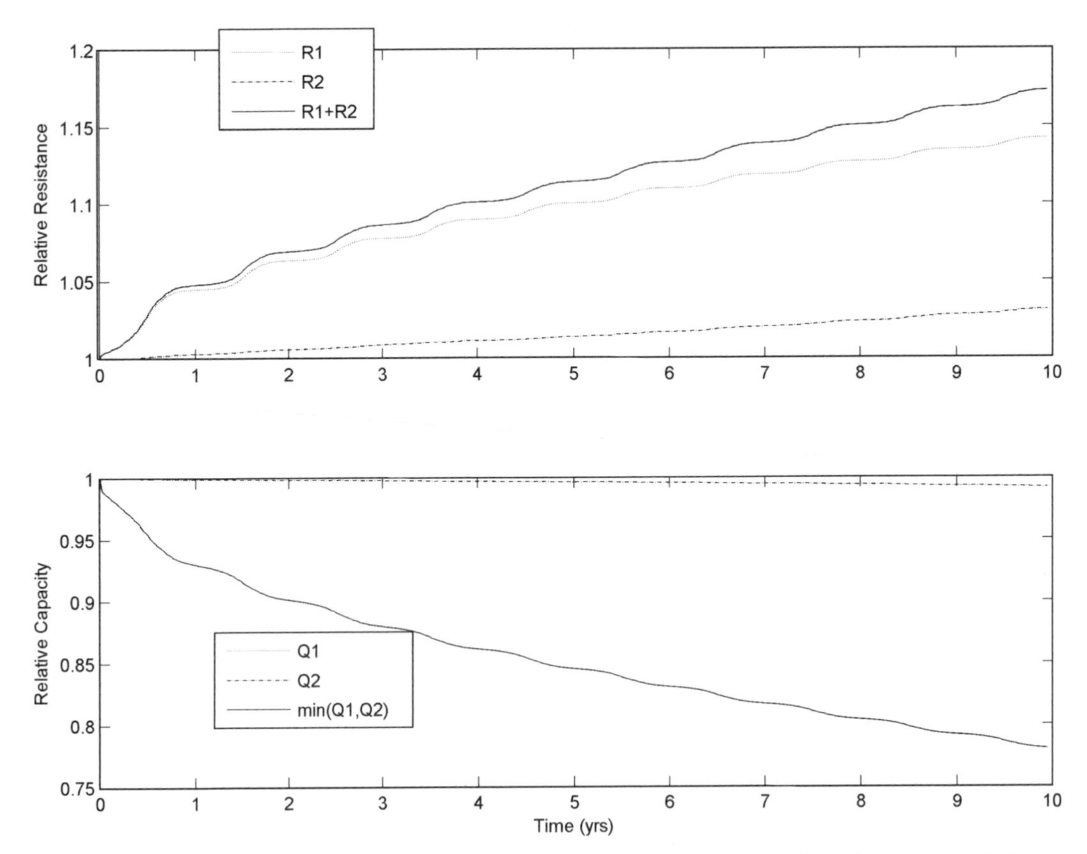

Figure 7.30 Baseline resistance growth and capacity fade. R1 and R2 represent the resistance growth due to calendar and cycling effects, respectively. Q1 and Q2 represent the capacity fade due to calendar and cycling effects, respectively.

year due to the increase in battery resistance, the random forecast error affects this parameter as well.

7.8.4 Increased Cooling

In this case we reduce the temperature at which the cooling system becomes active from 30°C to 20°C. We see that while peak temperatures are reduced slightly later in life (Figure 7.33), effects on both capacity fade and resistance growth are negligible. This analysis suggests that, for our assumed chemistry, increasing battery cooling beyond that necessary to constrain peak temperatures to operationally acceptable limits may not significantly benefit longevity. While energy costs associated with powering the more aggressive cooling system are increased, ultimately changes in average savings are within the noise of our random forecast error effect.

7.8.5 Reduced Target *SOC*

Here we reduce the target *SOC* of the peak shaving algorithm from 20% to 10%. This effectively increases the energy available for peak shaving, which could result

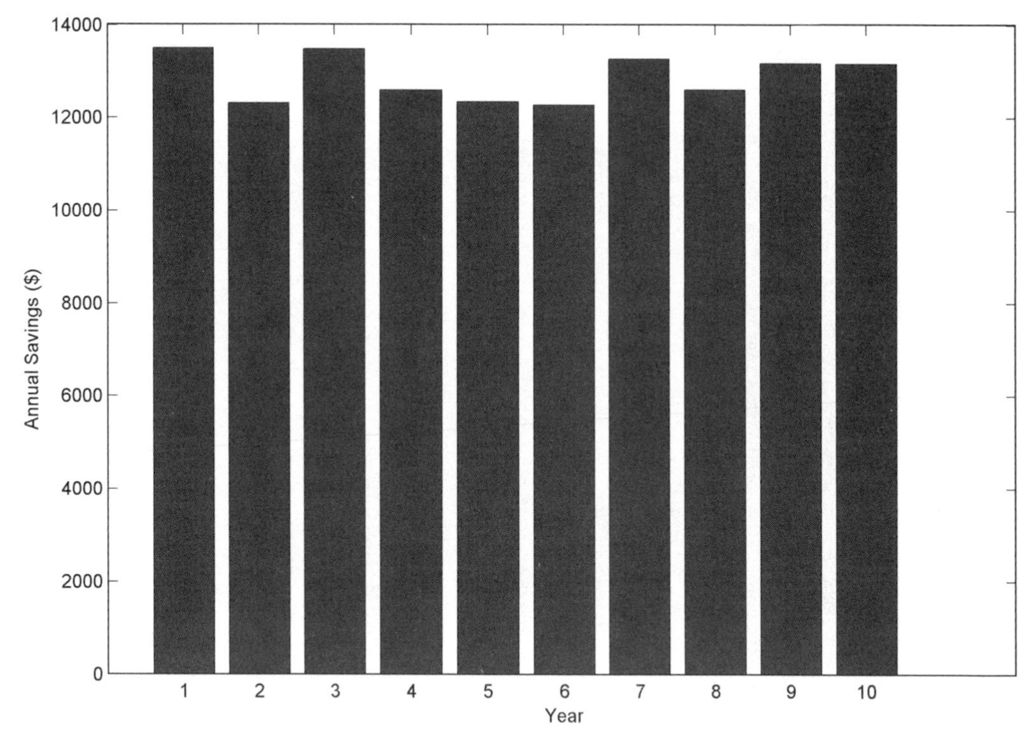

Figure 7.31 Baseline annual savings.

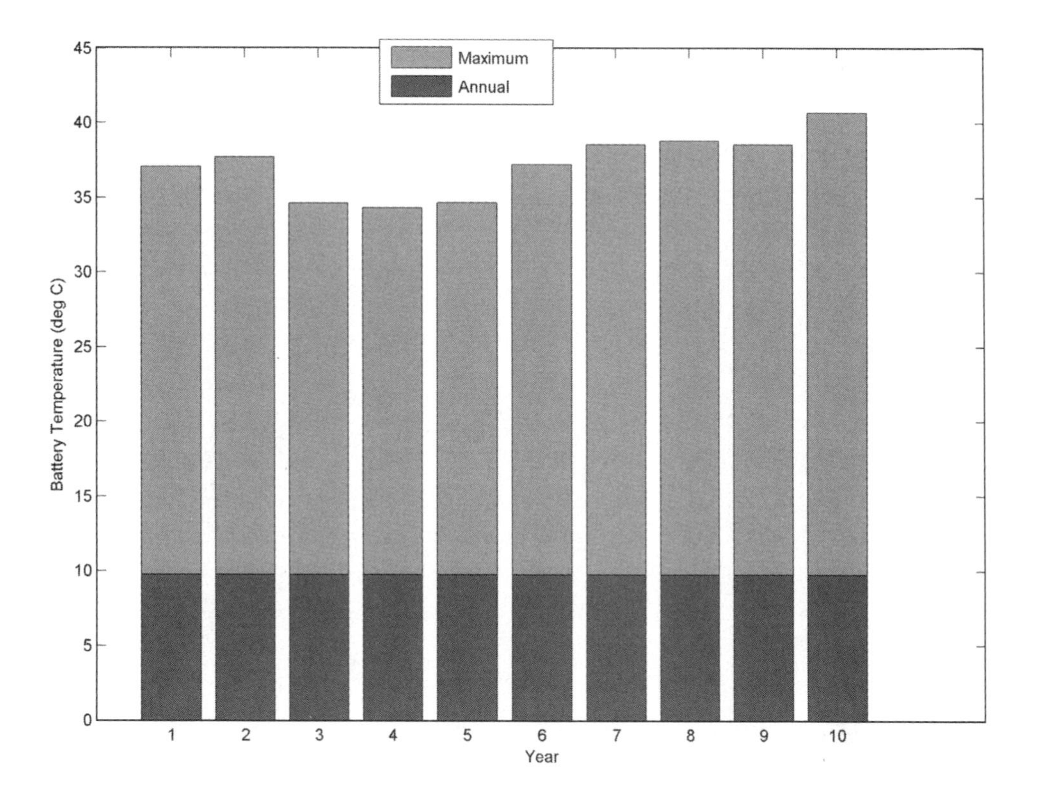

Figure 7.32 Baseline battery temperatures.

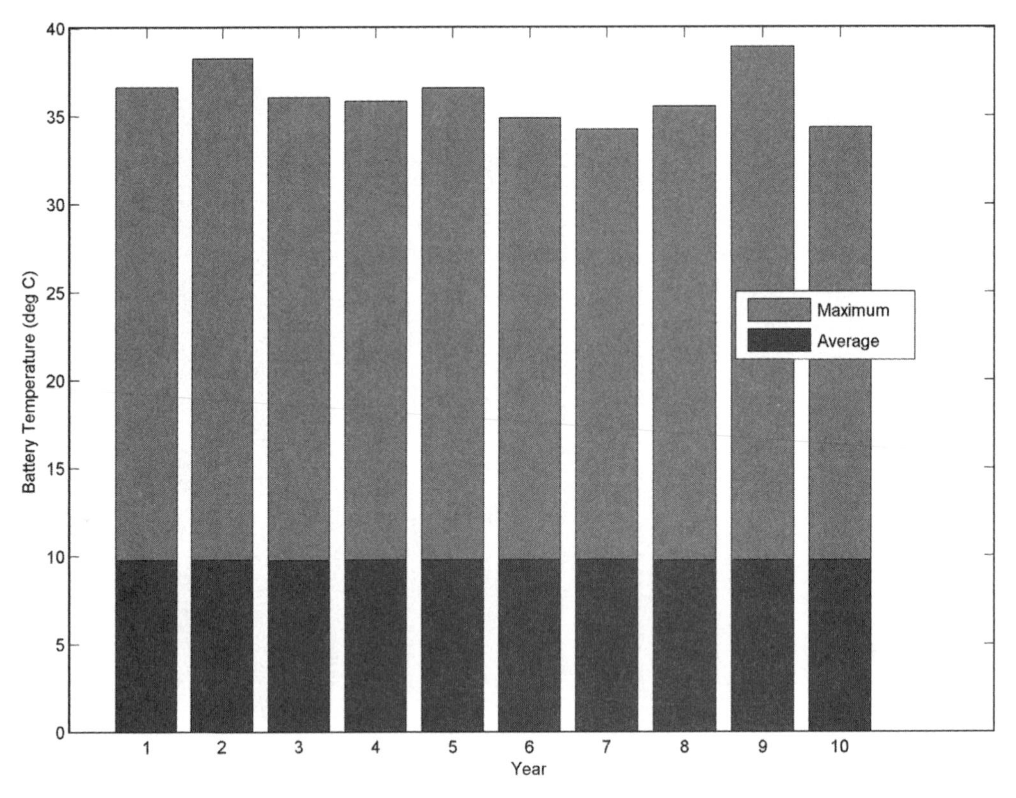

Figure 7.33 Battery temperatures for increased cooling scenario.

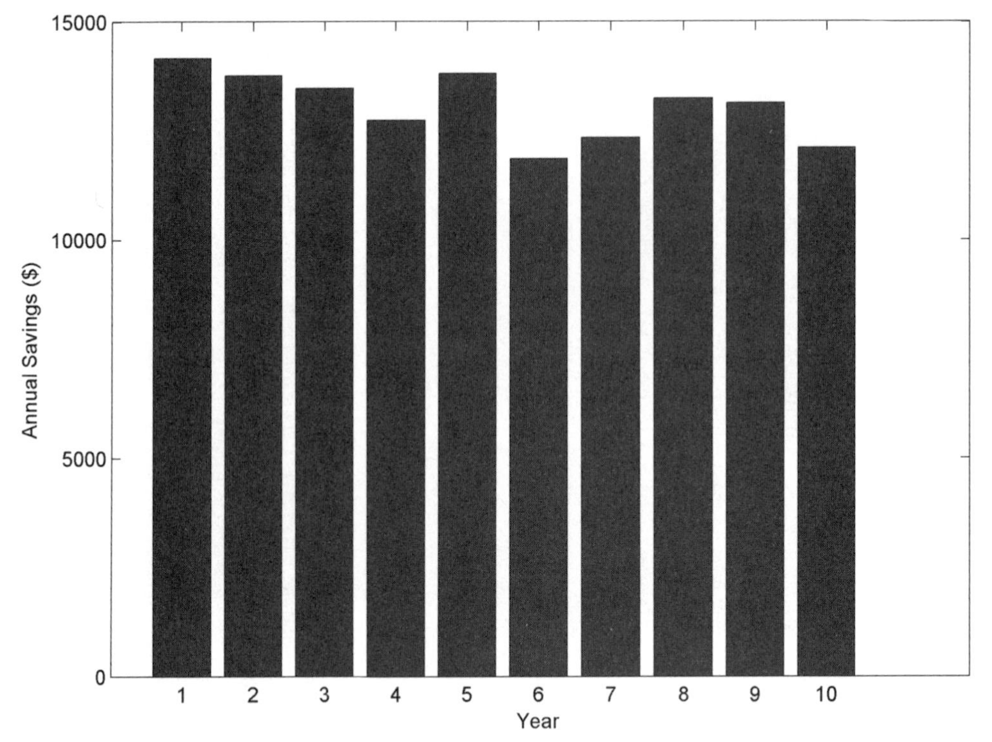

Figure 7.34 Annual savings for reduced target *SOC* scenario.

in increased savings. It also, however, decreases the margin for error associated with poor load forecasting and may affect the duty cycle in such a way as to increase max temperatures and degradation. Results of this simulation again show negligible impact on thermal response and wear. Further, annual savings (Figure 7.34) suggest that the decrease in target *SOC* has only a minor effect on total value, again on the scale of variation induced by our random forecast error.

7.8.6 Decreased Maximum *SOC*

Finally, we reduce the maximum *SOC* used for charging the battery from 100% to 90%. This directly decreases the energy available for peak shaving, which could result in decreased savings but will also decrease degradation rates. As seen in Figure 7.35, this strategy delivers as expected in terms of battery degradation: resistance change is decreased from +17% to +12%, and capacity change is decreased from –22% to –17%. This is due to the fact that the battery spends a large amount of time in the fully charged state awaiting an infrequent discharge order. Reducing the maximum *SOC* directly affects the wear induced during these times. However, there does appear to be a decrease in annual savings as well (Figure 7.36).

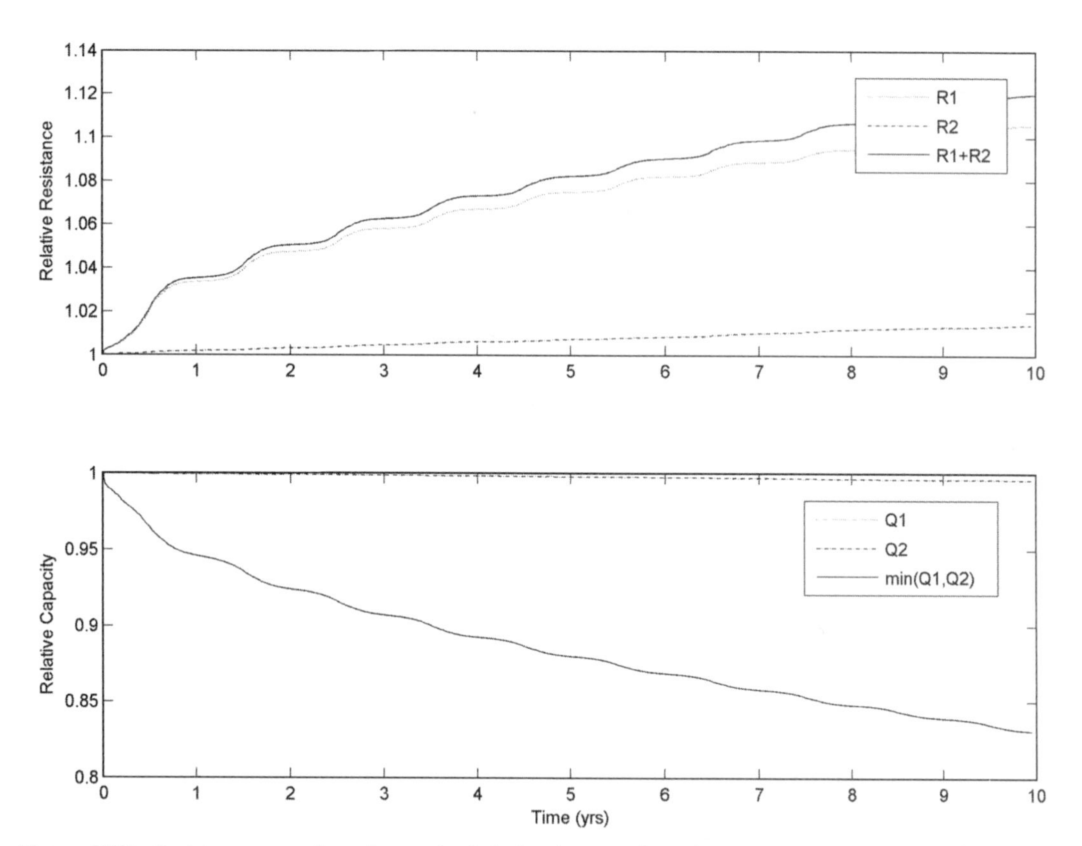

Figure 7.35 Resistance growth and capacity fade for decreased maximum *SOC* scenario. R1 and R2 represent the resistance growth due to calendar and cycling effects, respectively. Q1 and Q2 represent the capacity fade due to calendar and cycling effects, respectively.

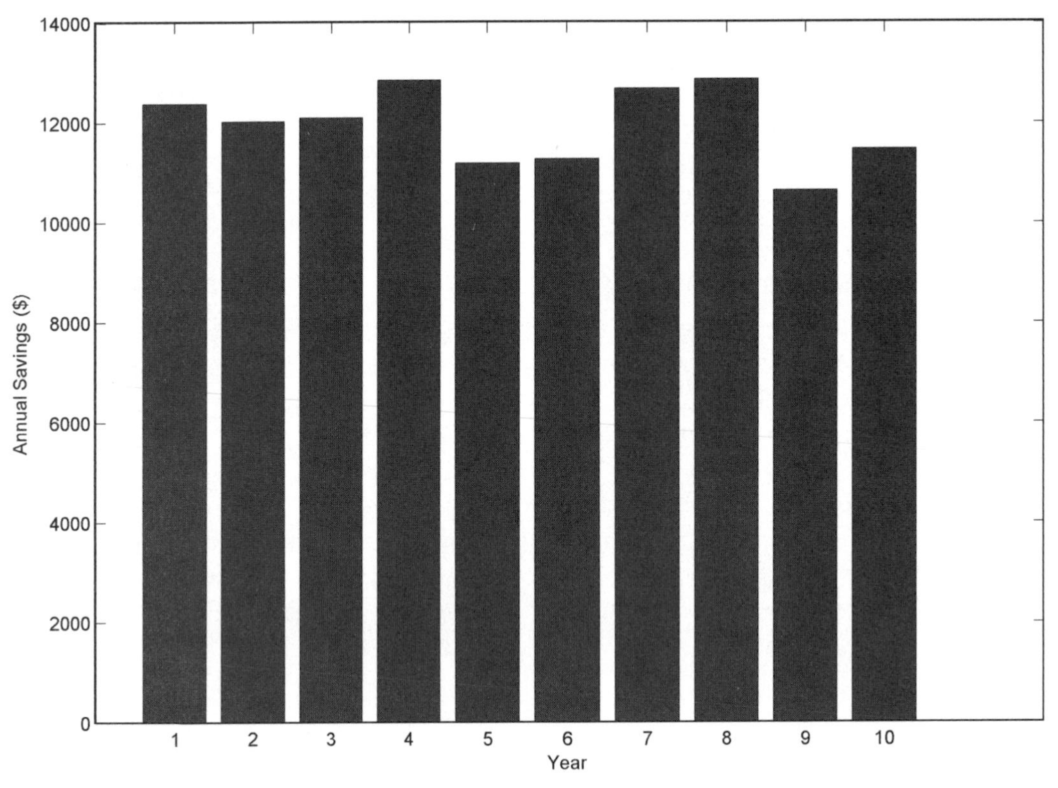

Figure 7.36 Annual savings for decreased maximum *SOC* scenario.

7.9 System Specification

The preceding system-level, long-duration analyses has begun to quantify the trades possible with a large energy storage system for behind-the-meter demand charge management. We have identified energy and power levels that offer the best annual savings within our allotted budget. We have also identified that an active direct-to-battery cooling system will be required to constrain maximum temperatures to acceptable levels. With respect to degradation, we have found that calendar fade at high *SOC* is the predominant factor in battery wear, as the battery spends a large amount of time resting in a fully charged state. While we have seen that reducing the maximum allowed *SOC* can offer gains in this area, more aggressive use of the cooling system does not.

Unfortunately, we have also seen that uncertainty in campus load forecasting affects the economics of the system more strongly than many of these factors. Running a large number of simulations at each condition and analyzing the results statistically is strongly recommended to address such factors. Barring such additional analysis, this factor encourages being conservative in setting our target *SOC*.

The identified system architecture is readily summarized in Table 7.6. It is important to note that we specify the 72.5 kWh as available energy because several simulations predict that all of this energy will be utilized. It is also important that we reference *SOC* to this available energy figure as well, as this aligns with our control assumptions and methods.

Table 7.6 System Architecture

Electrical	
Total available energy	72.5 kWh
Total power (AC)	145 kW
Control	
Maximum *SOC*	100%
Target *SOC*	20%
Thermal Management	
Architecture	Shaded container with direct active battery cooling
Max heat rejection	5,000W
Active temperature range	≥30 deg C

Table 7.7 Battery, Cell, and Module Worst-Case Protocol for Subsequent Development Activities

Step	*Description*	*Purpose*
1	Soak in a 50°C environment with battery cooling system inactive at 100% *SOC* for 1 day	Demonstrate the battery can survive maximum container temperatures.
2	Soak in a 50°C environment with battery cooling system active at 100% *SOC* for 1 day	Demonstrate ability of cooling system to pull down battery to 30°C and hold.
3	Discharge battery at a constant power 30-minute rate to 0% *SOC*, then charge battery at full power to 100% *SOC* in a 50°C environment with battery cooling system active.	Demonstrate ability to deliver rated power and energy without exceeding max temperature.
4	Soak in a –20°C environment at 100% *SOC* for 1 day	Demonstrate the battery can survive minimum container temperatures.
5	Discharge battery at a constant power 30 minute rate to 0% *SOC*, then charge battery at the same rate to 100% *SOC* in a –20°C environment	Demonstrate ability to deliver rated power and energy in minimum temperature environment.

To continue with more detailed development activities, such as the design of cells, modules, thermal management systems, and so forth, we must now define more concise worst-case scenarios and duty cycles.

Inspection of simulation results can reveal such worst-case power requirements and thermal environments. For example, container temperature simulations reveal that container temperatures can span –20°C to +50°C. Electrical duty cycles vary substantially but show occasional occurrence of max power and full discharge operation. Given these results, Table 7.7 defines a reasonable protocol to use for simulation and testing in developing hardware.

References

[1] National Transportation Safety Board, Hazardous Materials Accident Brief, Report NTSB/HZB-05/01, September 26, 2005.

[2] Kim, G. -H., K. Smith, J. Ireland, A. Pesaran, and J. Neubauer, "Fail-Safe Designs for Large Capacity Battery Systems," U.S. Patent Application, Number 13/628,208, September 27, 2012.

[3] Neubauer, J., and E. Wood, "Accounting for the Variation of Driver Aggression in the Simulation of Conventional and Advanced Vehicles," SAE World Congress, Detroit, MI, April 16–18, 2013.

[4] Smith, K., T. Markel, G.-H. Kim, and A. Pesaran, "Design of electric drive vehicle batteries for long life and low cost," IEEE Accelerated Stress Testing and Reliability Workshop, October 6–8, 2010, Denver, CO.

[5] Smith, K., E. Wood, S. Santhanagopalan, G. -H. Kim, and A. Pesaran, "Advanced Models and Controls for Prediction and Extension of Battery Lifetime," Large Lithium Ion Battery Technology & Application Symposia, Advanced Automotive Battery Conference, February 4, 2014, Atlanta, GA.

[6] Andrea, D., *Battery Management Systems for Large Lithium Ion Battery Packs,* Norwood, MA: Artech House, 2010.

[7] Baumhöfer, T., M. Brühl, S. Rothgang, and D. U. Sauer, "Production Caused Variation in Capacity Aging Trend and Correlation to Initial Cell Performance," *J. Power Sources,* Vol. 247, 2014, pp. 332–338.

[8] Verbrugge, M., and B. Koch, "Generalized Recursive Algorithm for Adaptive Multiparameter Regression. Application to Lead Acid, Nickel Metal Hydride, and Lithium-Ion Batteries," *J. Electrochem. Soc.,* Vol. 153, No. 1, 2006, pp. A187–A201.

.[9] Plett, G. L., "Extended Kalman Filtering for Battery Management Systems of LiPB-Based HEV Battery Packs: Part 3. State and Parameter Estimation," *J. Power Sources,* Vol. 134, No. 2, 2004, pp. 277–292.

[10] Smith, K., C. D. Rahn, and C. Y. Wang, "Control oriented 1D Electrochemical Model Of Lithium Ion Battery," Energy Conversion and Management, Vol. 48, No. 9, 2007, pp. 2565–2578.

[11] Smith, K., "Electrochemical Control of Li-ion Batteries," *IEEE Control Syst. Mag.* Vol. 30, No. 2, April 2010, pp. 18–25.

[12] Moura, S. J., N. A. Chaturvedi, and M. Krsti , "Adaptive Partial Differential Equation Observer for Battery State-of-Charge/State-of-Health Estimation via an Electrochemical Model," *J. Dyn. Sys. Meas. Control,* Vol. 136, No. 1, 2013, 011015.

[13] Northrop, P. W. C., B. Suthar, V. Ramadesigan, S. Santhanagopalan, R. D. Braatz, and V. R. Subramanian, "Efficient Simulation and Reformulation of Lithium-Ion Battery Models for Enabling Electric Transportation," *J. Electrochem. Soc.* Vol. 161, No. 8, 2014, pp. E3149–E3157.

[14] Santhanagopalan, S., and R. E. White, "State of Charge Estimation using an Unscented Filter for High Power Lithium Ion Cells," *Int. J. Energ. Res.* Vol. 34, No. 2, 2010, pp. 152–163.

[15] Clarus Concept of Operations, Publication No. FHWA-JPO-05-072, Federal Highway Administration (FHWA), 2005, http://ntl.bts.gov/lib/jpodocs/repts_te/14158.htm.

[16] Wikipedia article, "ISO 26262," http://en.wikipedia.org/wiki/ISO_26262, accessed July 21, 2014, and International Standards Organization, "ISO 26262-1:2011 Road Vehicles—Functional Safety," http://www.iso.org/iso/catalogue_detail?csnumber=43464.

[17] Doughty, D. H., and C. C. Crafts, "FreedomCAR Electrical Energy Storage System Abuse Test Manual for Electric and Hybrid Electric Vehicle Application," Sandia National Laboratory, SAND 2005-3123, 2005.

[18] Broussely, M., "Aging of Li-Ion Batteries and Life Prediction, an Update," 3rd International Symposium on Large Lithium-Ion Battery Technology and Application, Long Beach, CA, May 2007.

[19] Hall, J., T. Lin, G. Brown, P. Biensan, and F. Bonhomme, "Decay Processes and Life Predictions for Lithium Ion Satellite Cells," 4th International Energy Conversion Engineering Conf., San Diego, CA, June 2006.

[20] Smith, K., M. Earleywine, E. Wood, J. Neubauer, and A. Pesaran, "Comparison of Plug-In Hybrid Electric Vehicle Battery Life Across Geographies and Drive Cycles," SAE Technical Paper 2012-01-0666, 2012.

[21] Wood, E., J. Neubauer, A.D. Brooker, J. Gonder, and K. Smith, "Variability of Battery Wear in Light Duty Plug-In Electric Vehicles Subject to Ambient Temperature, Battery Size, and Consumer Usage," International Battery, Hybrid and Fuel Cell Electric Vehicle Symposium 26 (EVS 26), Los Angeles, CA, May 6–9, 2012.

[22] Smart, M., K. Chin, K., L. Whitcanack, and B. Ratnakumar, "Storage Characteristics of Li-Ion Batteries," NASA Battery Workshop, Huntsville, AL, November 2006.

[23] Belt, J. R., "Long Term Combined Cycle and Calendar Life Testing," 214th Meeting of the Electrochemical Society, INL/CON-08-14920, October 13–16, 2008.

[24] Gaillac, L. A., "Accelerated Testing of Advanced Battery Technologies in PHEV Applications," International Battery, Hybrid and Fuel Cell Electric Vehicle Symposium 23 (EVS 23), Los Angeles, CA, December 2–5, 2007.

[25] http://www.nrel.gov/about/overview.html, accessed June 30, 2014.

[26] http://www.nrel.gov/sustainable_nrel/sustainable_buildings.html, accessed July 24, 2014.

[27] http://www.xcelenergy.com/staticfiles/xe/Regulatory/Regulatory%20PDFs/rates/CO/psco_elec_entire_tariff.pdf , accessed October 1, 2013.

[28] http://rredc.nrel.gov/solar/old_data/nsrdb/1991-2005/tmy3/, accessed July 10, 2014.

[29] Jayaraman, S., G. Anderson, S. Kauschik, and P. Klaus, "Modeling of Battery Pack Thermal System for a Plug-in Hybrid Electric Vehicle," SAE Technical Paper 2011-01-0666, http://dx.doi.org/10.4271/2011-01-0666, 2011.

[30] Buford, K., J. Williams, and M. Simonini, "Determining Most Energy Efficient Cooling Control Strategy of a Rechargeable Energy Storage System," SAE Technical Paper 2011-01-0893, http://dx.doi.org/10.4271/2011-01-0893, 2011.

[31] Franklin, G. F., J. D. Powell, and A. Emami-Naeini, *Feedback Control of Dynamic Systems,* Addison-Wesley: Reading, MA, 1994.

[32] Mohammadpour, J., and C. W. Sahere, *Control of Linear Parameter Varying Systems with Applications,* London: Springer, 2012.

CHAPTER 8

Conclusion

Batteries have come a long way from the hobbyist voltaic pile to energy storage for entire communities. So have the processing and analysis tools used in the industry. Computational power available to design engineers has increased by several orders of magnitude, enabling the virtual design of entire battery packs. The theoretical bases for computational battery design are now sufficiently mature; the software tools have evolved from academic software to standard practice in the industry. Battery makers today have access to virtual design of cells of different formats based on newer and emerging chemistries. Optimization, which has traditionally been carried out by extensive design of experiments, is now performed by a series of computational case studies. Battery assembly traditionally performed in chemical plants has now been automated over clean rooms and robotic assembly lines. The technoeconomic evaluations have in turn adopted such sophisticated models and tools to provide users the ability to base their investment decisions on the actual physics of the batteries. Most importantly, the battery community has grown from the traditional group of materials chemists and solid-state physicists to encompass a variety of disciplines from environmental enthusiasts to industrial engineers. This has resulted in a paradigm shift in battery design, from the traditional wisdom in building batteries for consumer electronics applications, to assembling large battery packs. As outlined in Chapters 4 through 7, the contributions from the different groups enable the expedition of the technological maturity and deployment of batteries applications that demand several MWh of energy storage.

For instance, simple thermal management solutions utilized in single-cell applications such as cell phones are no longer effective in a battery pack several kWh in size. In fact, to date, there is very little guidance on thermal management for these large batteries. The calorimetric studies outlined in this book were built over the years evaluating different pack designs and have been shown to be effective indicators of battery performance under different load conditions. What remains to be implemented is the design of cells and packs that maximize utilization for specific applications, accounting for efficient heat distribution. For instance, affordable tuning of coolant flow rates in accordance with the response of the battery to ambient temperatures, and adaptive heat dissipation devices are currently being introduced in the market. Such implementations can build on the thermal evaluation of batteries outlined in Chapter 3.

As demonstrated Chapter 5, the linear extension of circuit-breakers and fuses used to protect small cells is not always a reliable means to ensure sequestration of a large-format cell in a vehicle battery pack. Even when identical cells are used, the implications on battery safety are widely different depending on the application at hand. The question of how well safety at the single-cell level translates to safety of the battery pack is raised more often in applications involving high-energy and power demands. The safe operating window for large format-cells is set based on different sets of metrics adopted by different organizations. The need for a sound understanding of these limits and a uniform set of test conditions that meet safety assessment under a variety of scenarios is yet to be developed. In part, this gap arises from the rapidly evolving nature of the technology. As addressed earlier, the understanding of the underlying factors by the diverse crew involved in battery manufacturing and utility will significantly help address this challenge.

On-board electronics available for battery management in the market today are sufficiently robust that we believe the tools outlined in the book will help control engineers and thermal management designs to take advantage of the insight the offline analyses provide, in order to design more reliable fuel gauges and battery life estimators for battery packs. Smart grid design based on physical state estimators is not far-fetched from the algorithms outlined in Chapter 4. Not only will such tools enable more efficient utilization of existing battery packs, they can promote the planning and distribution of load requirements across the day of the week or time of the day based on these routines. Field data on such topics is still sparse given the relatively recent usage of large-format Li-ion batteries in these applications. However, as the vehicle and grid-scale applications mature, readers will build more confidence in the utility of detailed analyses similar to the case studies outlined in Chapter 7.

In fact, among the first of assignments every prudent reader of this book should undertake would be to employ the tools outlined in this book for a design problem relevant to their immediate line of work. This effort will provide sufficient confidence in the methods and help to appreciate the multifaceted nature of battery technology.

About the Authors

Shriram Santhanagopalan is a senior engineer at the National Renewable Energy Laboratory (NREL). He develops novel design strategies to address the limitations of various electrochemical energy storage systems and to optimize their performance and safety. He has spent a good part of the last 10 years investigating several battery chemistries, building electrochemical models, developing test methodologies for lithium-ion batteries, and exploring materials for next generation batteries. He has worked extensively with several leading battery manufacturers and automakers around the world. Shriram has a Ph.D. in chemical engineering from the University of South Carolina. He is an active mem ber of the Electrochemical Society, the Society of Automotive Engineers, and the American Institute of Chemical Engineering. His work is recognized in the *Who's Who in Chemical Sciences* list by the American Chemical Society. He has over 100 publications, presentations, book chapters, patents, and invention-disclosures in this discipline.

Kandler Smith is a recognized expert in battery electrochemical modeling, control, and lifetime prediction. As a senior researcher at NREL since 2007, he leads projects in the areas of battery diagnostics and prognostics, life-extending controls and systems, and battery multiphysics modeling. Kandler holds a Ph.D. in mechanical engineering from the Pennsylvania State University in electrochemical control of Li-ion batteries and is an active member of IEEE, the Electrochemical Society, and the Society of Automotive Engineers. In addition to codeveloping several patents in battery safety and control, Kandler has licensed battery electrical, thermal, electrochemical, and lifetime software models to more than 10 external partners, helping to accelerate the battery development efforts in industry, academia, and other labs.

Jeremy Neubauer is a senior engineer with NREL's Transportation and Hydrogen Systems Center–Energy Storage Group. Dr. Neubauer specializes in evaluating and optimizing battery use strategies for automotive and grid applications. Past projects with NREL include evaluating lifetime of Li-ion batteries in community energy storage applications under different climates and thermal configurations, analysis and validation of plug-in electric vehicle battery second use strategies, and techno-economic assessment of battery electric vehicle fast-charging and battery-swapping business models. He has also worked closely with the United States Advanced Battery Consortium (USABC) to develop battery technical requirements for multiple electrified vehicle platforms. Recently, Dr. Neubauer has also ventured into the development of advanced battery chemistries with applicability to both automotive and grid markets under funding from an ARPA-E award. Prior to coming to NREL, Dr. Neubauer was chief engineer at ABSL Space Products, a leading manufacturer of Li-ion batteries for the space industry. There he developed energy storage solutions for long duration, high reliability, and manned space missions. Dr. Neubauer has bachelor's, master's, and doctorate degrees in mechanical engineering from Washington University in St. Louis.

Gi-Heon Kim leads the multiphysics lithium battery modeling task for the Transportation and Hydrogen Systems Center's Energy Storage Group. He serves as a technical monitor and technical advisor for the U.S. Department of Energy's Computer-Aided Engineering for Electric Drive Vehicle Batteries (CAEBAT) Program. He is the lead developer of NREL's pioneering multiscale, multiphysics battery model (multiscale, multidomain model), which helped the initiation of CAEBAT, resolving nonlinear interplay of lithium-ion battery physics in varied length scales. He is internationally recognized for his expertise in lithium-ion battery safety research and engineering, with multiple publications and patents.

Ahmad Pesaran is an internationally recognized expert on battery thermal management for hybrid and plug-in vehicles. He holds a Ph.D. in mechanical engineering from the University of California, Los Angeles. He has worked at the National Renewable Energy Laboratory on various energy-efficiency-related technologies including buildings, advanced air conditioning, and automotive batteries for over 30 years. He is currently the group manager for the Energy Storage group at NREL and leads several projects funded by the U.S. Department of Energy's Vehicle Technologies Office, in close collaboration with industrial partners. He is an active member of the FreedomCAR Electrochemical Energy Storage Technical Team and a member of the Society of Automotive Engineers and American Society of Mechanical Engineers. He has coauthored more than 100 papers and presentations on energy storage and electric drive vehicles.

Matt Keyser, a three-time R&D 100 Award Winner, currently leads the battery thermal characterization work at NREL, where he has spent the last 23 years exploring a wide range of transportation technology developments in areas such as battery optimization, thermal management, cold-start emissions reductions, and thermoacoustic cabin cooling. His work has been recognized through many awards and patents over the years. Among his accolades are the Colorado Governor's Award for Foundational Technology and the Top Young Innovator Award by the MIT Technology Review. He works closely with the United States Council for Automotive Research (USCAR) evaluating thermal efficiency of energy storage systems and thermal management strategies for vehicles. He has mentored numerous engineers and technicians in designing experiments and performing R&D in NREL's energy storage laboratories.

Index

A

Accelerating fade region, 84
Accelerating rate calorimeters (ARCs)
- defined, 128
- experiments, 131, 132
- heat/wait seek temperature profile, 129
- limitations, 130
- in measurement of reaction heats, 128–31
- in safety aspects of cell design, 131
- schematic of, 129
- testing batteries and, 130–31
- thermal response, 131
- thermal stability, 130

Adiabatic calorimeters. *See* Accelerating rate calorimeters (ARCs)
Air-cooled channel
- hydraulic radius and flow rate, 75
- optimal design, 75, 76

American Society for Testing and Materials (ASTM) methods, 132
Anodes
- lithium ion, 12
- lithium-sulfur, 16
- thermal stability of, 130

Applications
- automotive, 142–51
- battery requirements, 139–42
- grid, 151–62

Area regulation, 159–60
Atomic layer deposition (ALD), 61
Automatic generation control (AGC), 157, 158
Automotive applications
- battery electric vehicles (BEVs), 149–51
- drive cycles, 142–43
- plug-in hybrids, 145–49
- power assist hybrids, 144–45
- SLI, 143
- start-stop (micro) hybrids, 143–44
- *See also* Applications

Automotive battery design case study
- battery temperature prediction, 190–94
- control trade-offs versus lifetime, 194–96
- defined, 185–86
- fitting life parameters to cell aging data, 188–90
- life predictive model, 186–88

Automotive safety integrity level (ASIL), 185

B

Balance of plant, 169–70
Baseline
- annual savings, 209
- battery temperatures, 209
- resistance growth, 208
- scenario, 207

Batteries
- applications, 2
- defined, 1
- emerging chemistries, 19–20
- entropic heating of, 54–57
- flow, 8–9
- heat generation in, 47–51
- hybrid-flow, 9
- lead acid, 3–4
- Li-ion. *see* Li-ion batteries
- liquid metal, 20
- lithium-air, 18
- lithium ion anodes, 12
- lithium ion cathodes, 10–12
- lithium-sulfur, 15–16
- lithium-sulfur anodes, 16
- lithium-sulfur cathodes, 15–16
- metal-air, 17–19
- metal chloride, 7
- module and pack performance, 74
- nickel-based, 4–5
- operating principles, 21
- redox flow, 8–9

Batteries (continued)
relative performance of, 2–3
secondary, 1
sodium beta, 6–8
sodium-ion, 19–20
sodium sulfur, 6–7
thermodynamics inside, 21–24
types of, 1–20
zinc-air, 17
Battery electric vehicles (BEVs)
drive train, 149
HVAC and, 150
optimal range, 150
range, increasing, 150–51
range effect on vehicle utility, 151
short-term duty cycles, 149
USABC technology targets, 152
Battery electronic control module (BECM), 174
Battery life
beginning of life (BOL), 82
calendar life versus cycle life, 82–83
cell life, extending, 87–88
comparison for thermal management strategies, 172
degradation mechanisms, 89–99
end of life (EOL), 82, 86–87
fade mechanism, 88–89
failure modes, 81–82, 83
modeling, 99–108
overview, 81–99
physics, 81–82
regions of performance fade, 83–86
testing, 108–14
Battery Lifetime Analysis and Simulation Tool for Behind-the-Meter Lite (BLAST-BTM Lite), 199
Battery management
on-board electronics for, 218
roles of, 174
Battery management system (BMS)
arrangement, 175
hardware, 174–75
inputs and outputs, 175
roles of, 174
Battery packs
active cell balancing, 88
capacity, 88
enclosure design, 174
protective device effectiveness, 121–22
Battery reference model, 179–80
Battery requirements
electrical, 139–41
mechanical, 141–42
safety/abuse, 142
thermal, 141
Battery safety
challenges with localized failure, 121
chemical failure, 120–21
concerns, 117–21
electrical failure, 118
electrochemical failure, 119–20
evaluating, 128–37
mechanical considerations, 122–24
mechanical failure, 120
modeling insights on, 121–28
pressure buildup, 124–26
protective circuitry design, 126–28
thermal failure, 118–19
Battery temperature prediction
ambient conditions and, 190
passive thermal environment model, 193–94
solar radiation and, 190
thermal network model, 192–93
Battery thermal management system (BTMS)
design calculations for first-order evaluation, 70–74
designing, 68–75
designing, building, and testing, 74–75
design objective and constraints, 69
evaluation design calculations, 70–74
module properties, 69–70
net temperature difference, 71–72
performance prediction, 74
Beginning of life (BOL), 82
Behind-the-meter peak-shaving case study
auxiliary loads, 202
baseline, 207–8
control model, 200–202
decreased maximum *SOC*, 211–12
defined, 196–97
degradation analysis, 207
electrical analysis, 200–202
electrical model, 200

end user load profile and rate structure, 197–207
end user needs and constraints, 197
increased cooling, 208
preliminary techno-economic sizing analysis, 197–200
reduced target *SOC*, 208–11
thermal analysis, 202–7
thermal model, 200
Binder decomposition, 94
Boltzmann constant, 74
Break-in region, 84
Butler-Volmer kinetics, 102

C

Calendar life, 82–83
Calibrated thermocouples (TCs), 65
Capacity fade
for decreased maximum *SOC*, 211
global model for, 192
hot and cold climates, 193
semiempirical models, 105–6
Capacity measurement, 40
Cathodes
Li-ion cells, 24
lithium ion, 10–12
lithium-sulfur, 15–16
thermal stability of, 130
Cell balancing, 176–77
Cell formation cycles, 29
Cell-level testing
crush, 135
fire hazard, 136–37
hot-box, 135–36
impact, 136
overcharge, 136
overview, 133
pressure/humidity, 136
RMS, 135
short-circuit, 133–35
standards governing, 134
Cells
efficiency comparison between, 57–58
electrochemical models for, 31–39
fade mechanism, 88–89
general degradation mechanisms, 89
isothermal pouch, 64
stress buildup, 124
thermal stability of, 130
Charge buildup, 30
Charge/discharge, voltage dynamics during, 28–29
Charge transport
between electrodes and electrolyte, 36–37
within electrodes by electrons, 33–34
in electrolyte by ions, 34–35
Chemical failure, 120–21
Chevrolet Volt battery thermal system, 173
Circuit diagram
approach as simplistic, 31
for cell, 29–31
Community energy storage (CES), 161–62
Component characterization
conductivity measurements, 42–43
diffusivity measurements, 43–44
open circuit potentials, 42
reaction rate constants, 44–46
Compressed air energy storage (CAES), 153
Conductivity measurements, 42–43
Contactors, 170
Control trade-offs versus lifetime, 194–96
Cooling, increased, 208
Crush testing, 135
Current collector corrosion, 94
Current interrupt device (CID), 118
Current/power limits calculation, 182–83
Cycle life, 82–83
Cyclic fade, 107

D

Decelerating fade region, 84
Decreased maximum *SOC*, 211–12
Degradation
analysis, 207
key factors to adjust, 207
processes, 109
rate, 112
Degradation mechanisms
binder decomposition, 94
current collector corrosion, 94
in electrochemical cells, 89
electrochemical reactions, 90

Degradation mechanisms (continued)
electrode displacement and fracture, 98
electrolyte decomposition, 94–95
electromechanical processes, 90
gas buildup, electrode isolation, 98–99
in Li-ion cells, 89–99
lithium plating, 92–94
metal-oxide positive electrode decomposition, 95–97
particle facture, 97–98
SEI formation and growth, 91–92
SEI fracture and reformation, 97
separator viscoelastic creep, 98
Demand charge management (DCM)
discharges, 157
forecast, 157
implementation requirements, 155
UPS and, 154–57
Depth of charge, 195
Design of experiments, 110–11
Design process
component development, 183–84
management, 183
"V" design cycle, 183, 184
verification testing, 184
See also System design
Design standards, 184–85
Differential scanning calorimeters (DSCs)
batteries and, 60–62
defined, 58
heat capacity data, 60
heat flow measurement, 58, 59
heat flux, 58, 59
melting point and enthalpy changes, 61
safety features assessment, 62
typical heating for experiments, 59
use of, 60
Diffusivity measurements, 43–44
Discrete element method (DEM), 105
Disordering, 95
Dissolution, 96
Distribution of ions, 37–38
Drive cycles, 142–43
Duty cycle
area regulation, 159
BEVs, 149
peak shaving, 140
PHEVs, 147
Dynamic stress test (DST), 41

E

Electrical characterization
capacity measurement, 40
component characterization, 41–46
conductivity measurements, 42–43
diffusivity measurements, 43–44
Li-ion batteries, 39–46
open circuit potentials, 42
power measurement, 40–41
reaction rate constants, 44–46
Electrical design
balance of plant, 169–70
contactors, 170
ground default detection, 170
main pack fuse, 170
manual service disconnect (MSD), 169–70
module sizing, 169
power/energy ratio, 166–67
series/parallel topology, 167–69
See also System design
Electrical failure, 118
Electrical performance
circuit diagram, 29–31
electrical characterization, 39–46
electrochemical models for cell design, 31–39
Li-ion cell assembly, 24–28
thermodynamics, 21–24
voltage dynamics, 28–29
Electrical requirements, 139–41
Electric double-layer capacitors, 1
Electric drive vehicles (EDVs), 58
Electricity Storage Handbook, 152
Electrochemical capacitors, 1
Electrochemical dilatometry measurement, 112
Electrochemical failure, 119–20
Electrochemical models
for cell design, 31–39
charge transfer between electrodes and electrolyte, 36–37
charge transport in electrolyte, 34–35
charge transport within electrode, 33–34

comparison of simulation results, 39
distribution of ions, 37–38
predication versus experimental data, 32
Electrochemical potential, 23
Electrochemical reactions, 90
Electrodes
charge transport between electrolyte and, 36–37
charge transport within, 33–34
displacement and fracture, 98
isolation, 98–99
OCP of, 42
reactions, heat generation from, 49
Electrolytes
charge transport between electrons and, 36–37
charge transport by ions, 34–35
decomposition, 94–95
defined, 27
thermal stability of, 130
Electromechanical model properties, 100
Electromechanical processes, 90
Electronics and controls
battery management roles and, 174
battery reference model, 179–80
BMS hardware, 174–75
cell balancing, 176–77
current/power limits calculation, 182–83
state estimation algorithms, 177–79
state estimator, 180–82
Emissivity, thermal image and, 63–65
End of life (EOL)
criteria, 86
defined, 82
for energy applications, 86
Endothermic crystalline phase transition, 56
End users
load profile and rate structure, 197–207
needs and constraints, 197
Energy storage
community, 161–62
compressed air (CAES), 153
specifications, 153
system design, 165–213
systems, 165
upregulation requests and, 158
Energy Storage Handbook, 151–52
Entropic heat generation, 49–51, 54–57
Entropic measurements, 112
Extending cell life prediction, 87–88

F

Fade mechanisms, 88–89
Fade trajectories, 107
Failure
chemical, 120–21
electrical, 118
electrochemical, 119–20
localized, challenges with, 121
mechanical, 120
modes, 81–82, 83
thermal, 118–19
Fire hazard testing, 136–37
Flow batteries
challenges and future work, 9–10
hybrid, 9
performance improvement, 10
redox, 8–9
Fluid flow rates, 73
Fused busbar arrangements, 169

G

Galvanostatic intermittent titration technique (GITT), 44
Gas buildup, electrode isolation, 98–99
Grid applications
area regulation and transportable asset upgrade deferral, 157–60
community energy storage, 161–62
demand charge management (DCM), 154–57
energy storage specifications, 153
grid-connected applications, 162
overview of, 151–54
See also Applications
Ground default detection, 170

H

Heat generation
in batteries, 47–51
curves, 55
efficiency and heat generation, 54
from electrode reactions, 49

Heat generation (continued)
entropic, 49–51
from joule heating, 47–49
Heat transfer coefficient, 73–74, 77
Heat/wait seek temperature profile, 129
Hot-box test, 135–36
Hybrid electric vehicles (HEVs)
applications, 145
batteries, 145
power-assist, 144
Hybrid-flow-batteries, 9
Hybrid pulse power capacity (HPPC), 41

I

Imaging battery systems, 65–67
Impact and intrusion requirements, 142
Impact testing, 136
Infrared imaging
calibration and error, 65
imaging battery systems, 65–67
object emissivity and reflectivity and, 63–65
use of, 62
Intercalation, 27
Ions
charge transport in electrolyte by, 34–35
distribution of, 37–38
ISO 26262, 184–85
Isothermal battery calorimeters (IBCs)
applications, 54–58
basic operation, 51–54
defined, 51
efficiency and heat generation, 54
efficiency comparison between cells, 57–58
entropic heating of batteries, 54–57
heat conduction of, 52
schematic of, 52
temperature-controlled, 52
Isothermal pouch cell, 64

J

Jellyroll, 28, 127
Joule heating, 47–49

K

Kinetics, 102–3
Kirchhoff's loop rule, 30

L

Lattice instability, 95–96
Layered-transmission-metal oxides, 11
Lead acid batteries
during charging process, 3–4
cost of, 4
defined, 3
disadvantages of, 4
short cycle life, 4
Life parameters, fitting to cell aging data, 188–90
Life predictive model, 186–88
Li-ion batteries
anodes, 12
capacity measurement, 40
cathodes, 10–12
challenges and future work, 13–15
component measurement, 41–46
cost, 14
degradation processes, 109
electrical characterization, 39–46
electrochemical operating window, 90
electrolytes, 13
fire hazard, 136–37
mechanical stress sources, 104
modeling insights on safety, 121–28
new material chemistries, 14–15
performance, 14
power measurement, 40–41
safety, 13–14, 117–21
Li-ion capacitors (LICs), 56, 57
Li-ion cells
anodes, 24
assembling, 24–28
cathodes, 24
change in OCPs, 26
circuit diagram, 29–31
common degradation mechanisms, 89
electrolytes, 27
intercalation, 27
ion movement, 25
jellyroll, 28
voltage difference, 25
See also Cells

Li-ion electrolytes, 13
Li loss, 112
Linear parameter varying (LPV) system, 182
Liquid metal batteries, 20
Lithium-air batteries, 18
Lithium cobalt oxide ($LiCoO_2$), 11
Lithium ion anodes, 12
Lithium ion cathodes
  classes, 10
  layered, 11
  olivines, 12
  spinels, 11–12
Lithium plating, 92–94
Lithium-sulfur batteries
  anodes, 16
  cathodes, 15–16
  challenges and future work, 16
  defined, 15

M

Main pack fuse, 170
Manual service disconnect (MSD), 169–70
Mass conservation, 102
Mechanical design, 173–74
Mechanical failure, 120
Mechanical requirements, 141–42
Mechanical stress modeling, 103–5
Metal-air batteries
  challenges and future work, 19
  defined, 17
  lithium-air, 18
  research on, 19
  zinc-air, 17
Metal chloride batteries, 7
Metal-oxide positive electrode decomposition, 95–97
Modeling, battery life
  mechanical stress, 103–5
  overview, 99
  physics-based, 99–105
  reaction/transport models, 99–103
  semiempirical models, 105–8
  See also Battery life
Model predictive controls (MPC), 183
Module sizing, 169
Multireaction transport model, 101

N

Newton's law of cooling, 73
Nickel-based batteries, 4–5
Nickel metal hydride (NiMH) systems, 5
Nickel-zinc batteries, 5
Nonlinear system, 182
Nusselt number, 77

O

Observability of the system, 181
OCV behavior, 113
Olivines, 12
On-board electronics, 218
Online parameter identification, 178
Open circuit potential (OCP)
  change in, 26
  of individual electrodes, 42
  values, 24
Optimization, thermal management system, 75–79

P

Pack capacity, 88
Particle facture, 97–98
Passive thermal environment model, 193–94
Peak shaving duty cycle, 140
Performance fade
  accelerating fade region, 84
  break-in region, 84
  decelerating fade region, 84
  example during lifetime, 85
  nonlinear, 83
  regions of, 83–86
  trajectory equations, 106
Physics-based models
  mechanical stress, 103–5
  reaction/transport, 99–103
  See also Modeling, battery life
Plotting dQ/dV, 113
Plug-in hybrid electric vehicles (PHEVs)
  applications, 53, 145–49
  battery duty cycles, 147
  battery response example, 146
  battery technology targets, 148
  control trade-offs versus lifetime, 194–96

Plug-in hybrid electric vehicles (continued)
parallel, 145–46, 147
porosity, 48
powertrain, 145
series, 146–47
through-the-road, 147
Polyvinylidene-difluoride (PVdF), 94
Positive temperature coefficient (PTC), 118, 122
Potential
concept illustration, 22
electrochemical, 23
open circuit (OCP), 24, 26
Potentiostatic intermittent titration technique (PITT), 44
Power-assist hybrids, 144–45
Power/energy ratio, 166–67
Power measurement, 40–41
Pressure buildup
debate, 124
factors contributing to, 125
weld strengths and, 125
See also Battery safety
Pressure/humidity testing, 136
Protective circuitry design, 126–28
Protective devices, in multicell packs, 121–22
Pseudocapacitors, 1
Pumped-hydro storage (PHS), 152–53

R

Rate model equation, 108
Reaction rate constants, 44–46
Reaction/transport models
film growth, 100–102
kinetics, 102–3
mass conservation, 102
overall total reaction, 100
side reactions, 99–100
Tafel kinetic equations, 103
Redox battery operation, 2
Redox flow batteries, 8–9
Reduced target *SOC*, 208–11
Reference diode, 113
Reference governor, 178
Reference model, 178
Reference performance tests (RPTs)
in benchmarking performance, 111
common temperature, 111
defined, 109
degradation rate and, 112
design, 109
objective, 86
Reflectivity, thermal image and, 63–65
Resistance, 33
Root mean square (RMS), 135, 139

S

Safety/abuse requirements, 142
Safety evaluation
ARCs, 128–31
cell-level testing, 133–37
thermochemical characterization, 131–33
See also Battery safety
Screening/benchmarking tests, 110
SEI formation and growth, 91–92
SEI fracture and reformation, 97
Semiempirical models
capacity fade data, 105–6
cyclic fade, 107
defined, 105
fade trajectories, 107
performance fade trajectory equations, 106
rate model equation, 108
regression steps, 107–8
See also Modeling, battery life
Separator
defined, 27
viscoelastic creep, 98
Series/parallel topologies
benefit of P before S, 167
benefit of S before P, 167
defined, 167
example, 168
fused busbar arrangements, 169
safety and, 168
Short circuits
combating, 126–28
components leading to, 126
as major concern, 126
permutations of, 126
testing, 133–35
types of, 127

SLI batteries, 143
Sodium beta batteries
  challenges and future work, 7
  defined, 6
  issues impeding, 8
  metal chloride, 7
  sodium sulfur, 6–7
Sodium-ion batteries, 19–20
Spinels, 11–12
Standards, design, 184–85
Start-stop (micro) hybrids, 143–44
State estimation algorithms, 177–79, 182
State estimator
  continuous time linear, 180
  defined, 178
  observability matrix, 181
State of charge (SOC)
  calculation, 177
  decreased maximum, 211–12
  estimate adjustment, 178
  reduced target, 208–11
State of health (SOH), 177–78
State of power (SOP), 177–78
Stress buildup, 124
Surface effects, 97
System architecture, 213
System design
  aspects, 165–66
  automotive battery design case study, 185–96
  behind-the-meter peak-shaving case study, 196–212
  electrical design, 166–70
  electronics and controls, 174–83
  mechanical design, 173–74
  process, 183–84
  standards, 184–85
  system specification, 212–13
  thermal design, 171–73
System specification, 212–13

T

Techno-economic sizing analysis, 197–200
Temperature differences, 71–72
Temperature distribution, 123
Testing
  ARCs and, 130–31
  cell-level, 133–37
  crush, 135
  design of experiments, 110–11
  electrochemical dilatometry measurement, 112
  entropic measurements, 112
  hot-box, 135–36
  impact, 136
  life, 109–10
  Li loss, 112
  OCV behavior, 113
  overcharge, 136
  plotting dQ/dV, 113
  pressure/humidity, 136
  reference diode, 113
  results, 114
  RPTs, 111–12
  screening/benchmark tests, 110
  standards governing, 134
  steps, 109–10
  verification, 184
  See also Battery life
Thermal behavior
  conclusions, 79
  differential scanning calorimeters (DSC), 58–62
  experimental measurement of parameters, 51–58
  heat generation, 47–51
  infrared imaging, 62–67
  system attributes, 67–79
Thermal design, 171–73
Thermal electric devices (TEDs)
  defined, 51
  heat flow measurement, 53
  heat sink measurement, 53
  power requirement, 52
  use of, 51–52
Thermal energy, 62–65
Thermal failure, 118–19
Thermal imaging
  calibration and error, 65
  imaging battery systems, 65–67
  object emissivity and reflectivity and, 63–65
  use of, 62

Thermal management
  behind-the-meter peak-shaving case study, auxiliary loads, 202
  solutions, 217
Thermal management system
  designing, 68–75
  designing, building, and testing, 74–75
  design objective and constraints, 69
  design trade-offs comparison, 78
  desired attributes, 67–79
  evaluation design calculations, 70–74
  module properties, 69–70
  optimization, 75–79
  performance prediction, 74
Thermal network model, 192–93
Thermal parameters
  applications of, 54–58
  experimental measurement of, 51–58
  isothermal battery calorimeters (IBCs), 51–58
Thermal requirements, 141
Thermal stability, 130
Thermodynamics, 21–24
Thermomechanical characterization, 131–33
Transmission and distribution upgrade deferral, 158
Transportable asset upgrade deferral, 157–60

U

Uninterruptible power supply (UPS), 155–56
United States Advanced Battery Consortium (USABC), 148

V

"V" design cycle, 183, 184
Verification testing, 184
Vibration requirements, 142
Voltage dynamics, 28–29

Z

Zero emissions vehicle (ZEV) program, 150
Zinc-air batteries, 17

RECEIVED
FEB - 2 2015